Advanced Health Technology

Everything worth winning in life boils down to teamwork and leadership. In my positions as a businessman, athlete, community leader, and University trustee, there are tremendous parallels between all of these endeavors that mirror an extreme team sport such as medical technology. Understanding the game, defining the game, playing your position at your highest performance, and helping others play their best game. *Advanced Health Technology* represents an incredible opportunity to level up the game of healthcare and highlights the multiple disciplines – or positions to be mastered – while laying out winning plays to make that next level happen.

– Ronnie Lott, Managing Member, Lott Investments; Member, Pro Football Hall of Fame, and Trustee, Santa Clara University

Healthcare stakeholders are paralyzed from making progress as risks explode in volume and complexity. This book will help readers understand how to manage and transcend risks to drive the quadruple aim of improved patient experiences, better patient and business outcomes, improved clinician experience, and lower healthcare costs, and also learn from working successful examples across projects, programs, and careers to get ahead of these multidisciplinary healthcare risks.

Everything worth winning in life boils down to teamwork and leadership. In my positions as a businessman, athlete, community leader, and University trustee; there are tremendous parallels between all of these endeavors that mirror an extreme team sport such as medical technology. Understanding the game, defining the game, playing your position at your highest performance, and helping others play their best game. *Advanced Health Technology* represents an incredible opportunity to level up the game of healthcare and highlights the multiple disciplines, or positions to be mastered while laying out winning plays to make that next level happen.

Ronnie Lott, Managing Member, Lott Investments; Member, Pro Football Hall of Fame, and Trustee, Santa Clara University

In my role as SVP Chief Digital and Information Officer, I am passionate about "seeking first to understand." The most relevant solutions come from extreme teamwork beyond our internal teams in IT, and partnering with physicians and administration to make certain that we bring real value to addressing opportunities and challenges to drive optimal physician and patient experience. It all starts with empathy which requires foundational knowledge across the work that folks are doing on the full technical stack as well as physicians, administration, and the board, while building and developing those dynamic teams of diverse skills. *Advanced Health Technology: Managing Risk While Tackling Barriers to Rapid Acceleration* gives you a head start on that knowledge foundation.

Aaron Miri, SVP, Chief Digital & Information Officer at Baptist Health; Board Director at College of Healthcare Information Executives, CHIME; Co-Chair for the U.S. Department of Health and Human Services' Federal Health IT Advisory Committee

Sherri Douville has brought together impressive teams of respected engineers, IT practitioners, CISOs, physician executives, and more to develop this unique book, *Advanced Health Technology* that's both broad and deep. These experts have collaborated to bring you insights brimming with experience and know-how that comes from their years of combined experiences. This will enable any enterprise to accelerate knowledge and preparation around governance obligations for identifying and mitigating emerging risks, including cybersecurity.

Aimee Cardwell, SVP & CISO, UnitedHealthcare Group

As an entrepreneur, educator, and global health leader; I am honored to be called upon by top academic institutions, industry, and governments to advise on the future of connected care. I often do that by convening top experts for insightful debate and analysis. I was happy to have Sherri Douville participate at one such roundtable and know that this unique book – *Advanced Health Technology* – will make those conversations much more efficient and actionable.

Anurag Mairal, PhD
Director, Global Outreach Programs, Stanford Byers Center for Biodesign
Lead Faculty, Technology Innovation and Impact,
Center for Innovation in Global Health
Adjunct Professor, Stanford School of Medicine
Executive Vice President, Orbees Medical

Great corporate directors and boardrooms play a wide range of critical roles that include monitoring functions and advisory functions to understand and manage emerging risk. Leading practices in digital and cybersecurity oversight are emerging. And this next threshold of corporate governance is engaging CIOs CISOs directors and other areas to make sure the boardroom is a high-performing part of every company's digital future. As leaders of this charge at Digital Directors Network, we recognize the criticality of domain expertise and the valuable richness of this resource, *Advanced Health Technology: Managing Risk While Tackling Barriers to Rapid Acceleration* book in addressing the majority of emerging risks in healthcare including cybersecurity.

Bob Zukis, CEO Digital Directors Network, USC
Marshall Business School Professor

In my role as CEO for AACSB, the largest accreditation body recognizing the world's top business schools, we focus on building risk literate leaders with the needed competencies of holding paradox, compassion, and depolarization to drive the necessary multi-disciplinary solutions to today's extraordinary business challenges. Through learning from this book, *Advanced Health Technology: Managing Risk While Tackling Barriers to Rapid Acceleration*, you can quickly and efficiently digest the landscape of emerging risks across clinical, operational, and technology domains in healthcare which allows you to lead through the challenging paradox of innovation and risk from a lens of empathy for all the major functions. It's also very important for business leaders to be able to transcend politics with data and facts in the medical technology sector and this book demonstrates an example of and helps the reader with that.

Caryn Beck-Dudley, President and CEO at AACSB International

What I love about my CIO role at Renown includes strategy, leadership, turnarounds and transformation, all facilitated through coaching and mentoring. On the cusp of an exciting revolution of Connected Care, tomorrow's top CIOs and executive teams need to tackle a new, diverse set of risks. Taking in the lessons in *Advanced Health Technology: Managing Risk While Tackling Barriers to Rapid Acceleration* will put the reader ahead of the game and help make it all more fun too.

Charles Podesta, Chief Information Officer at Renown
Health, Serial CIO, & CIO Advisor

Throughout my career in multiple leadership roles spanning mobile, e-Commerce, software, energy, finance, semiconductor, and industrial sector companies, I've been honored to be reputationally and officially recognized for having expertise in and championing safety and technology. That's why I resonate with this book, *Advanced Health Technology* which is a force multiplier for strategic leaders who recognize that tackling emerging risks as a critical competency for tomorrow's leaders to drive effective, competitive governance.

Cheemin Bo-Linn, CEO and Corporate Board Director, Top 50 Board of Directors NACD 2019, Lead Independent Director, Serial Audit Chair, Chair of Compensation, Chair of Nom/Gov/ESG/Technology, and Cyber Security Lead

As technology rapidly advances, healthcare systems are highly focused on care quality, efficiency and leveraging the most out of their digital assets. Rapid digital transformation is crucial for meeting these needs, but many risks and barriers can impede the best of efforts. *Advanced Health Technology: Managing Risk While Tackling Barriers to Rapid Acceleration* picks up right where its companion *Mobile Medicine* left off. Editor Sherri Douville and the fabulous cohort of multidisciplinary minds and book contributors again share their proven in-the-trenches successes, guidance, and best practices to create value while managing the complexities of risk in an expanding digital ecosystem.

Dan Howard, MBA, CHCIO, PMP, CIO, San Ysidro Health

Healthcare has labored for decades under the belief that innovative technology is innovation. Technology, alone, is not innovation, it is only a tool. Digital transformation will only result in incremental change until all the things that go along with technology - - the people and the processes - - security, engineering, privacy, medicine, governance, IT, patients, and others work together. Healthcare has long been one of the most risk averse sectors that must change to do transformation. The risks must be understood and managed and that will require a new way of making change and managing risk. This book enables you to transform your thoughts, strategies, vision, and actions. The days of silos within and across healthcare, technology, care sites, payers, and everything else are over.

David Finn, CISA, CISM, CRISC, CDPSE, Vice President, CHIME, for AEHIS, AEHIT, AEHIA

Over the years we've known each other and worked together, Sherri and I have shared a passion for harm reduction. Whether that is due to failure of effective uses and deployment of technology or harm by one of today's biggest threats, drug overdose. We both recognize the need to strategically address the root causes of threats to health and wellbeing. Given that *Mobile Medicine* and related works are changing the way a whole generation of CTOs and CIOs are thinking about healthcare technology; this new book *Advanced Health Technology: Managing Risk While Tackling Barriers to Rapid Acceleration* takes the reader deeper into their leadership journey of transforming medicine with technology by understanding and addressing diverse risks rather than ignoring them.

Dean Shold, Co-Founder FentCheck, Former Partner at Accenture, Former CTO Stanford Healthcare and Alameda Health System, Former CTO at Medigram and current advisor.

I am thrilled to see this definitive guide to overcoming the large number of heterogeneous risks facing healthcare and all mission critical industries. Brave, sensible and intelligent teams will see these challenges as opportunities to accelerate the transformation of medicine using *Advanced Health Technology: Managing Risk While Tackling Barriers to Rapid Acceleration* as their guide.

Duy-Loan Le, Texas Instruments Senior Fellow (ret.), Board of Director – Wolfspeed, National Instruments, Ballard Power Systems, Atomera, Medigram

In my roles as a CIO, Board Director, Advisor and Investor passionate about driving innovation forward; I was focused on cybersecurity as critical to our success in leveraging health information technology across research, teaching, and patient care. As the risk environment for healthcare organizations has exploded and to be able to continue innovating; we need to stay ahead of the multidisciplinary risk landscape, which this book *Advanced Health Technology: Managing Risk While Tackling Barriers to Rapid Acceleration* allows us to do efficiently and effectively.

Eric Yablonka, Advisor, Investor, Board Director and former CIO and Associate Dean at Stanford Medicine

In an era where emerging risks are facing many boards of directors, senior management, and front line healthcare providers, *Advanced Health Technology: Managing Risk While Tackling Barriers to Rapid Acceleration* takes a fresh approach in helping the reader understand the evolving risk landscape, how these risks can significantly disrupt patient care and healthcare operations, and what steps leadership can take to ensure organizational risk is minimized.

James Brady, PhD, VP and CISO at M Health Fairview

Since the introduction of the mobile phone, technology has rapidly developed in both the tremendous increase in processing power, with equally great reduction in size. Coupled with the explosion of high-speed wireless Internet access, this technological revolution has changed the global face of communication and human interaction. However, these massive improvements in computing power and information come at a tremendous risk when not properly managed. Sherri and her team of expert authors capture the critical aspects of technology and risk, explained in straightforward terms, and help to prepare the audience to understand and manage cut-edge trends in new technology.

Jim St.Clair, Executive Director, Linux Foundation Public Health

As a 25-year leader in the life sciences industry, I have helped to pioneer IT commercialization capabilities leveraging innovative strategies that were implemented leveraging state of the art program management; with the goal of ensuring operational success and excellence for the enterprise and all business partners. Critical to winning in a competitive, world class capacity is the ability to accurately assess and manage multiple dimensions of risk, which is why *Advanced Health Technology: Managing Risk While Tackling Barriers to Rapid Acceleration* is a must-read for any executive that recognizes the need for innovation and transformation in medicine.

Joe Mulhearn, Head of Revenue, Medigram Inc. and former Senior Director of IT at Merck & Co.

As a seasoned specialist physician CEO and now board director; it's obvious to me that safely integrating advanced technologies into healthcare will radically improve our productivity and experience for both patients and clinicians. In order to reap the rewards, we have to identify and tackle emerging risks to healthcare systems. Amazingly, solving for most emerging risks can be done by focusing on advanced technologies in the right way with cybersecurity and privacy. This book, *Advanced Health Technology* shows you how.

Kathy Garrett, MD, Board Director, Chair, Quality Committee at Orlando Health & Strategic Advisor Committee on Governance, American Hospital Association

It is widely recognized that healthcare is at an inflection point that requires transformation. *Advanced Health Technology: Managing Risk While Tackling Barriers to Rapid Acceleration* is the perfect tool for stakeholders who seek to take meaningful action and deliver results.

Laura Huang, Associate Professor at Harvard Business School; Author *of Edge: Turning Adversity into Advantage*; best 40 business school professors under 40 by Poets & Quants

I have 20+ years of providing counsel and industry leadership for connected healthcare across the full technical stack; I am excited to see this team decode and efficiently solve for the landscape of risks that prevent the innovation so critically needed now through this book *Advanced Health Technology*.

Noel Gillespie, Partner, Procopio, Cory, Hargreaves & Savitch LLP

As someone who works extensively in Healthcare IT and digital health, I'm painfully aware that medicine desperately needs to innovate with technology. At the same time, the risk landscape is exploding and mutating, thus freezing many from taking action. This book will help master your understanding of emerging risks and overcome the challenges by applying the lessons described from *Advanced Health Technology: Managing Risk While Tackling Barriers to Rapid Acceleration*.

Paddy Padmanabhan, Founder & CEO Damo Consulting, Co-author of Healthcare Digital Transformation and host of The Big Unlock podcast

Throughout my over 20 year career practicing law across health systems, life sciences, and technology; my passion has driven me to lead the necessary balancing of risk, innovation, and ensuring regulatory compliance. Healthcare has an urgent need to innovate now in the face of an exploding risk landscape and that's why this book *Advanced Health Technology* is essential reading for critical guidance to executives working in healthcare.

Rosie Goddard, Chief Legal Officer and Chief Compliance Officer, Medigram Legal & Regulatory Advisor

With many parallels in critical safety, highly regulated industries; it's my responsibility as both a CFO and risk strategy leader as serial board director to learn from other industries such as healthcare. While Advanced Health Technology was written for healthcare leaders to overcome the majority of their emerging risks by focusing on *Advanced Health Technology* capabilities; through my earlier career experience in healthcare/biotech at Amgen, I can envision how this book will help all of us with fiduciary obligations to predict and manage emergent risks for smarter, more efficient governance and planning of our technology ecosystem and business.

Shannon Nash, Esq., CPA, Chief Financial Officer at Wing (an Alphabet Company) Serial CFO, Lead Independent Director, Nominating & Governance Chair, Board Member User Testing (NYSE: USER), Audit Committee Chair and Board Director --Lazy Dog Restaurant & Bar.

In my role as one of the few physician CIO's and combined CIO and CMIO; this resource supports my vision for an integrated technical strategy that can truly enable clinicians. That can only be done by building bridges and partnerships across all the necessary functions. To benefit from what advanced health technology has to offer us in our drive to transform care; we have to develop new models for understanding, building, managing, and leading very diversely skilled teams that can effectively manage risk that allows advanced technology to be a true asset in our clinical goals. This book is the prescription for that.

Stephanie Lahr, MD, CHCIO, Monument Health, Chief Information Officer and Chief Medical Information Officer

In my career as an operator, investor, and board director, I have leveraged STEM innovation to drive shareholder value, on both strategic and finance fronts. I've had the privilege to do this working with and leading outstanding teams. Today's risks and challenges are exponentially more complex. They require sophisticated teamwork, cross-functional literacy, and depth of experience to scale solutions to intractable problems like those in healthcare. Use this terrific resource, *Advanced Health Technology*, to jumpstart your transformation journey.

Sue Siegel, Serial Corporate Board Director, Advisor, Former CEO & VC, Board of Director – Illumina, Align Technology, Nevro, The Engine, Kaiser Family Foundation

As a veteran of innovation finance for hundreds of startup financing deals; the one ingredient contributing to winning has been an understanding of risks. There's tremendous parallels of mastering risk between medical technology and early stage finance. To get in on the wave of the next generation of medical technology, give yourself a head start by reading *Advanced Health Technology: Managing Risk While Tackling Barriers to Rapid Acceleration.*

Tom Bondi, Tax Partner, Armanino LLP

My roles as an executive and educator require continually learning new material from the most innovative minds in healthcare; that material needs to cover professional education while addressing academic requirements. *Advanced Health Technology: Managing Risk While Tackling Barriers to Rapid Acceleration* does all these and much more by comprehensively covering culture, infrastructure, engineering, cyber, and product management topics. Whether you're already a healthcare executive or strive to be one, this book provides executive guidance and practical knowledge you can immediately apply.

Will Conaway, Chief Growth Officer, The HCI Group/ a Tech Mahindra Company

The most satisfying part of my career has been driving repeatable hard technical innovation and commercializing that innovation which changed organizations and lives. No other sector needs innovation more than healthcare. With *Advanced Health Technology: Managing Risk While Tackling Barriers to Rapid Acceleration*, real and motivated leaders will be able to overcome the obstacles to sustainable, transformative innovation.

Wim Roelandts, Retired Chairman Applied Materials, Retired CEO Xilinx, Board Chair at Medigram, Serial Board Director

Advanced Health Technology

Managing Risk While Tackling
Barriers to Rapid Acceleration

Edited by
Sherri Douville

Foreword by Edward W. Marx, Leading Serial Top CIO, and CEO at Divurgent

Routledge
Taylor & Francis Group

A PRODUCTIVITY PRESS BOOK

First published 2023
by Routledge
605 Third Avenue, New York, NY 10158

and by Routledge
4 Park Square, Milton Park, Abingdon, Oxon, OX14 4RN

Routledge is an imprint of the Taylor & Francis Group, an informa business

ISBN: 9781032391496 (hbk)
ISBN: 9781032391489 (pbk)
ISBN: 9781003348603 (ebk)

DOI: 10.4324/9781003348603

Typeset in Garamond
by Deanta Global Publishing Services, Chennai, India

Dedication

I dedicate this book to my constant muse in this work, my dear husband, Art (Arthur W. Douville, Jr. MD). The first time I met my better half Art was close to 20 years ago. I was invited to a talk he was giving at a hospital for continuing medical education by a mutual friend and physician, where Art, always way ahead of his time, had led the design and delivery for a physicians' symposium focused on multicultural literacy as a matter of professional survival for the modern physician. The past few years have really amplified how prescient Art really was at the time. Our work and lives together have sparked a shared passion for discovering how to bring forward the best that advanced technologies have to offer to optimize and enhance medicine as quickly as possible. It turns out there are many technical challenges and even more people challenges, both of which we explore with their solutions in this book, from multiple and necessary functional perspectives.

We further dedicate this work to all Art's physician colleagues, healthcare workers around the world, and public health officials who have been doing the impossible in the face of extraordinary challenges. This includes, though is not limited to, the incredible levels of medical misinformation and politicization of their work, and occupational threats of violence they face. We all owe them a huge debt of gratitude for persisting in their work to take care of us despite such incredible challenges that would indicate widespread lack of empathy and understanding for what they face every day.

– Sherri Douville, CEO at Medigram, and Editor, *Advanced Health Technology*

Contents

Foreword

Advanced Health Technology
Managing Risk While Tackling Barriers to Rapid Acceleration

When Sherri Douville asked me to write the foreword for her first book, *Mobile Medicine: Overcoming People, Culture, and Governance*, I was naturally delighted and honored to oblige. When Sherri asked me to write the foreword for her new book, *Advanced Health Technology: Managing Risk While Tackling Barriers to Rapid Adoption*, I was downright humbled. With the success of Sherri's first book and her reputation as an industry mover and shaker, I felt perhaps the student has become the teacher!

I have been blessed to have authored and coauthored a few books. I was fortunate to serve with some amazing teams and innovative organizations along the way. These experiences helped shape and prepare me for my current role as chief executive officer for Divurgent, a mid-size healthcare consulting, design, and services firm. Along this broad path, I have met many fantastic leaders who are truly transforming healthcare, perhaps none more poignant than Sherri Douville.

We first met in Denver, Colorado, almost ten years ago where we discussed healthcare transformation and leadership at length. While I remained on the East Coast and Sherri on the West, our paths would cross many times. Her work as an influencer, leader, pioneer, and CEO of Medigram crisscrossed not just our country, but far beyond our borders. Sherri's influence continues today. *Mobile Medicine* went on to debut as the #1 release in its categories of medical technology and medical informatics while remaining a longer-term bestseller and to sit at the top of a very competitive field of quality published works. We can now say Sherri is a worldwide influencer of all things in healthcare transformation and leadership.

Advanced Health Technology shows how preparing for innovative technology can equally equip readers to tackle the majority of emerging risks affecting healthcare systems overall. This knowledge set positions readers to drive a profound and distinctive legacy of leadership by more safely and successfully transforming healthcare with technology. Technical innovation is urgently needed in healthcare and medicine, a clinical environment facing enormous workforce shortages and challenges. However, the path to success is not clear or simple by nature.

The goal of our new book on risk management for advanced technology in medicine is to provide expert guidance for overcoming approximately 75% of the top emerging risks facing healthcare and medicine (Douville, 2022). Sherri once again displays her collaborative style by leveraging the same diverse team of respected authors to demystify and bring clarity to the complex of risk management by bridging engineering, cybersecurity, privacy, medicine, leadership, and legal perspectives, together with advanced clinical technology development, deployment, and implementation experts.

Some of the ways the book differentiates itself include:

- Features all the "hottest" topics requested for speaking engagements by leading industry conferences, such as staff burnout, information blocking and privacy, cybersecurity, interoperability, data transformation, modern leadership, workforce development, and clinical product design considerations
- Written by a multidisciplinary community of engineers, IT practitioners, information security officers, physician executives, engineering faculty, physician educators, entrepreneurs, and lawyers
- Provides totally unique content written from the team's individual and collective applied professional experiences combined with research
- Most of the chapters are drafted by teams of experts working together
- Coauthor expert team is diverse, expanding perspectives and knowledge sets beyond points of view traditionally present in medical technology literature

Perhaps my favorite aspect of the book that is tackled head-on and then woven throughout is its emphasis on human-centered design. It took us a generation to hone the important tool of process improvement into manufacturing and engineering. It took us another generation to learn that while important, these same tools do not lend themselves well to people-intensive industries like healthcare. Human-centered design is quickly emerging as the right tool set for healthcare transformation and Sherri once again is on the leading edge with this concept and book.

Listen, you have finite time and books take a while to digest and to reflect upon. There are many communication modalities vying for your time. I get it. I myself love podcasts and social media for my go to resources. Trust me on this one: buy the book. Reflect and share and then help advance healthcare technology and transform healthcare.

Edward W. Marx
CEO
Divurgent

Preface

This has been developed for you, the healthcare leader, and is written for you to write your story and path to success. We offer this resource to enable you to simply and elegantly elevate your technical strategy while comprehensively driving down risk. It may sound harder than it needs to be. But it's not. Extract yourself from the noise and take advantage of what we aim to do. That is to provide a rich set of not just signals but guideposts. We present you with the tools to think about transcending the crowd on the field so that you can focus on delivering advanced technologies for your patients and clinicians while addressing and reducing substantial risk emergent in your enterprise at the same time; in fact, the majority of emerging risks many stakeholders can perceive but don't truly understand. This may act to freeze them in fear from moving forward. Not you though. The problem actually is the apparent size and siloed nature of the problem space. Our expert coauthors tackle advanced healthcare technology in the clinical business-to-business (B2B) healthcare provider use cases. There is also a review of evidence-based patient engagement models. We provide program and risk management tools for adoption of advanced technology.

This new book shows how preparing for advanced healthcare technology can equally equip readers to tackle the majority of emerging risks affecting healthcare systems overall. This knowledge set will position readers to drive a profound and distinctive legacy of leadership by more safely and successfully transforming healthcare with technology as quickly as possible. When Becker's declares that future healthcare CEOs will come from CIO ranks, whether you are a technical or general management leader in healthcare, you will dramatically benefit from these insights.

An evolved and augmented coauthor team from our internationally acclaimed, best-selling book, *Mobile Medicine: Overcoming People, Culture, and Governance*, has now brought their individual and collective knowledge to this new second book specific to managing risks to advanced healthcare technology design, development, deployment, and implementation. In *Advanced Health Technology*, the team addresses the top concerns and barriers to innovation for executives leading in healthcare systems, the life sciences, and medical device technology companies. As with all complex, cross-system challenges, the solution requires education in the latest thinking across fields, including engineering, medicine, technology, and leadership. The chapters provide advice on tackling root causes to tech hype and medical misinformation, cybersecurity, common technical knowledge gaps, effective leadership frameworks, addressing technical security staff burnout, and the importance of allyship for slashing execution risk in clinical technology development and deployment. When recent research shows that physicians believe that technical skills will be more critical than clinical skills within the decade (Herzhoff 2022), this book provides important direction and insights with practical solutions for executives seeking to develop, deploy, and drive clinical, financial, and other benefits from medical work augmented with devices, data, and other advanced technologies.

An additional objective of *Advanced Health Technology* is to accelerate the cross-stakeholder collaboration required as healthcare and technology interact. Executives can better lead and initiate efforts that *build effective strategies with stakeholders at the same time as they tactically focus on specific and clearly identified actions.* This book provides clarity by framing knowledge and solutions in order for leaders and investors to prioritize, roadmap, and accurately resource healthcare IT innovation from technical, cybersecurity, legal, clinical, financial, and management leadership capabilities perspectives. Considering it is common for ecosystem parties to evade managing privacy and security, the book aims to provide direction for readers to provide and participate in informed leadership, ultimately leading to more safe, successful, and ultimately faster development and deployment of advanced technologies in medicine. Curated real-world expertise and research-led expertise combine in *Advanced Health Technology* to significantly prepare and accelerate leaders on this mission.

Note: To you, active and future board of director and executive candidates. Understanding these risks depicted in this table represents an incredible, massive unfair advantage for you to convey in your current or future candidacy. Competent risk understanding and management will continue to soar in demand for rare competence.

In the face of ongoing cybersecurity threats and infection threat from a pandemic, more and more stakeholders are centering on the language of risk, in particular boards.

Provides Solutions to More than ¾ New and Emerging Risks

Clinical Risk	Financial and Operational Risks	Technology Risks	Emerging Risks
Patient safety	Workforce planning	Cybersecurity and ransomware preparedness	Telemedicine
**Physician technology–related burnout	Revenue cycle management	Data governance	Social, governance and diversity, equity and inclusion
Acute care at home	Vendor management	Biomedical devices	New regulations
	Joint venture management		Disinformation

Source: *Tackling the Majority of Emergent Risks in Healthcare via Advanced Technology.* Douville, Medigram, Inc. (2022).

Our book does not dive into the following risks, deferring instead to several existing healthcare law texts that do this well.

Existing/Legacy Legal and Regulatory Compliance Risks
Pharmaceuticals
Physician Financial Transactions
Emergency Medical Treatment and Labor Act

Source: *Classic Risks That Should Be Well Understood by Legal.* Douville, Medigram, Inc. (2022).

You will notice that this book has only a small section on evidence-based consumerism. Other than that, we encourage the deepest possible separation between medical technology and consumer technology. There are several reasons for this, the most important one being the barriers that will remain intractable gulfs for the foreseeable future. These include but are not limited to:

1. Privacy and cybersecurity protections which are questionably existent and, some argue, nonexistent
2. Competency gaps that are needed for training and onboarding patients and their care providers to software solutions
3. Medical Evidence requirements that the medical community demands for adoption of any care intervention.

We show in the following two tables successively:

1. What to expect from the awareness and evolution of national privacy policy
2. How to differentiate the capabilities needed in medical grade technology with the gap that compares to consumer grade technology even in healthcare.

Many Chief Information Security Officers have been calling to delete health apps. But have you heard there's recently been bipartisan agreement on a national privacy bill? (Kern 2022). Not so fast. Any law has a long road from being passed to being implemented and effective and we present this table to help you understand why we're years, potentially decades, from meaningful enforcement for digital health privacy. That is why we believe that direct-to-consumer products will present the most risky and least understood risk profiles in the digital health ecosystem and encourage the reader to focus on advanced technology in the context of clinical operations. This book also does not focus as much on healthcare administrative applications, many of them which can be performed in a web app context. The latter not being advanced technology would also not be a focus of this book.

Tremendous Noise, Competing Interests, Some Progress, with Misalignment – Actual Change

Policy Intention	Laissez-faire/Do Nothing	State Laws	Pass National Privacy Law	Create Regulation	Enforce Regulation Lightly or No Enforcement = Symbolic	Meaningfully Enforce = Structural Change
Policy status	National privacy	Patchwork	Possible	TBD	TBD	TBD
Policy purpose	Intention	Patchwork intentions	New intention	Meaning	Signal	Impact
The bottom line: actual results	No	**Variable Mostly No**	No	No	No	**Yes**

Source: "The Messy Road to Regulatory Clarity," Douville, Medigram Inc. (2022).

The following table is about what we'd expect successful solutions in a healthcare IT context to address. Looking at several clinical AI programs, we noticed a number of them that would score close to zero on all the following elements. We hope that health services researchers and implementation science experts will evolve this potential scoring model. This table helps to visualize what's needed and expected in healthcare IT and what we typically see with consumer grade versus medical grade products and solutions.

Capabilities: Technology Lifecycle Management Specific to Healthcare IT Market	Consumer Health App, B2C, Consumer Data Is the Product	Enterprise Health App, B2B, Model Does Not Monetize Data	Mobile Medicine/ Advanced Technology, B2B, Requirements for MD Adoption
Risk of Criminalization Related to Health Data	Highest	Lower	Lowest
Advanced technical performance	N/A or low	Lower	Highest
Proven clinical outcomes	N/A or low	Lower	Highest
User education program	N/A or low	Lower	Highest
Operator education program	N/A or low	Lower	Highest
User support	N/A or low	Lower	Highest
Technical support	N/A or low	Lower	Highest
Implementation program	N/A or low	Lower	Highest
Privacy and compliance	N/A or low	Lower	Highest
Cybersecurity controls and auditing	N/A or low	Lower	Highest
Testing/QA	N/A or low	Lower	Highest
Total score (0–100)			

Source: Distinguishing between Consumer Grade App Requirements and Medical Grade Apps. Douville & Ng Medigram, Inc. (2022).

Scoring Table

Score	Result	Call to Action
90–100	Clearly defined and standardized	Operational
70–90	Credibly emerging	Leadership
0–70	Nascent	Research

Source: Scoring Table. Douville, Medigram, Inc. (2022).

Intended Audience

This book was written for our friends and colleagues: visionary, gifted, connected, and passionate leaders across healthcare and medical technology. Common roles would be the following in no particular order, and not limited to:

- Chief Information Officer
- Chief Digital Officer
- Boards of Directors
- Chief Medical Informatics Officer
- VP Engineering
- Chief Information Security Officer
- Chief Medical Officer
- Senior Technical Fellow
- Head of Regulatory
- Head of Quality
- Head of Human Capital
- Chief Strategy Officer
- Chief Technology Officer
- General Counsel
- Chief Privacy Officer/Head of Regulatory
- Chief Financial Officer
- Chief Operations Officer
- Chief Executive Officer
- Medical School Dean
- Dean of Engineering
- Dean of Business
- Education Accreditation Bodies
- Policymakers
- IT executives and business relationship managers
- Medical Device Company executives
- IT and Leadership Researchers
- Strategic Clinical Service Line Owners
- Biotech executives
- Organizational psychologists

How Is This Book Organized?

This book is comprised of 16 chapters under the following four parts.

Part I: Tackling Barriers to Rapid, Exponential Acceleration of Advanced Technologies in Medicine

In Chapters 1–3, we cover data transformation, how to close common technical knowledge gaps for devices, data, and applications, and how to lead with the goal of slashing medical and technical misinformation.

Part II: Management and Leadership Competencies and Objectives for Driving the Science, Medicine, and Engineering of Advanced Medical Technology Forward

In Chapters 4–8, we address the importance of allyship in effective teamwork, the necessity and managing against the pitfalls of transformational leadership, the dangers of hubris in medical technology, how humility is necessary for development and adoption of medical technology, and what you need to know about the root causes of burnout facing cybersecurity professionals.

Part III: How Cybersecurity Enables Deployment of Advanced Technologies
 In Chapters 9–11, the contributors reveal how to solve for gaps in today's security frameworks in a medical technology context, how to manage third-party risk, and what to know and prioritize about Hospital at Home.

Part IV: The Practical Technical, Legal, Management, and Leadership Steps to a More Interactive Health System
 In Chapters 12–16, the contributors reveal lessons they've learned over decades and thousands of experiences in mitigating innovation risk, what to know about the latest rules for healthcare interoperability, what our next steps should be toward engaging patients effectively with technology, how to comply with newly enforceable information blocking rules, and what you need to understand about the continued rise of technical standards in medical technology leadership.

What's Next

We imagine you finding flow and joy in your work as you confidently engage your team in designing your strategy. It will be so clear that it allows you to easily and quickly explain to all your stakeholders how your advanced technology strategy enables and empowers your organization to competently address the majority of what would otherwise be emerging risks. This is how you can win through innovation.

Sherri Douville

Acknowledgments

I'm incredibly honored to be able to partner again with the team that brought you the longer-term best-selling book on Amazon, *Mobile Medicine: Overcoming People, Culture, and Governance*, which debuted as a #1 new release in both medical technology and medical informatics categories. I'm deeply grateful to the team for sharing their knowledge and expertise into such a rewarding body of work that will continue to have an incredible impact on the trajectory of the field of Mobile Medicine. In this new book, our team is able to go deeper into key areas critical to teamwork for advanced technologies and we also aimed to lead by example in the way we built this book. Several team members stepped up additional efforts to enable this depth to come to life. First and foremost, my muse in our work to eradicate this leading cause of preventable death, a delay in information, my better half, Dr. Art Douville, we have to thank for his steadfast support, constant insight and feedback, and literal work on the book as well as the many forfeited nights and weekends it takes to tie together the many moving parts that make up a highly regulated project which is a book such as this. In terms of teamwork, we're grateful to the partnership and leadership of several team members, in particular: William C. Harding, who acts as spiritual CTO figure of the overall book effort. As a passionate expert across bioengineering, medical devices, networking, data, applications, and interoperability, William brought technical leadership and depth forward into many of the chapters and across many of them as connective tissue. Our de facto Chief Information Security Officer voice of the book, Mitch Parker brings his vision, experience, and passion for a strengthened and more effective healthcare cybersecurity ecosystem to several chapters of the book. Further, our legal experts Lucia C. Savage and Peter McLaughlin have been critical thought partners as we "bring compliance to everything we do" in order to best manage risk through this book. We also had much help with project and program management by the masterful hands of Brittany Partridge and Mike Ng. Another instrumental teammate in assisting us in providing thematic refinement of our cybersecurity sections as well as providing overall product marketing expertise is book contributor, Allison J. Taylor. Further, we want to recognize our chapter contributing leaders beyond the aforementioned who also include the incredibly talented physician leaders Dr. Brian McBeth and Dr. Joshua Tamayo-Sarver, and our amazing talent, culture, and leadership coach at Medigram, Karen Jaw-Madson. Last but certainly not least, we thank all contributors listed and whose bios we encourage you to read next. Learn with and from these extraordinary contributors whom you will meet by way of their bios and whose insight you should seek out when they do interviews, write articles, through their next books, and for you to consider hiring their companies to partner with. There are many to thank for the support that enabled this work; the management and leadership talent and infrastructure provided by the Medigram team and supported by our board Wim Roelandts, and Tom Bondi who coach me biweekly and weekly, respectively, in their areas of expertise in business, technology, ecosystem, general management, legal, and finance skills. Our board at Medigram recognizes these books as an ecosystem development pillar indicative of a true learning organization, the only kind of organization that can be truly successful in the future of advanced healthcare technologies. We also have much gratitude for the editorial reviewers who are spotlighted with their contributions of their observations on the value of this work in the front matter of this book. Critically, I learned much of what I know about academic publishing by way of learning from my coauthors on the chapter we worked on together in a prior published book – "How Can We Trust in IoT? The Role of Engineers in Ensuring Trust in the Clinical IoT Ecosystem" – lead author and eminent trust in medical data scholar, Jodyn Platt, PhD, MPH, and our chapter coauthor, passionate public health bioethicist, Ann Mongoven, PhD, MPH, in the book *Women Securing the Future with TIPPSS Trust, Identity, Privacy, Protection, Safety, and Security for Connected Healthcare* (Springer 2022), led and edited by Florence Hudson.

Many physicians, technology, and other healthcare leaders have provided inspiration and insights beyond those we had recognized in *Mobile Medicine*; they further include Dr. Ryan Collins, Dr. Lisa Shieh, Dr. Vanila Singh, Dr. Matt Gillette, Dr. J. Kersten Kraft, Dr. David Noller, Srini Madala, Subhash Tantry, Dipty Desai, Dr. Malathi Srinivasan, Bill Russell, Scott Becker, Tom Andriola, Nancy Hall, Adele Allison, Jody Tropeano, Julie Bernicker, Rachel Fredman, Will Conaway, Ramana Annamraju, and Anurag Mairal.

Ed and Simarn Marx are the uber mentors. There have been many friends and supporters, as well as those who provided intellectual, investment, and emotional support to our work beyond those mentioned in Mobile Medicine such as Frank Marshall, Dan Warmenhoven, Rich Parenteau, Beth Perrell, Maria Roelandts, Michele Kirsch, Kristi Markkula Bowers, Stefanie Lingle Beasley, David Finn, Carolyn McCusker, Dennis Lanham, Thane Kreiner, Sarah Granger, Susan Solinsky, Jerry Antimano, Sharyn Bires, Michael McCarthy, Theresa Palmer, Lisa Magneson, and Joy Ajlouny.

We especially want to thank the reviewers of *Mobile Medicine* who are all listed in *Mobile Medicine: Overcoming People, Culture, and Governance*. We can't thank the publisher's editor, Kristine Mednansky, enough for all of her patient coaching again.

Sherri Douville

About the Editor

Photo credit to Hillary Ungson.

Sherri Douville

Sherri Douville is CEO and board member at Medigram, the Mobile Medicine company. She is recognized on top U.S. CEO lists in eight categories of technology and healthcare by CEO, Boardroom Media as one of the highest-ranking tech executives on Crunchbase globally. She is a best-selling, repeat editor, lead author, and contributor to this 3rd forthcoming book, *Advanced Health Technology* (Taylor & Francis 2023). Sherri is the co-chair of the IEEE/UL JV for the technical trust standard SG project for Clinical IoT in medicine, P2933. She is passionate about redefining technology, software and data for medicine and advanced health technologies in a way that's worth the trust of clinicians, our family, and friends. Ms. Douville leverages her books to inform her work on the CHIME CDH security specialization certification. She advises and co-founded the Cybersecurity curriculum for the Black Corporate Board Readiness and Women's Corporate Board Readiness programs at Santa Clara University. She serves as series editor for Trustworthy Technology & Innovation and Trustworthy Technology & Innovation in Healthcare (Taylor & Francis).

About the Editor

About the Contributors

Felix Ankel

Felix Ankel is Medical Director for health professions education at the HealthPartners Institute. He is an attending physician at Regions Hospital in Saint Paul, MN. He is Professor of Emergency Medicine at the University of Minnesota. He is a former residency director, Accreditation Council of Graduate Medical Education (ACGME) designated institutional official (DIO), and Council of Emergency Medicine Residency Directors (CORD) board member. He currently serves as a Director for the American Board of Emergency Medicine. He is a contributor to icenetblog.royal-college.ca on the future of health professions education.

Shantanu Chakrabartty

Shantanu Chakrabartty is Clifford Murphy Professor and Vice-Dean for Research and Graduate Education in the McKelvey School of Engineering at Washington University in St. Louis. His research covers different aspects of analog computing, and in particular, self-powered sensing, quantum sensing, and neuromorphic computing systems. Chakrabartty is a fellow of the American Institute of Medical and Biological Engineering (AIMBE), a recipient of National Science Foundation's CAREER award, the University Teacher-Scholar Award from Michigan State University (MSU), and the 2012 Technology of the Year Award from MSU Technologies. Chakrabartty has published over 190 journal and conference publications along with ten issued U.S. and international patents. He has served on the editorial board for the *IEEE Transactions on Biomedical Circuits and Systems* and the *Frontiers of Neuromorphic Engineering* journals. Chakrabartty holds a PhD from the Johns Hopkins University and a Bachelor of Technology from the Indian Institute of Technology Delhi, India. He previously was on the faculty at MSU, and he is a co-founder of several startups in the area of self-powered monitoring technologies.

Michael DeKort

Michael DeKort is a systems engineer, engineering and program manager. He has worked in areas spanning the Department of Defense (DoD), Aerospace, IT, Autonomous Systems Air and Ground, focusing on FAA Level D Simulation, Smart Cities, Unmanned Aerial Urban Air Mobility, UAM, and Cybersecurity. He is a recipient of IEEE Barus Ethics Award. He formerly worked for Lockheed Martin on the Aegis Weapon System in the Department of Homeland Security. He serves as a Society of Automotive Engineers, SAE subject matter expert for Ground and Air Autonomy Testing and Simulation. He participates in multiple air and ground transportation industry initiatives and has written several articles. Michael is passionate about safety in engineering.

Arthur W. Douville

Arthur W. Douville is Chief Medical Officer at Medigram and an Attending Neurologist. Douville has held numerous leadership and administrative positions in healthcare, including Chief of Staff and Chief Medical Officer in two separate health systems and Regional Vice President and Chief Medical Officer at Verity Health System in Northern California. In these roles, he oversaw infection control and biohazard assessment in hospital environments, as well as physician relations, including clinical integration, patient safety and quality, regulatory compliance, and the development of innovative clinical programs, and physician technology deployment and adoption plans. He was part of the leadership team charged with bundled payment and Hospital Consumer Assessment of Healthcare Providers and Systems (HCAHPS) initiatives. As Associate Medical Director of the Crimson Analytic program for the Advisory Board (Washington, DC), his role was helping physicians understand and leverage the data by which they are being measured. Douville has over two decades of experience in executive physician leadership. He has published work in managing change in physician communication, culture, and the adoption of medical technologies, including as a contributor to *Mobile Medicine: Overcoming People, Culture, and Governance* (Taylor & Francis 2021). He is in the active practice of neurology in Los Gatos, CA, and acts as Stroke Medical Director in the Santa Clara County Health System based in San Jose, CA.

Ken Fuchs

Ken Fuchs is Sr. Standards Consultant at Draeger Medical Systems, Inc., a manufacturer of electronic medical devices, including patient monitoring, ventilation, anesthesia, and warming therapy systems. He is responsible for coordination of Drager Medical Systems' participation in the United States and international standards development activities.

He currently serves as the Chair of the IEEE 11073 Standards Committee (SC) for healthcare device interoperability and Secretary of the IEEE/UL P2933 Standards Committee. He co-chairs the AAMI MP working group for multi-parameter patient monitors. He is also an IHE DEV co-chair. He also participates in a number of other standards development efforts in AAMI, ISO, IHE, and HL7.

He was recently awarded the 2020 HIMSS-ACCE Excellence in Clinical Engineering and Information Technology Synergies Award and is a recipient of the ACCE Professional Achievement in Technology Award. He holds a number of patents related to patient monitoring systems technology.

Ken's career has focused on networking, connectivity, and system architectures at various point of care medical device companies, including Draeger Medical Systems, Siemens Medical Solutions, Mindray Medical, and the non-profit Center for Medical Interoperability. He holds a MEng. in Bio-Medical Engineering from Rensselaer Polytechnic Institute in Troy, NY, as well as an MBA from Babson College, Wellesley, MA.

His professional affiliations include Institute of Electrical and Electronic Engineers (IEEE), Association for the Advancement of Medical Instrumentation (AAMI), Integrating the Healthcare Enterprise (IHE), American College of Clinical Engineering (ACCE), and Devices (DEV).

William C. Harding

William C. Harding is Distinguished Technical Fellow with 41 years of industry experience, including 24+ years at Medtronic. He has a bachelor's degree in Computer Science emphasizing Electrical Engineering, a master's degree in Information Systems, and a PhD in Technology Integration. William has had a very successful career in missile launch/tracking systems, drug interdiction systems, rechargeable cell manufacturing, and medical device manufacturing. Recognized as a leader and go-to person in extended reality (XR), vision trace/recognition technology, and process automation, William has initiated and championed innovative manufacturing solutions and patented medical product designs that continue to have significant impacts across business units in the areas of process improvement, manufacturing automation, product development/ traceability, and FDA validation. With more than 80 technical conferences, symposium presentations, lectures, seminars, and workshops under his belt, William continues to establish the highest level of standards through his professionalism, ethical behavior, mentoring, and guidance both internal and external to Medtronic. Lastly, as Emeritus Chair of the Technical Fellows, Emeritus Chair of the Tempe Technical Guild, and Chair of the Tempe ABLED employee resource group, William is dedicated to improving Medtronic and the world's vision of diversity and innovation.

Florence D. Hudson
Florence D. Hudson is Founder and CEO of FDHint, LLC, a global advanced technology and diversity and inclusion consulting firm, and Executive Director of the NSF Northeast Big Data Innovation Hub at Columbia University. She is Principal Investigator for the COVID Information Commons (https://covidinfocommons.net) funded by NSF, providing an open resource to enable global researcher collaboration to address the COVID-19 pandemic, and Founder of the National Student Data Corps (https://nebigdatahub.org/nsdc), which was created to teach data science fundamentals to students across the United States and around the world, with a special focus on underserved institutions and students. A former IBM Vice President and Chief Technology Officer, Internet2 Senior Vice President and Chief Innovation Officer, Special Advisor for the NSF Cybersecurity Center of Excellence, and aerospace engineer at the NASA Jet Propulsion Laboratory and Grumman Aerospace Corporation, she Chairs the global IEEE/UL Working Group on Clinical Internet of Things (IoT) Data and Device Interoperability with TIPPSS – Trust, Identity, Privacy, Protection, Safety and Security, and has published books on TIPPSS. She is a published Springer editor and author for their Women in Engineering and Science series, including *Women Securing the Future with TIPPSS for Connected Healthcare: Trust, Identity, Privacy, Protection, Safety, Security* (ISBN: 978-3-030-93591-7). She is an experienced board director, including for IEC Electronics (NASDAQ: IEC) as an Independent Board Director and Compensation Committee member. She has served on non-profit and Academic Boards, including for the Society of Women Engineers, Princeton University, Cal Poly San Luis Obispo, Stony Brook University, Blockchain in Healthcare Today, Neuroscience Outreach Network, Union County College Cyber Service, and the IEEE Engineering in Medicine and Biology Society. She earned her Mechanical and Aerospace Engineering degree from Princeton University, and executive education diplomas from Harvard Business School and Columbia University.

Craig Hyps
As a distinguished engineer and solutions architect with Ordr, an IoT and connected device cybersecurity company focused on healthcare and mission critical industries, Craig drives next-generation solutions that allow organizations to keep pace with the hyper-connected Internet of Things through ML/AI on big data platforms for securing devices. Prior to joining Ordr in 2018, Craig brings a 20+ year track record recognized with unique deep technical acumen, domain experience with Cisco Systems defining policy and access control solutions, including Cisco Identity Services Engine (ISE) and Software-Defined Access (SDA). Craig was a leading force behind Cisco Medical NAC and the advancement of network access control (NAC) for IoT. Craig is an active member of the Institute of Electrical and Electronics Engineers IEEE/Underwriter Labs UL Joint Venture P2933 Working Group. This standard is focused on delivering actionable best practices for

Clinical IoT with TIPPSS (Trust, Identity, Privacy, Protection, Safety, Security). Craig was also a collaborator and National Institute of Standards and Technology (NIST) speaker for the Internet Engineering Task Force (IETF) and the Manufacturing Usage Description (MUD) specification which provides a framework for securely implementing Operational Technology devices. Craig is the author of numerous articles and frequent presenter on the topics of establishing trust, network access control design, medical device security, and Zero Trust cybersecurity. Craig holds an AB degree from Dartmouth College plus leading cybersecurity certifications that include CISSP, CCSP, and CCSI. Craig's professional goals are to have a material ongoing impact leveraging his influence in the industry for advancing protection, privacy, security, and the resilience of critical services. His passion is to drive innovation through the development of technical standards and solution design to deliver effectively against cybersecurity challenges. He enjoys providing critical thinking and best practice guidance to design solutions that uniquely solve today's and tomorrow's challenges.

Karen Jaw-Madson

Karen Jaw-Madson is Principal of Co.-Design of Work Experience. She is the author of *Culture Your Culture: Innovating Experiences @Work* (Emerald Group Publishing, 2018), founder of Future of Work platform A New HR, executive coach, and instructor at Stanford University's Continuing Studies Program. She enables decision-makers to address organizational challenges that affect business performance through coaching and developing leadership; enabling organizations to leverage culture, diversity, and employee experience; optimizing talent by aligning people with strategy; and driving change management and transformation.

A former corporate executive, Karen is known as a versatile leader across multiple industries with experience in developing, leading, and implementing numerous organizational initiatives around the globe. She has been featured in Inc., Fast Company, Fortune, Thrive Global, and Protocol, as well as written for publications such as Forbes, Greenbiz, SHRM's HR People+Strategy, TLNT.com, HR.com's *HR Strategy & Planning Excellence* magazine, and *HR Professional* magazine. Other publications where she appears as a contributor include *Mobile Medicine: Overcoming People, Culture, and Governance* (Taylor & Francis 2021), *Punk XL* (Experience Leadership), and *The Secret Sauce for Leading Transformational Change*. Karen has a BA in Ethnic and Cultural Studies from Bryn Mawr College and an MA in Social-Organizational Psychology from Columbia University.

Kate Liebelt

Kate Liebelt is a life sciences and healthcare leader who is passionate about strategy, technology, and innovation. Kate serves as Executive Director and Chief of Staff to the Head of Research & Development at Organon & Co., a pharmaceutical company dedicated to women's health. Prior to Organon, Kate held a variety of roles in both consulting and industry. Kate was a management consultant and Chief of Staff at Deloitte Consulting, LLP, and PricewaterhouseCoopers, serving a variety of clients across the healthcare ecosystem transforming supply chains and other enterprise workflows. Kate transitioned from consulting into a Chief of Staff role at AbbVie in the global pharmacovigilance division and previously held R&D project leadership roles at Takeda Global R&D, Inc., Baxter Healthcare Corp., TAP Pharmaceutical Products, Inc., and the Washington University in Saint Louis Office of Technology Management/Technology Transfer. Kate earned a BA from Washington University in Saint Louis and a Certificate in Health Innovation from the University of Pennsylvania. Kate is based in Chicago. She is a contributor to *Mobile Medicine: Overcoming People, Culture, and Governance* (Taylor & Francis 2021) and is a champion for diversity, equity, and inclusion and volunteers with a variety of civic and industry organizations that align with her personal mission to create a more equitable world for women and minorities.

Brian D. McBeth

Brian D. McBeth, an attending emergency physician, has worked in clinical medicine and hospital administration for more than 20 years. His degrees are from Stanford University (BA) and the University of Michigan (MD), where he also completed a residency and served as chief resident in emergency medicine. He has had past academic appointments as Assistant Professor of Emergency Medicine at the University of Minnesota and the University of California, San Francisco. He has served as Medical Director and Chair for the Department of Emergency Medicine at O'Connor Hospital (San Jose, CA), as well as the Chair of O'Connor Hospital's Quality Improvement Committee. He recently completed an educational fellowship, teaching medical students and residents in the Department of Emergency Medicine at Tel Aviv Sourasky Medical Center (Ichilov Hospital) in Israel, and has taught and lectured around the world on quality in medicine and patient safety. He completed the American Association of Physician Leadership's program as a Certified Physician Executive (CPE) and is currently serving as the Chief Quality Officer (Interim), County of Santa Clara Health System and most recently as Physician Executive at O'Connor Hospital in the Santa Clara County Health and Hospital System, where he oversees patient safety, hospital quality programs, and physician performance and professionalism. He has published in the areas of physician culture and administrative communication, physician impairment and wellness, infection prevention, and medical ethics.

Peter McLaughlin

Peter McLaughlin is a partner in the Boston office of Armstrong Teasdale LLP and a member of the firm's Data Innovation, Privacy & Security team. Peter guides clients through complex technology transactions and data privacy issues. Since 2005, he has developed a particular focus in healthcare technology and the international movement of data.

When not doing client work, he can be found procrastinating on book chapters he has committed to writing. He contributed two chapters to *Mobile Medicine: Overcoming People, Culture, and Governance* (2021), two chapters to *Advanced Health Technology: Managing Risk While Tackling Barriers to Rapid Acceleration* (2022), and he is editor of an upcoming ABA Business Law book on healthcare technology. He is also currently serving as an adjunct professor on privacy and data regulation at Northeastern University School of Law.

Peter graduated from Georgetown University Law Center in 1993. After graduation he worked in Europe for five years and then served as assistant general counsel to companies, including Sun Microsystems and Cardinal Health. At Cardinal, he was the firm's first privacy director, responsible for global privacy strategy and operations.

Mike Ng

As Medigram's Head of Operations, Mike utilizes his strong organizational expertise and mindset, so the team is aligned and succeeding. Extreme teamwork is critical in this environment and requires someone with Mike's background and temperament to drive organizational connectivity and engagement. Mike prioritizes creating the kind of clarity needed from a broadly diverse stakeholder setting in a polarizing time to drive the best and most efficient results through true teamwork.

He will not settle until useful products are delivered. Mike drove the manuscript production for the first *Mobile Medicine* book working with luminaries across medicine, law, cybersecurity, and technology. Now as a coauthor to this second book publication, Mike is contributing his voice and impact to making advanced technology in healthcare real. He supervised this book's manuscript production which was produced twice as fast as the prior project.

In addition, Mike has had an instrumental role, as thought partner, to Medigram's whole executive team, including the CEO and CTO/CSO, both being distinguished by merit-based recognition on top US CEO and CTO lists and both ranked highly on Crunchbase.

Mike previously developed his project management and operational skills across multiple industries. He's led multimillion dollar mission-critical construction projects for the US military. Prior, Mike also held key leadership roles in a family business. Mike holds a BS in Construction Management from California Polytechnic State University, San Luis Obispo, CA.

Mike is passionate about partnering with transformational leaders and driving critical, yet misunderstood aspects for delivering organizational outcomes.

Mitch Parker

Mitch Parker is Chief Information Security Officer/Executive Director, Indiana University Health.

He has 11 years of experience in this role, having established effective organization-wide programs at multiple organizations. He is responsible for providing policy and governance oversight and research, third-party vendor guidance, proactive vulnerability research and threat modeling services, payment card and financial systems security, and security research to IU Health and IU School of Medicine. In this role, Mitch collaborates across the organization and with multiple third parties to improve the people, processes, and technologies used to facilitate security and privacy for the benefit of IU Health's patients and team members.

He also publishes in multiple publications, including CSO Magazine, Healthcare IT News, HealthsystemCIO.com, Security Current, Healthcare Scene, and HIMSS's blog. He has also contributed a chapter for *Cybersecurity in Healthcare*, an essay to *Voices of Innovation* (which was published in March 2019 by HIMSS), has a chapter in the upcoming book, *Healthcare Cybersecurity*, for the American Bar Association's Health Law section, and is a contributor to *Mobile Medicine: Overcoming People, Culture, and Governance* (Taylor & Francis 2021). Mitch has also been quoted in numerous publications, including the *Wall Street Journal*, *ISMG*, *HealthITSecurity*, and *Becker's Hospital Review*.

Mitch is also a co-vice chair of the IEEE P2933 working group, Trust, Identity, Privacy, Protection, Safety, and Security of the Internet of Things (IoT), and a co-subgroup chair of the P2418.6, Blockchain in Healthcare and Life Sciences Cybersecurity and IoT subgroups. Mitch also participates in other IEEE working groups related to security of the Internet of Things and collaborates with researchers and professionals worldwide on establishing and understanding standards for cybersecurity.

Brittany Partridge

Brittany Partridge is a leader in technology for Clinical Communications and Virtual Care. Currently, she is the UC San Diego Health Virtual Care Technical Lead. In this role, some highlights include designing and developing cross-functional workshops, and rolling out 500+ iPhones, increasingMyChart Video Visits over 1000% and integrated interpreters into Telemedicine offerings to assist with the COVID response. She obtained a BS in Health Care Administration from CSU-Sacramento, while serving in Fire/EMS. After graduation, she worked for the Emergency Medical Services Authority automating their Emergency Resource Reporting, and fell in love with the intersection between clinical and technical. Brittany then moved to Austin Tx where she earned her Health IT certificate from UT and an MBA in Healthcare from UT-Tyler. Brittany worked at Seton (later grouped under the Ascension Umbrella) for seven years. She began as a physician educator for the Electronic Health Record and brought three hospitals live from paper to electronic. From there she was promoted to Clinical Informatics Specialist, where she ran many deployments in the EHR. Some of her favorites included ePrescribing and Merge Hemo. She has also served as Informatics Department portfolio manager, handling intake, resourcing, and governance of EHR optimization requests. She moved to Seton's Good Health Solutions Center to run CI projects for Virtual Care, Innovation, and the Patient Logistics Departments. Prior to moving to San Diego, she transitioned to Ascension Connect, expanding her Virtual Care projects from Texas to all of Ascension's health ministries. Brittany has been a long time member of AMIA and has served on the Public Policy Committee. She also participates in the Women in AMIA working group, and is very excited to be joining the WIA Leadership Program. She has attended the National Library of Medicine Informatics Course, and is involved in HIMSS education initiatives. When she is not working, Brittany enjoys programming, volunteering with Remote Area Medical, and the outdoors, she is happiest on, in, or by the water.

Neil Petroff

Neil Petroff received a doctorate in Mechanical Engineering from the University of Notre Dame in 2006, and then served a postdoctoral position in the hand rehabilitation lab at the Rehabilitation Institute of Chicago (now Shirley Ryan Ability Lab) for one year. He has more than ten years of industrial experience in steelmaking and processing and hip implant development. Neil has been in academia since 2013, first with Purdue Polytechnic South Bend, and currently as Assistant Professor and Department Head in the Department of Engineering Technology at Tarleton State University. He also serves as the Mechanical Engineering Technology program director, and recently achieved tenure and promotion to Associate Professor, effective from September 2022. Neil works closely with the laboratory for wellness and motor behavior in the school of Kinesiology to develop smart devices for its clients. Neil's research interests are in physical medicine and rehabilitation and health and human performance. Neil has published with IEEE – Engineering in Medicine and Biology Society (EMBC), Society for the Advancement of Material and Process Engineering (SAMPE), and was a chapter coauthor on the book *Mobile Medicine: Overcoming People, Culture, and Governance*, published by Taylor and Francis in 2021. He is also an associate editor for the *Neural and Rehabilitation Engineering* theme for EMBC.

David Rotenberg

David Rotenberg is a veteran engineering leader of the Israeli technology industry, passionate about innovation and excellence. He brings more than 20 years of hands-on experience in designing and building complex hardware and software multidisciplinary systems. David specializes in IoT and cybersecurity.

David is currently a senior VP of R&D and Engineering in Telefire, a leading manufacturer of mission-critical, high-compliance, and high-risk systems for smart buildings management and fire safety. In addition, David serves as a board member and advisor for several technology companies in the fields of software development and cybersecurity. Prior to his current work, David spent a decade working with medical device companies and developing cutting-edge technologies to save and improve lives.

David is an essential part of the P2933 IEEE and UL joint standard team, contributing technical thought leadership and practical direction for system of systems design, trust, and identity elements. He also contributed to the UL2900 standard and is coauthor of several technical papers.

David holds a BSc in Electronics Engineering from the Tel Aviv university and an MBA from the Technion, as well as certifications in cybersecurity and for serving on Boards of Directors.

Lucia C. Savage

Lucia C. Savage is a nationally recognized expert on healthcare regulation, digital health, and health information privacy. Using strategic advice to advance digital technology to deliver healthcare, she drives the Omada Health's privacy, regulatory, and public policy strategies. Founded in 2011, Omada Health is one of the oldest virtual-first digital healthcare providers in the United States and has provided well-established cardio-metabolic and physical therapy protocols via a virtual-first approach to over 500,000 individuals.

Ms. Savage is also an advisor to Evidation Health, a digital health research platform that measures health in everyday life and enables anyone to participate in ground-breaking research and health programs; a member of the Board of Directors of Tidepool, a 501(c)(3) that is developing an open-source, fully interoperable closed-loop insulin pump (clearance pending at FDA) that allows people with diabetes to see and understand their own data; and a member of the Board of Directors of Academy Health, where she is chair of its Committee on Advocacy and Public Policy.

Lucia is Rock Health's 2021 "Top 50 in Digital Health" for her policy acumen and digital health advocacy, is a contributing author to Amazon bestseller *Mobile Medicine: Overcoming People, Culture and Governance*, and has testified before the Senate Committee on Health, Education, Labor and Pensions on health information interoperability and digital health.

Prior to joining Omada, she served the Obama Administration as Chief Privacy Officer at the US Department of Health and Human Services Office of the National Coordinator for Health IT. Lucia has a BA with honors from Mills College and received her Juris Doctor summa cum laude from New York University School of Law.

Viktor Sinzig

Viktor Sinzig is an executive and customer success professional with 12 years of experience leading onboarding and customer success initiatives with new technology in Fortune 500 companies. His expertise includes building solution implementation best practices, leading customer and partnership onboarding, solutions design, business process design, and establishing KPIs to monitor customer health and technology adoption. He has worked with a broad range of commercial and government organizations, including ABB, Boehringer Ingelheim, Bosch, Dematic, Medtronic, Porsche, Roche Diagnostics, Software AG, and Siemens – he has spent his career working with technologies and companies that are ready to break away from "Pilot Purgatory."

Currently, Viktor is Director of the Program Management and Customer Success at OSARO, where he builds programs that streamline complex deployment of robotics automation software. He leads a global team of project managers, solutions engineers, and customer success executives.

Outside of OSARO, Viktor regularly speaks and lectures about augmented reality, artificial intelligence, financial technology, and other innovative technology sectors. Prior to OSARO, Viktor was Management Consultant at KPMG and PwC, where he provided strategy and technology transformation consulting services. At PwC, Viktor was responsible for building their FinTech investment practice. His team identified startups to partner with, and then deployed the startup's technology to PwC's clients.

To stay on the cutting edge of artificial intelligence, extended reality, and financial technology, Viktor continues to be an active advisor and technology investor.

Eric Svetcov

Eric Svetcov, Information Security Leader with International Experience and Deep Cloud Computing Knowledge, is the CTO/CSO for Medigram. He has published multiple articles and books, built IT and security from scratch four times, including for Intuitive Surgical. He is recognized on several top CTO lists by Startups.US as one of the top ranked technology executives on Crunchbase.

Eric is the coauthor of the *CCISO Body of Knowledge* and is one of the original CCISO certification trainers. He is an Advisory Council member for the CISO Executive Network, and has led the first global Cloud Computing Company (Salesforce) through ISO 27001 Certification and did it again with MedeAnalytics. He led MedeAnalytics through HITRUST Certification, and has been a Caldicott Guardian, Chief Privacy Officer, and Data Protection Officer. Eric brings deep international experience – Europe, Middle East, North America, APAC, and ANZ – to the problems facing global-scale medical information risks and security requirements. Previously, his work informed federal-level HIPAA audit processes. Eric's work has supported more than 24 billion patient encounters in the United States and the United Kingdom with records of more than 100 million unique patient lives from more than 900 healthcare organizations.

He has taught professional training classes internationally on Data Security and Cloud Computing, including one certified for reimbursement by the Singapore Government. He is a contributor to *Mobile Medicine: Overcoming People, Culture, and Governance* (Taylor & Francis 2021). Eric has been a board member (and former Chair) of the American Board of Cybersecurity and Information Assurance (ABCIA), as well as a former board member of ISACA (Auckland Chapter).

In his off hours, Eric enjoys sports with his kids and spends time on his passion for bringing up standards for cybersecurity nationally.

Joshua Tamayo-Sarver

Joshua Tamayo-Sarver, MD, PhD, FACEP, is Vice President of Innovation at Vituity and Inflect Health, overseeing the innovation efforts that leverage technology through strategic partnership, investment, or internal incubation. In addition to being the VP of Innovation, he works as a staff physician in the Emergency Department at Good Samaritan Hospital in San Jose, CA, where he is the past Department Vice-Chair and Director of Quality Improvement. He holds a bachelor's degree with honors in Biochemistry from Harvard University, a certificate in negotiation from the School of Law and Harvard University, a medical degree from Case Western Reserve University, and a 10 × 10 certificate in medical informatics from Oregon Health Sciences University. Dr. Tamayo-Sarver completed his PhD at Case Western Reserve University in Epidemiology and Biostatistics. He completed his residency in emergency medicine at Harbor-University of California, Los Angeles. He also spent a year working as an EMT and health educator in El Salvador. He is board certified in Emergency Medicine and Clinical Informatics.

Dr. Tamayo-Sarver's vision is to help transform healthcare so that every encounter between the patient and the healthcare system is rewarding, productive, and positive for the patient, the provider, and the system. He has developed a deep understanding of the problem spaces in healthcare through (1) extensive clinical practice as an emergency physician; (2) as a physician executive at a national physician staffing company by managing a wide range of providers; (3) as a technology executive leader providing software, technology, and analytical solutions to over 100 hospitals; and (4) as an executive leading the product development, testing, deployment, go-to-market strategy, and scaling of multiple technology solutions in the healthcare ecosystem. His goal is to work within a dynamic team environment to leverage this understanding to produce solutions that will allow patients to get direct access to the appropriate care, allow providers to give the quality care they want to deliver, and enable an efficient system with minimal waste.

Allison J. Taylor

Allison J. Taylor is a Silicon Valley technology go-to-market strategist and entrepreneur who has brought over 20 software solutions and services to market across 35 countries, representing over $3 billion in revenue. She founded her first company in Israel at age 24 and is a former Middle East journalist and New York City medical trade press editor. An award-winning cybersecurity veteran, Allison consults senior leadership and their diverse cross-gen teams on growth and transformation strategies as Founder and CEO of consulting firm Thought Marketing LLC. Clients have included Cisco Systems, EMC/Dell, Honeywell, and Juniper Networks in addition to many early and mid-stage tech startups worldwide. Allison's executive operating experience includes cybersecurity product leadership positions at McAfee and Nokia, strategist roles at Sun Microsystems (now Oracle), and corporate communications leadership at Check Point Software Ltd. She contributed to the cybersecurity chapters for the Taylor & Francis book, *Mobile Medicine: Overcoming Culture, People, and Governance Challenges.* Allison speaks multiple languages and holds an MS in Communications from San Jose State University in California and a BA in Spanish and Journalism from the University of Richmond. She regularly hosts her LIFT podcast for thought-provoking conversations with industry leaders.

Introduction

Technical innovation is urgently needed in healthcare and medicine, a clinical environment facing enormous workforce shortages and challenges. However, the path to success is not clear or simple by nature. The goal of our new book on risk management for advanced technology in medicine, published by top academic book publisher Taylor & Francis, is to provide expert guidance for overcoming approximately 75% of the top emerging risks facing healthcare and medicine. A diverse set of respected authors demystify and bring clarity to the complex relative to risk management by bridging engineering, cybersecurity, privacy, medicine, leadership, and legal perspectives, together with advanced clinical technology development, deployment, and implementation insights.

This book squarely focuses on advanced technology in a medical enterprise context to help with timing, scope, and focus.

There are enormous challenges on the consumer side of healthcare due to lack of practical privacy protections today. One huge part of the challenge is that there are application software–specific (McKeown 2022) as well as hardware and device–specific privacy gaps (McKeown 2022). There are also connectivity-related privacy issues. At the core of the matter, consumers will have to accept that with the confluence of devices, data, and connectivity gaps in privacy, there is little real privacy in practical terms of our information being actually private. This is when reidentification risk is so high at 81%. Further, data brokers sell linkages to physical devices and names and addresses. Knowing they don't really have privacy, perhaps consumers will choose to use few and better apps that come with greater levels of privacy and security. This might be only for instances and locations where they don't mind most of the information being ultimately available in the public. Though this is a real safety and well-being set of issues for patients and clinicians.

Advanced technologies for medicine require you, our reader, to accept and prepare for multiple layers of complexity and risk to master illustrated by data, networking, and hardware, which were all mentioned in the above example. Advanced technology for healthcare is a lot like aiming to personally manage the risk of the pandemic competently for one's team, family, and organization. If you really care about the well-being and productivity of your teams, you are certainly likely to be looking at pandemic risk management as a multiple layer risk management enterprise. Advanced healthcare technology demands the same level of depth and complexity mastery; that's part of what makes it fun for the right teams and people (Figure 0.1).

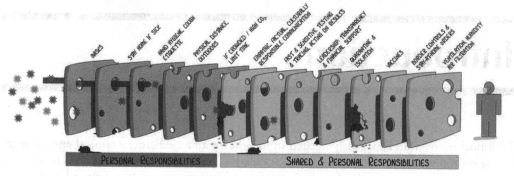

Swiss Cheese Words. Source: Used with permission from Ian M. Mackay, University of Queensland.

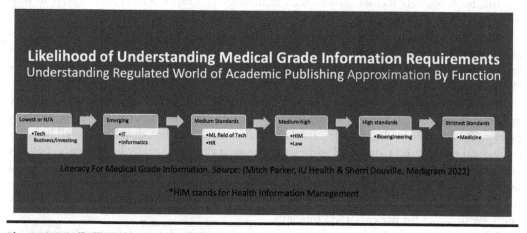

Figure 0.1 Likelihood of understanding medical-grade information requirements. Source: With permission from Ian M. Mackay, University of Queensland.

In early June, a group of 40 Congressional Democrats wrote a letter to one company asking them to stop collecting and retaining location information. Lawmakers urged the company to stop collecting location data for fear it could be used to criminalize those seeking healthcare.

The Senators also sent a letter to the FTC asking it to investigate "unfair and deceptive" privacy practices, alleging the "enabling the collection and sale of hundreds of millions of mobile phone users' personal data."

The Senators concluded that there had been little done to inform consumers of the harms of the privacy of their data and are acting as lawmakers to bring a stop to the practices being called out. The purpose of this section isn't to denigrate these companies as this has worked in a consumer-grade context to fuel rapid scale and profits; the goal is to help you, the reader, to understand what the status quo is for privacy in the consumer technology market today and to assist you to reflect for yourself on how that strongly diverges, even opposes strongly the goals and ethics of medicine. It will take years for this consumer-grade privacy situation to get sorted out as depicted in the "Messy Road to Regulatory Quality" table above. I lived through regulatory inflection in BioPharma; though the function at BioPharma 2000's was different – restrictions were on marketing) as opposed to restrictions on data practices for privacy – the road appears to me to be an exactly parallel one. It will be the story of a public sentiment battle very similar to the path and history of tobacco. Traveler beware as we explained in our blog on Medium (Douville 2022).

There was a time when focusing on patient engagement and consumerism looked like the easier path to digitization in healthcare. Though no one could have predicted the challenges and potential "side effects" as data privacy harms (Tangari 2021). By transcending the noise and challenges of consumer-grade technology – in particular, privacy and focusing on advanced technologies in a clinical context– you can also tackle the majority of emerging healthcare risks at the same time. This suggestion applies whether the technology is a consumer use case or sold as "consumerization of IT" or even as sold as business to business software as a service; though in each case it has been built in a consumer-grade methodology with the likely business goal to traffic in user data. We do not believe that the consumerization of medical technology is beneficial for reasons of privacy, security, and quality; all risks that are exploding in significance in your world.

To conclude this introduction, we leave you with a "recipe" for success in advanced healthcare technology to keep in mind as you read this book. In order for success to manifest, there needs to be capabilities for the following ten items. Presence or absence of any of these competencies helps to indicate whether the technology in question is a consumer-grade technology where the consumer owns most of the responsibilities for privacy, security, and implementation; or whether it is more advanced and something your ecosystem will be involved in helping to manage.

1. Technical performance
2. Clinical evidence
3. User education
4. Operator education
5. User support
6. Technical support
7. Implementation program
8. Privacy and compliance
9. Cybersecurity controls and auditing
10. Testing/QA

We also propose a way to think about levels of risk and "classes" of software in medicine. Counterintuitively, as a healthcare executive, we propose that if you leapfrog the following three listed risk levels (Class 0, Class 1, Class 2) and focus primarily on Class 3, you can expect faster, better results in terms of clinical outcomes, workforce development, and rapid risk reduction. Focusing in this way gives the consumer technology ecosystem the time it needs to mature. In the meantime, you can focus on your need to protect patients and clinicians from harm and liability respectively.

Proposed Classifying Software and Who Is Responsible in the US Example

Risk level	Example	Privacy	Security	Implementation	Sustainability
Class 0	Fitness app	Consumer	Consumer	Consumer	Consumer
Class 1 Low risk	Sensors Pulse OX EKG, EEG software	Vendor	Vendor	Patient/BYOD Bring your own device	Patient/BYOD Bring your own device
Class 2 Moderate risk	Telemedicine MD visit Linq II (patient data recorder)	Vendor	Vendor	Regional Consortium *MUSC exemplary telehealth model	Regional Consortium *MUSC exemplary telehealth model
Class 3 Highest risk	Pillcam software Implantable device software	Vendor	Vendor	Vendor	Vendor

Source: *Proposal of Risk Levels & Responsibilities for Medical Technology*. Douville, Medigram, Inc. & Harding, Medtronic, plc (2022).

Part I: Tackling Barriers to Rapid, Exponential Acceleration of Advanced Technologies in Medicine

This part comprises Chapters 1–3. In these chapters, we characterize the need described as the ability for solutions to securely exchange data between all elements of healthcare environments, facilities, and organizations using standards and practices that promote seamless error-free data transformation. Data transformation is a process that would ideally occur throughout the healthcare environment. The goal would be for data to be received, shared, and processed in a future seamless state. The point of addressing knowledge gaps is that they are the root of all fear, uncertainty, doubt, hype, misinformation, and abuse. Understand as well that advanced technology development is an extreme team sport. The social media era allowed for massive value creation based on the top layer of technology, user experience. However, progress in medical technology and advanced technology demands looking at the whole body of knowledge, multiple layers deep of technology, and how we build in things such as privacy and security. Many business leaders want to generalize and group all technologists together. However, there is evolving complexity and need for precision in developing multi-component systems and products. The fact is that different kinds of technologists have just as much distance between their perspectives and skill sets as different functions do (e.g., engineering, IT, and marketing or clinical medicine). In order to get individuals to work together productively, we need to identify and close knowledge gaps that create misalignment, drive waste, and increase failure. Like the pandemic, the tech industry's introduction into medicine has been fraught with challenges, including misalignment on what's considered technology misinformation. There are common widespread gaps to developing and deploying effective solutions that physicians trust. We identify the most common sources of mistrust and misinformation for technology and medicine in a tech sector specific to medicine context.

Part II: Management and Leadership Competencies and Objectives for Driving the Science, Medicine, and Engineering of Medical Technology Forward

This part comprises Chapters 4–8. Today, boards and investors are measuring and assessing for effectiveness with building a diverse workforce. The problem right now is that segments of the market and the workforce as well as specific industries are unaware of this mandate. Most importantly, talent demands this and cites a lack of diversity as a top cause of turnover making it a talent retention requirement. We'll discuss the signs and evidence to both look for and strive for to embrace the identity of being a great ally. We aspire for our readers to make a real difference in their organizations and communities, and beyond. The contributors combine personal stories, lessons, and evidence. We also explore transformational leadership; although traditionally the focus has been on enthusiasm and inspiration, the practice requires exponentially greater operating, management, and execution, coordination, and communication acumen than traditional management does. In absence of discipline, a transformational leadership style can contribute to burnout through two of the largest causes of burnout stemming from the lack of key management and execution competencies that will be explored by students of executive leadership and its required competencies. In medical technology, behaviors that are the opposite of HUBRIS are necessary for success. HUBRIS in technology holds back the true potential of technology from impacting industries that critically need digital transformation such as medicine. The coauthors review what's perhaps accepted in the technology world, but not in medicine–personal conduct and narcissism, individualism. Technologists may often bring a "cult of personality" or contrarian personality style to a medical technology interaction which guarantees a clash with a physician's tendency to think critically. If the former is compounded by a lack of demonstrating literacy for a medical and clinical context by the technologist in the eyes of the physician, then that's a recipe for potentially irreconcilable conflict between the technologist and the physician, hindering medical technology success. Humility is separately investigated for leaders who want to understand the critical role of humility with regard to technology in a healthcare setting. The authors present an argument for its importance in educational systems – in both medical and technical education – from bedside teaching to integration of technical instruction in clinical settings to mentoring and systems-based education models. Humility is examined in the context of professional communication, considering traditional models and changing leadership structures and hierarchies – using the lenses of patient safety, professionalism, and organizational vision to consider humility's role in shaping a healthcare's organization relationships with patients and its staff. In order to enable the medical profession with advanced technologies, we must drive cybersecurity as the enabler. Cybersecurity capabilities are based on having and retaining the right people. Providing deep insight in the light of the "Great Resignation," a guide in this section is designed for leaders who want to understand more about burnout and its root causes through the lens of Information Security.

Part III: How Cybersecurity Enables Deployment of Advanced Technologies

This section includes Chapters 9–11. We open this part by familiarizing the reader with the importance of understanding organizational and professional subcultures, their Practices and Characteristics (P's and C's), structures and histories, and standards for evidence and quality used to make decisions. It discusses the challenges that IT and technology companies have in understanding these attributes unique to the field of medicine. The goal is to provide communication plans that deliver change. Digital Transformation and Mobile Applications have left all businesses, especially Healthcare Delivery Organizations, reliant upon multiple third parties to deliver solutions by leveraging technology to customers and patients. This requires organizations to build and acquire specialized knowledge, specifically in the security and interoperability requirements of systems, to better deliver solutions. The goal is to provide healthcare executives with the knowledge needed to understand regulations and requirements, and the processes by which leaders can apply them to implement proper risk management. Cybersecurity concerns need to be front and center for new care models such as Health at Home or Hospital at Home. These programs present challenges, however, beyond the logistics of sending a provider to the patient. Chief amongst these hurdles is the security of patient information. Medical devices, patient monitoring tools, and remote access to patient files are significantly more complicated to use outside the familiar healthcare facility because the patient data does not benefit from the layers of security within the traditional U.S. hospital environment. We demystify the path forward.

Part IV: The Practical Technical, Legal, and Management Steps to a More Interactive Health System

This part comprises Chapters 12–16. In this part, we aim to help leaders understand how to manage innovation in a way that enables high-risk ideas and minimizes investment risk. Whether incubating in-house or implementing an innovative commercial solution, having a well-tested framework helps determine the resources you need and maximizes your likelihood of a positive return. Through this work, you will better understand how to use a proposed framework along with an in-depth explanation of the rationale for the framework so that those responsible for leading innovation can modify it to suit the unique aspects of your organization. Regardless of how many bright and energetic people you have in your organization, good or even great ideas will get caught in a quagmire of uncertainty without process, leading to frustrations that can lead to an abandonment of great innovative thinking. The goal is not to rehash all the challenges of innovation or describe the next big thing, but rather to provide a reasonable process that can apply to an innovation program so that great ideas can be nurtured and unsuccessful ideas can be terminated quickly. We also summarize the latest on healthcare data interoperability, namely open-specification Application Programming Interfaces, and the prohibitions against information blocking in the context of the 25-year effort of US policymakers to digitize health information and improve interoperability. There are practical tips about implementing these new standards. Given the context, readers will also receive some insights about how full interoperability can benefit health systems. Also highlighted are peer-reviewed real-world examples of hospitals or health systems using technology to engage and partner with patients in our call for more research on patient engagement. HHS issued regulations prohibiting information blocking practices and mandating standardized ways to enable data sharing among different platforms and systems. These rules have come into effect in 2022 and so now constitute an essential component of certain systems or applications holding electronic protected health information. This chapter walks through the interoperability and data blocking rules and offers practical steps on how to implement these mandates within a health information system a company is developing or considering acquiring. In medical software, you must understand how regulators perceive the field today and how they currently and plan to regulate, for example, software as a medical device. The coauthors of the last chapter have experience uniquely helping to build and lead a working group body as technical and industry leaders, thereby bringing a perspective of practical and applied understanding around the standards process that often underpins regulatory processes. The problem right now is that multiple segments of the tech sector are illiterate in regulatory processes as well as the standards that can and need to precede them. However, in order to lead in medical software, there must exist the capability to convene and drive the evolution of an ecosystem of informed, contributing partners to consensus. When this is done well, progress and the introduction and successful adoption of advanced technologies in medicine can be realized. The coauthors of the final chapter will explore the current successful advancement of a technical standards working group as a model.

Sherri Douville

References

Douville, Sherri. "Confusion, Strife, Challenge: What to Expect with Tech Sector Regulatory Change Related to Privacy Practices." *Medium* (July 5, 2022). https://sherridouville.medium.com/confusion-strife -challenge-what-to-expect-with-tech-sector-regulatory-change-related-to-privacy-1674cfb41864

Gooch, Kelly. "21 Top Management Risks in Healthcare for 2022." *Becker's Hospital Review* (April 18, 2022). https://www.beckershospitalreview.com/hospital-management-administration/21-top-man agement-risks-in-healthcare-for-2022.html?origin=BHRSUN&utm_source=BHRSUN&utm _medium=email&utm_content=newsletter&oly_enc_id=6333D8773701G4Z

Herzhoff, Jeff. "Clinician of the Future: A 2022 Report." *Elsevier* (March 15, 2022). https://www.elsevier .com/connect/clinician-of-the-future

Jee, Charlotte. "You're Very Easy to Track Down, Even When Your Data Has Been Anonymized." *MIT Technology Review* (July 23, 2019). https://www.technologyreview.com/2019/07/23/134090/youre -very-easy-to-track-down-even-when-your-data-has-been-anonymized/

Kern, Rebecca. "Bipartisan Draft Bill Breaks Stalemate on Federal Data Privacy Negotiations." *Politico* (March 6, 2022). https://www.politico.com/news/2022/06/03/bipartisan-draft-bill-breaks-stalemate -on-federal-privacy-bill-negotiations-00037092

McKeon, Jill. "Senators Call on FTC to Investigate Apple, Google's "Deceptive" Data Privacy Practices." *Health IT Security* (June 28, 2022). https://healthitsecurity.com/news/senators-call-on-ftc-to-investi gate-apple-googles-deceptive-data-privacy-practices

McKeon, Jill. "Senators Question Talkspace, BetterHelp on Patient Data Privacy Practices." *Health IT Security* (June 30, 2022). https://healthitsecurity.com/news/senators-question-talkspace-betterhelp -on-patient-data-privacy-practices

Tangari, Gioacchino. "Mobile Health and Privacy: Cross Sectional Study." *BMJ* 373 (2021): n1248.
*https://muschealth.org/medical-services/telehealth

TACKLING BARRIERS TO RAPID, EXPONENTIAL ACCELERATION OF ADVANCED TECHNOLOGIES IN MEDICINE

I

1

Chapter 1

What to Know about Data Transformation for Advanced Technologies in Medicine

William C. Harding, Florence D. Hudson,
Shantanu Chakrabartty, and Sherri Douville

Contents

Introduction

Data transformation means different things to different people and if you aren't aware of what data transformation means in your business, then continue reading as we seek to characterize it in relatable terms. For example, transforming data can be related to applying additions, subtractions, encryption, compression, interpretation, masking, and even the existence of corruption. Data transformation is also a key point to keep in mind as you think along the lines of technology integration, interoperability, and data aggregation. But even more importantly, when you think about data transformation, you must be looking through the lens of data integrity and data security. That said, this chapter will provide you, the reader, with a solid overview of what is data transformation, with an emphasis on considering data transformation within medicine/healthcare/life sciences.

DOI: 10.4324/9781003348603-2

What Is Data Transformation?

According to Augustyn (2017), the transformation of data occurs between a multitude of dissimilar solutions. With that in mind and in consideration of the issue of disconnected healthcare technology, it is proposed that there is a need within healthcare communication solutions (i.e., interactions between humans, hardware, and software) to function and exchange data *interoperably* across dissimilar models/foundations, where, according to Venegas, Fernández, and Maldonado (2016), interoperability represents securely transforming data using standardized interfaces (Cook et al. 2016; Janaswamy and Kent 2016; Kohn, Topaloglu, Kirkendall, Dharod, Wells, and Gurcan 2022; Stegemann and Gersch 2021; Wager, Lee, and Glaser 2021). Specifically, the need that is being described is the ability for solutions to securely exchange data between all elements of healthcare environments, facilities, and organizations using standards and practices that promote seamless error-free data transformation. Whereas, the transformation of data is a process that should occur between all components of a healthcare environment, which involves the interconnection of dissimilar elements (i.e., people and technology) such that data is received, shared, and processed (Theorin et al. 2017).

Data Transformation in Healthcare

According to Cook et al., the issues associated with healthcare solutions not being able to seamlessly transform data between dissimilar technology is rooted in the processes associated with assessing, integrating, and sustaining new technology into healthcare environments. Accordingly, the issue of needing seamless data transformation represents a gap in existing technology adoption models to examine the perceptions of all relevant stakeholders as their technical experiences potentially influence assessment, decision-making, adoption, and sustainability of transformative technology (Cook et al. 2016; Daghfous, Belkhodja, and Ahmad, 2018; Ingebrigtsen et al. 2014; Liebe, Hüsers, and Hübner 2015; Petersen 2018; van Oorschot, Hofman, and Halman 2018).

Specifically, the ability to transform data is presently accomplished using methods such as non-standardized electronic healthcare/medical records (EHR/EMR) and data protocols that connect the vast array of dissimilar healthcare components. Transforming the data from the point of collection to the point where it is used to make health-related decisions requires that human errors are not present and where solution response time could literally mean a patient's life or death (Konnoth 2016). Furthermore, and according to Mandl, Mandel, and Kohane (2015), variability between solutions is an issue where the absence of a standard communications method for transforming data will result in data instability across the multitude of dissimilar technology.

EMRs and EHRs in Transforming Data

In using EHRs/EMRs to solve the issues of data transformation between dissimilar technology, the healthcare industry is attempting to connect technology using a solution that was created for the purpose of addressing billing issues (Zheng, Abraham, Novak, Reynolds, and Gettinger 2016). That approach compounds the problems, where healthcare professionals act as the interface between dissimilar solutions, creating a greater potential for human error as they manually collect and transform/transcribe the data into a usable format. As suggested by Mandl, Mandel, and Kohane, the inability to establish the basic foundational element of what is described as a transformational engine, using standardized input, output, and functional modules, exemplifies the conditions that will result in continued increases in healthcare-related costs and decreased patient care.

Kraus, Schiavone, Pluzhnikova, and Invernizzi (2021) suggested that healthcare environments that can successfully integrate transformative technology will be able to seamlessly transform data across dissimilar technology and ultimately provide improved patient care. The point of data transformation emphasizes that there is a need for the healthcare industry to consider technology integration from a higher level, where the adoption of transformative technology represents a potential for improved therapeutic outcomes, reduced human error, and sustainability within the scope of adaptive solutions.

Definitions and Terms

Transformational Engines: A transformation engine is a principal element that encapsulates other elements (e.g., input and output connectors) that are used to transform data within a technology solution from one format to another format using specific data mapping rules that are defined in the engine's configuration module.

Input Connectors: An input connector takes data that enters a transformational engine through a data queue and stores the data in a data stream to be processed separately by the execution layer of the engine. Within the scope of this discussion, an input connector is an element of a transformational engine, which facilitates the transfer of data between dissimilar technological solutions.

Interoperative: It is the ability for humans, hardware, and software technology to connect and communicate using a standard protocol, where data may be transformed across dissimilar technology (Wager, Lee, and Glaser 2021). Technology that is considered interoperable represents solutions that are able to connect with other technology and exchange data using standardized protocols through a common interface such as is characterized by a transformational engine. That proposal is supported by the material presented by Evans et al. (2017) and Slight et al. (2015) that showed technology cannot be considered as successfully integrated within an environment if that technology is incompatible with existing technology.

Output Connectors: Output connectors take data that is changed into a different format by the transformational engine and pushes that data into queues that pass the data to designated target locations. Like the input connectors, an output connector is an element within the concept of a transformational engine, which facilitates the transfer of data between dissimilar technological solutions.

Input/Output Connectors and Transformative Engines: Input and output connectors were designed around an internally developed transformative engine that is adaptive in its ability to connect dissimilar technology and transform data using a defined set of configurable rules. Data and solution control within the transformative engine are then securely managed through a standard interface layer where it is received, processed, and transmitted in accordance with the established rules. Furthermore, the transformative engine is designed such that it can use instructions that are carried with the data to adjust any predefined rules and influence the processing of data through a method described as bidirectional metadata encapsulated token authentication (Paffel and Harding 2018). The data specifically acts as the controlling element (e.g., a virtual messaging bus) that determines how it will interface and transform between dissimilar technology and across a standardized interface layer (Macarulla, Albano, Ferreira, and Teixeira 2016).

Need for Data Transformation in Healthcare

There is a need to examine more than just the potential benefits associated with adopting new technology, where technology adopters must consider the risks of misuse as well as the ability to secure any and all data that is transformed across new technology and existing technology. Accordingly, Konnoth proposed that the added burden associated with collecting, controlling, and distributing data must be factored into the assessment and decision-making phases associated with technology adoption. That said, and within the scope of manufacturing environments, there is a need to deploy technology solutions that work interoperably with existing systems, such that seamless transformation of data is needed. Thus, the results of this discussion can be used to reinforce the need to consider technology interoperability in earlier phases of project planning and not just at the point when technology is being deployed.

Where Quantum Computing Can Make a Difference

Quantum computing has real-life applications in that it is exponentially faster than traditional computers when working through algorithms. This means it can tackle big data problems with large probability sets more efficiently, like effects of chemicals in drugs, financial trading, and analysis in quantum chemistry, to name but a few. The use cases of quantum computing in these problems are still under development, but new algorithms yielded by quantum computing will accelerate the rate at which these big data problems are resolved.

In quantum networking, the goal of post-quantum cryptography is to develop systems that are secure against both quantum and classical computers. Financial institutions are leading the way for secure quantum networking, because of the clear importance of secure transactions at both ends and even in between. International aerospace teams are conducting quantum experiments dedicated to quantum use in satellites, enabling faster data transfer and better data encryption. And some countries, such as Switzerland, are even using quantum networking to secure voting and protect elections.

The bottom line is that quantum computing is the next big thing in computing; we just have a few hurdles to surmount before it becomes any better than traditional computing. Quantum networking already has such high demand because current encryption methods are always under active threat. Specifically, the need for quantum computing is likely to grow in the coming years, but more as a complement to investments in things like exascale computing – or computing that performs at least one quintillion (one billion billion) calculations per second. Bringing those technologies together will unlock tremendous computational potential.

Quantum's power also allows us to process imagery – which requires significantly more processing power than traditional datasets – at scale. This ability could allow clinicians to more quickly analyze images, such as CT scans, and identify any anomalies, resulting in a faster diagnosis and improved patient care. For example, "Multi-factor optimization problems – those are the types of things that quantum will be good at," said Matt Kinsella, managing director at Maverick, the VC arm of $8 billion hedge fund Maverick Capital that has invested in quantum computing technology.

In terms of the future of data transformation through the lens of quantum computing, here are a few questions to consider and possibly answer:

■ If data can exist in two or more locations at the same time and that data state is mirrored in all locations, how do we apply an equal amount of protection for that data in all locations? We explain this in terms of a prospective IEEE standard definition, TIPPSS.

   TIPPSS and Quantum opportunities
   – Can we trust more in the information and recommendations based on quantum computing to improve responsiveness to a challenging situation, improving trust in the recommendation, which could improve outcomes?
   – Could we improve safety in medical procedures by using quantum computing in real-time healthcare, in an OR, leveraging real-time information with predictive analytics and very high-performance computing; can recommendations be made which could improve outcomes, leveraging more context and experience, to be a more expert system to deliver real-time recommendations?
   – Can the ability of quantum computing to assess many images and data very quickly improve predictive analytics for the TIPPSS elements?
   – Can distributed big data and analytics, injecting quantum, locally or centrally, improve health outcomes?
■ Can different efforts to transform data occurring at the same time but in different locations cancel each other out?
■ Can a quantum bit/byte/word be locked in one location such that no transformation in another location can occur?
■ Can normal data contain qubits?

   Note: *A practical example of qubits:* An example of single state versus multi-state data might be characterized through an examination of a mouse running a maze trying to find a piece of cheese. However, in our traditional and current world of binary data, the mouse would systematically check each path for the cheese and would probably continue checking each path, even after finding the cheese. In contrast, a mouse working in a quantum state would be able to check each path of the maze at the same time.

Cybersecurity and Quantum Technologies for Healthcare

Ensuring data, devices, and humans remain safe in the increasingly digitally transforming world of healthcare is of the utmost importance. While the leverage of advanced technologies can bring great potential value, these technology innovations can introduce increased risk of digital attacks which could compromise privacy, protection, safety and security of data, devices, and humans.

We must design security into healthcare devices and solutions from the beginning. As noted in the KPMG report on Security and the IoT Ecosystem (Hudson, Laplante, and Amaba 2018):

> Security really needs to be designed into IoT solutions right at the start. You need to think about it at the hardware level, the firmware level, the software level and the service level. And you need to continuously monitor it and stay ahead of the threat.

The cybersecurity community has been increasing its focus and communications regarding medical devices. On January 19, 2016, an "Open Letter to the Healthcare Community Leaders, from the Security Research Community" was published (Woods, Coravos, and Corman 2019). This letter urges stakeholders to:

- Acknowledge that patient safety issues can be caused by cybersecurity issues.
- Embrace security researchers as willing allies to preserve safety and trust.
- Attest to these five foundational capabilities to improve visibility of their Cyber Safety programs.
- Collaborate now to avert negative consequences in the future.

This letter is available on the website for the organization I Am The Cavalry, along with a Hippocratic Oath for Connected Medical Devices which includes the following statement:

> I will revere and protect human life, and act always for the benefit of my patients. I recognize that all systems fail; inherent defects and adverse conditions are inevitable. Capabilities meant to improve or save life, may also harm or end life. Where failure impacts patient safety, care delivery must be resilient against both indiscriminate accidents and intentional adversaries. Each of the roles in a diverse care delivery ecosystem shares a common responsibility: As one who seeks to preserve and improve life, I must first do no harm.

There are many cybersecurity tools and techniques that can be deployed in connected healthcare, including many types and levels of encryption as explained in the book *Women Securing the Future with TIPPSS for IoT – Trust, Identity, Privacy, Protection, Safety, Security for the Internet of Things* (Hudson 2019). As detailed in this book in Chapter 11, "Securing IoT Data with Pervasive Encryption," pervasive encryption is a strategy in which data is encrypted everywhere it travels and/or resides, including data in transit and data at rest, at multiple layers of the computing environment. It is a crucial strategy to enhance privacy and security in our increasingly connected world.

There are many forms of encryption:

- Network encryption
- Application-level encryption
- Database-level encryption
- File- and data set-level encryption
- Disk- and tape-level encryption

There are new technologies on the horizon to secure the clinical Internet of Things (IoT) data and devices, including homomorphic encryption, crypto anchors, and quantum technology. However, homomorphic encryption is designed to provide a means to perform mathematical operations on encrypted data, which when decrypted provides the correct result (Hudson 2019). Just like other forms of encryption, homomorphic encryption uses a public key to encrypt the data. Unlike other forms of encryption, it uses an algebraic system to allow functions to be performed on the data while it is still encrypted (Marr 2019). Only the individual with the matching private key can access the result.

Crypto anchors are a blockchain-based technology designed to validate the authenticity of objects, such as pharmaceuticals in the medical supply chain (Hudson 2019; Balagurusamy et al. 2019). Ensuring the authenticity of pharmaceuticals and medical devices is of critical importance in delivering safe healthcare. This type of data or metadata about pharmaceuticals and other elements of the medical supply chain can use technologies such as crypto anchors to add Internet of Things technologies to medical resources such as pharmaceuticals to increase trust, identity, privacy, protection, safety, and security in patient care.

Quantum technology is an emerging field of physics and engineering, encompassing technologies that rely on the properties of quantum mechanics, including quantum entanglement, quantum superposition, and quantum tunneling. Researchers are building low-cost quantum-based security solutions which can be applied to improve cybersecurity for healthcare from devices to data and the medical supply chain. One such team from Washington University in St. Louis is developing quantum tunneling–based timer technology (Hudson and Chakrabortty 2021; Chakrabartty and Zhou 2021; Chakrabartty and Zhou 2019) to create self-powered timers with no electromagnetic (EM) signature and therefore no side channels, eliminating risks of intrusion through a power-side channel or EM-side channel. It is also tamper-proof, whereas if someone tries to manipulate the device physically or digitally, it changes the functionality of the device due to its quantum nature. Uses include creation of a Trusted Platform Module (TPM), key distribution, access control, authentication, and IP protection. The self-powered timer hardware works in conjunction with a dynamic authentication software stack to provide an end-to-end security solution. The team hopes to someday use the timers for securing sensor data in a variety of applications, like monitoring neural activity or monitoring glucose level inside the human body.

Advanced technologies can improve the TIPPSS – Trust, Identity, Privacy, Protection, Safety and Security – of data and devices in connected healthcare and mobile medicine, but there is more to do. The IEEE/UL P2933 standards working group on clinical IoT data and device interoperability with TIPPSS is building TIPPSS frameworks with technical and process elements to further the state of the art in this area, for today and the future. Professionals from medical device manufacturers, healthcare providers, technology, cybersecurity, academia, and research are working together in the P2933 working group, and are publishing recommendations to improve TIPPSS. The book *Women Securing the Future with TIPPSS for Connected Healthcare* showcases the work of 17 women from technology, cybersecurity, medicine, law, and other domains pertinent to improving TIPPSS for connected healthcare and mobile medicine (Hudson 2019). Considerations and recommendations are included related to cybersecurity frameworks, reference models for information assurance and security, and adaptive cybersecurity for existing and future healthcare and organizational needs. Chapter topics include:

- Health Data Management for Internet of Medical Things, by Oshani Seneviratne
- Architecting and Evaluating Cybersecurity in Clinical IoT, by Tanja Pavleska
- Do No Harm: Medical Device and Connected Hospital Security, by Gabrielle E. Hempel
- The Case for a Security Metric Framework to Rate Cyber Security Effectiveness for Internet of Medical Things (IoMT), by Zulema Belyeu Caldwell
- How Can We Trust in IoT? The Role of Engineers in Ensuring Trust in the Clinical IoT Ecosystem, by Jodyn Platt, Sherri Douville, and Ann Mongoven
- The Hospital of the Future and Security: An Arranged Marriage, by Alexis Diamond and Joanna Lyn Grama
- The Right Not to Share: Weighing Personal Privacy Threat vs. Promises of Connected Health Devices, by Jenny Colgate and Jennifer Maisel
- Ransomware: To Pay, or Not to Pay – That Is (One) Question, by Melissa Markey
- TAP and Intelligent Technology for Connected Lifestyles: Trust, Accessibility, and Privacy, by Katherine (Kit) Grace August, Mathini Sellathurai, and Paula Muller

▪ The Rise of IoMT: Leveraging a Polycentric Approach to Network-Connected Medical Device Management, by Cory Brennan and Emily Dillon
▪ Editor in Chief, Florence D. Hudson

References

Augustyn, Jason. "Emerging science and technology trends: 2017–2047." (2017).

Balagurusamy, Venkat S. K., Cyril Cabral, Srikumar Coomaraswamy, Emmanuel Delamarche, Donna N. Dillenberger, Gero Dittmann, Daniel Friedman et al. "Crypto anchors." *IBM Journal of Research and Development* 63, no. 2/3 (2019): 4–1.

Chakrabartty and Zhou. 2019. Self-powered timers and methods of use. US Patent 10,446,234, filed October 24, 2017, and issued October 15, 2019.

Chakrabartty and Zhou. 2021. Self-powered sensors for long-term monitoring. US Patent 11,041,764, filed February 28, 2017, and issued June 22, 2021.

Cook, Erica J., Gurch Randhawa, Chloe Sharp, Nasreen Ali, Andy Guppy, Garry Barton, Andrew Bateman, and Jane Crawford-White. "Exploring the factors that influence the decision to adopt and engage with an integrated assistive telehealth and telecare service in Cambridgeshire, UK: A nested qualitative study of patient 'users' and 'non-users'." *BMC Health Services Research* 16, no. 1 (2016): 1–20.

Daghfous, Abdelkader, Omar Belkhodja, and Norita Ahmad. "Understanding and managing knowledge transfer for customers in IT adoption." *Information Technology & People* 31 (2018): 428–454.

Evans, Steve, Doroteya Vladimirova, Maria Holgado, Kirsten Van Fossen, Miying Yang, Elisabete A. Silva, and Claire Y. Barlow. "Business model innovation for sustainability: Towards a unified perspective for creation of sustainable business models." *Business Strategy and the Environment* 26, no. 5 (2017): 597–608.

Hudson, Florence D., ed. *Women Securing the Future with TIPPSS for IoT: Trust, Identity, Privacy, Protection, Safety, Security for the Internet of Things.* Springer, 2019.

Hudson, F.D. and Chakrabortty, S. "Quantum techniques & technologies for cybersecurity in healthcare, Converge2Xcelerate." YouTube video, 2021. https://www.youtube.com/watch?v=gdthGzzTUYg.

Hudson, Florence D., Phillip A. Laplante, and Ben Amaba. "Enabling trust and security: TIPPSS for IoT." *IT Professional* 20, no. 2 (2018): 15–18.

Ingebrigtsen, Tor, Andrew Georgiou, Robyn Clay-Williams, Farah Magrabi, Antonia Hordern, Mirela Prgomet, Julie Li, Johanna Westbrook, and Jeffrey Braithwaite. "The impact of clinical leadership on health information technology adoption: Systematic review." *International Journal of Medical Informatics* 83, no. 6 (2014): 393–405.

Janaswamy, Sreya, and Robert D. Kent. "Semantic interoperability and data mapping in EHR systems." In *2016 IEEE 6th International Conference on Advanced Computing (IACC)*, pp. 117–122. IEEE, 2016.

Kohn, Martin S., Umit Topaloglu, Eric S. Kirkendall, Ajay Dharod, Brian J. Wells, and Metin Gurcan. "Creating learning health systems and the emerging role of biomedical informatics." *Learning Health Systems* 6, no. 1 (2022): e10259.

Konnoth, Craig. "Governing health information." *University of Pennsylvania Law Review* 165 (2016): 1317.

Kraus, Sascha, Francesco Schiavone, Anna Pluzhnikova, and Anna Chiara Invernizzi. "Digital transformation in healthcare: Analyzing the current state-of-research." *Journal of Business Research* 123 (2021): 557–567.

Liebe, Jan-David, Jens Hüsers, and Ursula Hübner. "Investigating the roots of successful IT adoption processes-an empirical study exploring the shared awareness-knowledge of directors of nursing and chief information officers." *BMC Medical Informatics and Decision Making* 16, no. 1 (2015): 1–13.

Macarulla, Marcel, Michele Albano, Luis Lino Ferreira, and César Teixeira. "Lessons learned in building a middleware for smart grids." *Journal of Green Engineering* 6, no. 1 (2016): 1–26.

Mandl, Kenneth D., Joshua C. Mandel, and Isaac S. Kohane. "Driving innovation in health systems through an apps-based information economy." *Cell systems* 1, no. 1 (2015): 8–13.

Marr, Bernard. "What is homomorphic encryption? And why is it so transformative." 15th November, 2019. Available at: https://www.forbes.com/sites/bernardmarr/2019/11/15/what-ishomomorphic-encryption-and-why-is-it-sotransformative.

Paffel and Harding. 2018. Data driven schema for patient data exchange system. US Patent 10,147,502, filed October 10, 2016, and issued December 4, 2018.

Petersen, Carolyn. "Patient informaticians: Turning patient voice into patient action." *JAMIA Open* 1, no. 2 (2018): 130–135.

Slight, Sarah Patricia, Eta S. Berner, William Galanter, Stanley Huff, Bruce L. Lambert, Carole Lannon, Christoph U. Lehmann et al. "Meaningful use of electronic health records: Experiences from the field and future opportunities." *JMIR Medical Informatics* 3, no. 3 (2015): e4457.

Stegemann, Lars, and Martin Gersch. "The emergence and dynamics of electronic health records–A longitudinal case analysis of multi-sided platforms from an interoperability perspective." (2021). https://doi.org/10.24251/HICSS.2021.746

Theorin, Alfred, Kristofer Bengtsson, Julien Provost, Michael Lieder, Charlotta Johnsson, Thomas Lundholm, and Bengt Lennartson. "An event-driven manufacturing information system architecture for industry 4.0." *International Journal of Production Research* 55, no. 5 (2017): 1297–1311.

Van Oorschot, Johannes A. W. H., Erwin Hofman, and Johannes I. M. Halman. "A bibliometric review of the innovation adoption literature." *Technological Forecasting and Social Change* 134 (2018): 1–21.

Venegas, Roberto Rodríguez, Emigdio Archundia Fernández, and Ma Guadalupe Olvera Maldonado. "The diffusion of an authoritarian innovation in the implementation of the e-accounting in Mexico." *European Journal of Business and Social Sciences* 5, no. 7 (2016): 51–65.

Wager, Karen A., Frances W. Lee, and John P. Glaser. *Health Care Information Systems: A Practical Approach for Health Care Management.* John Wiley & Sons, New York, 2021.

Woods, Beau, Andrea Coravos, and Joshua David Corman. "The case for a Hippocratic oath for connected medical devices." *Journal of Medical Internet Research* 21, no. 3 (2019): e12568.

Zheng, K., J. Abraham, L. L. Novak, T. L. Reynolds, and A. Gettinger. "A survey of the literature on unintended consequences associated with health information technology: 2014–2015." *Yearbook of Medical Informatics* 25, no. 1 (2016): 13–29.

Chapter 2

Closing Knowledge Gaps, Critical in Advanced Technology for Medicine: Building Shared Knowledge to Drive Down Risks in Medical Technology

William C. Harding, Viktor Sinzig, Craig Hyps, Neil Petroff, David Rotenberg, Brittany Partridge, and Sherri Douville

Contents

DOI: 10.4324/9781003348603-3

Knowledge Gaps

Knowledge gaps between stakeholders can be costly in business, specifically as those gaps represent misunderstandings or even a lack of understanding of how things might work together. For example, without an understanding of how healthcare data could be used, it is possible that important data is discarded or ignored (Rodrigues et al. 2022). Whereas, by listening to the needs of end users (i.e., relevant stakeholders), we gain an understanding of how important all data is and how that data might teach us something that we didn't already know. Accordingly, it could be gleaned from those statements that without the perspectives of all relevant stakeholders, we might miss something important, ultimately widening gaps between what is known and unknown. That said, a similar conclusion can be drawn regarding how gaps in knowledge associated with cybersecurity can result in corrupted or compromised data. Thus, a need to reduce knowledge gaps and reduce risks to a business or even a patient could represent the difference between life and death. In considering what data is "important," it is also possible for the converse to be true; data, or at least data relationships, can be considered important, but only leads to further obfuscation. The classic case of this is the confusion between correlation and causation. For example, as cheese consumption increases, more people die by entanglement in their bedsheets (Schwartz and Stanovsky 2022). Finally, we can create new data by fusing information from multiple sensors or building highly accurate user profiles from disparate and unstructured pieces of information. This is a huge business trying to take advantage of every type of footprint we leave in all forms of media. It has been estimated that 90% of all data in the digital universe is unstructured (Cahyanti and Permana 2022). The latter is notorious for purposely fomenting misunderstanding, and, therefore, increasing knowledge gaps. Data itself can embed bias and as we massage it more and integrate it into algorithms, which may also embed bias, the overall system bias becomes ever more entangled. For these reasons, practices of data and text mining for prescriptive modeling can be seen as an issue of cybersecurity.

Technology Integration and Interoperability

With a clearer grasp of what is meant by knowledge gaps and the risks associated with knowledge gaps, we examine some of the areas and concepts where knowledge gaps can have undesired consequences. One of those first concepts that receive our focus is technology integration and the subset of interoperability. That said, in this section, we present and discuss the importance of reducing knowledge gaps to successfully deploy interoperable solutions.

Intro: Technology integration for the purpose of exchanging and transforming data represents the central theme associated with interoperability. It also represents a concept or model that is not well understood and thus results in considerable knowledge gaps across developers and users that, if not resolved, will lead to failure. That said, this discussion will highlight the current state of interoperability within healthcare environments, the risks associated with current embodiments of proposed interoperable solutions, and things that need to be considered when building an interoperable solution.

Definition: Interoperability is the ability for humans, hardware, and software technology to connect and communicate using a standard protocol, where data may be transformed across dissimilar technology (Wager, Lee, and Glaser 2017). Technology that is considered interoperable represents solutions that are able to connect with other technology and exchange data using standardized protocols through a common interface such as is characterized by a transformational engine. That proposal is supported by material presented by the Evans et al. (2017) and Slight et al. (2015) that showed technology cannot be considered as successfully integrated within an environment if that technology is incompatible with existing technology.

Summary – Technology Integration and Interoperability: Interoperability may not be an obvious point to consider when we speak in terms of reducing knowledge gaps; but in the healthcare and life sciences world, the issues associated with ill-conceived transfer of knowledge/data/information do in fact represent a major barrier to successfully reducing knowledge gaps. So, as we think about the elements, behaviors, and methods associated with creating successful interoperability, we need to consider how a lack of vision toward data and technology interoperability represents one of the biggest barriers to the successful transfer of knowledge.

Interoperability example: As the design engineer for a steel producer, part of the The job of one coauthor of this chapter was to approve orders based on production capabilities and assign a predefined process flow. Orders entered into the fulfillment system were planned out by when the product would be at a particular process by calendar-year week. This is a long process, about 12 weeks or so, including several process steps that may take place as far away as on different ends of the country. Raw materials must be converted to steel, the liquid steel must be cast, the cast slab must be rolled, and so on until the final product (output from the steelmaking process) is delivered to the customer (an input to the customer's process, e.g., a panel to make a car door) for further value adding. The problem is that once the order is entered, these process steps are fixed and typically could only be changed manually. Some of these processes have very limited locations in which the process can take place; however, not all. As an example, it is not uncommon to have to side trim a steel coil to its proper width. This can occur at several locations, both in house and at an outside processor (OP). Next, consider all possible issues that could occur in the process over the course of those three months to cause delays, putting the supplier at risk of not delivering on time. For the sake of argument, let's assume the originally scheduled side trim process could not be done at the scheduled processor. A smart, interoperable system could parse this updated information and automatically reroute the side trim process to an OP with excess capacity that is on the way to the final customer to make up time. Our forecasting gets better the closer we get to the final event, e.g., weather prediction. Now, imagine an analogous case of a critical care situation with a compressed timeline – medical transport. Is it possible, based instead of following the predefined procedure – a particular ambulance service going to a particular hospital – we attempt to make better decisions based on the most currently available information? None of this can happen, however, without the entire supply chain exhibiting a certain level of interoperability. The transfer of information must also be secure. As a worst case, we don't want a malevolent agent using that same information to purposely reroute the ambulance for the purpose of putting the patient at risk.

Knowledge Gaps across Cultural and Geographical Diversity

In this section, we examine the widening knowledge gap between areas that have highly skilled healthcare professionals and those areas that are underserved or are in need of improved healthcare solutions as well as equally skilled professionals.

Healthcare resources are inadequate, not well-coordinated, nor allocated equitably, especially for people living in rural portions of the United States, which accounted for about 19% of the US population in 2016. Access to programs and resources, including network connectivity, are exacerbated in rural areas due to geographic, demographic, and population density disparities. However, what is truly lacking is the data synthesis needed to improve outcomes to the single-patient resolution.

It is ironic that while we may be drowning in data, a "smart" hospital generates an estimated 5 TB of data/day, we have been unable to leverage that data to improve outcomes. Like any service/resource, rural areas are not good investments because of sparse population densities, higher poverty versus metropolitan areas, and greater number of uninsured. And because rural towns offer relatively fewer amenities in return, they are less likely to attract top talent. During practice, it can be difficult for healthcare providers to remain current and "sharp" since they are less likely to be exposed to the variety of cases seen from a larger population.

It seems technology will be vital in closing gaps in these cases. Data-driven initiatives in healthcare are Healthy People 2030 and the National Academy of Engineering's grand challenge to advance health informatics (Brach and Harris 2021). The former has a workgroup in Health Communication and Health Information Technology (HC/HIT). One objective of the HC/HIT workgroup is to increase the proportion of adults who use IT to track healthcare data or communicate with providers. The latter states health informatics "should enhance the medical system's ability to … analyze the comparative effectiveness of different approaches to … therapy." In the UK, the Engineering and Sciences Research Council has a grand challenge – Expanding Frontiers of Physical Intervention – which promotes effective interventions through the use of physical and digital devices for targeted approaches (Rew 2020).

Ultimately, if the clients cannot get to the healthcare, then the healthcare must be brought to the client. This will increasingly be through a "tele-" mode. Addressing accessibility, however, is only half the problem. A second issue is the cost, even if it were conventionally available, of delivering healthcare. For example, the state of Texas in the past decade has led the nation in the number of rural hospital closures, making up nearly one-fifth of all closures across the United States. This means patients must drive further to receive care. Those who are unable to drive themselves must rely on others for transportation. Ultimately, if clients can be better served, they will have increased knowledge and understanding of the benefits of healthcare, leading to behavioral changes that support long-term health, and also drive broader changes in policies and regulations related to decreasing healthcare disparities originating from knowledge gaps.

Reduced Collaboration

Knowledge transfer plays a major role in mitigating risks related to reduced collaboration and slowing innovation. Specifically, it has been well documented that collaboration and engagement with diverse perspectives enriches our results (Jardi, Webster, Petrenas, and Puigdellívol 2022), whereas a lack of collaboration often leads to reduced knowledge sharing and even a widening of knowledge gaps.

Analogy/Example: Sports teams are also required to mitigate these types of risks. Team collaboration and upfront preparation are essential in team sports like Formula 1 – if the team manager, engineers, tire changers, and/or technicians are not trained and lack the proper knowledge in their specific function, this will adversely affect the overall team performance and will prevent the team from achieving strong results and ongoing best-in-class innovation. One mishap may lead to others, which eventually slows team competitiveness. A lack of communication and training also has the potential to destroy the overall team culture, therefore eliminating any chances of enduring, strong organization values and greatness. This, in return, may result in reduced collaboration within the team, and degradation in performance.

Just like in Formula 1, ensuring strong company values and adequate knowledge throughout a team is critical. There are several key elements that organizations should consider:

Work with the Right Partner(s): An organization's culture and values should be the number one selection criterion. This may seem obvious, but it is often overlooked. It's important that organizations work with partners externally and internally that follow shared guiding principles and common core values. Working with an organization with the experience and required skills will also be essential.

As an example, I've worked with organizations looking to adopt a company-wide healthcare system, and change management always seems to get deprioritized – this is an immediate red flag, and is almost always the wrong approach. Change management requires close interaction with the end users, and this is how organizations can really understand the Voice-of-Customer (VoC), and to make sure that the system meets the end users' ultimate needs (Parker and Partridge 2021). If the system is unable to meet the needs of the end users, then the system has a high potential for failure – even if the technology is stable and works perfectly! If there is not enough preparation and upfront scoping work completed in the early stages of an implementation, the technology may not be tailored to meet the needs of the end users – and end users are our most important customers!

During the "Onboarding" phase of the customer journey, it's critical that the deployment of advanced technology follows a methodological framework to achieve the desired results. Throughout the onboarding phase, working with a partner who acts as a "trusted advisor" can guide organizations looking to adopt company-wide systems through the end-to-end implementation with proper guidance and best practice knowledge. Almost always, these types of partners will leverage a proven methodology that has worked in the past, and can be applied to other organizations looking to accomplish similar business needs.

Follow a Proven Methodology or Framework: Firms like BCG (Boston Consulting Group), KPMG (Klynveld Peat Marwick Goerdeler), McKinsey, and PwC (PricewaterhouseCoopers) have gold standard frameworks and best practices that have worked for their clients in the past – these types of firms follow best practice guidelines that have been built over time (through many lessons learned!). Typically, these types of implementations follow key phases with specified tasks and deliverables – for the purpose of this section, we will focus on the "Scope" phase identified in the "Implementation Framework" (Figure 2.1). One caveat is that with new elements or change to context; best practices have to often be augmented or refined to address a new or expanded scope. At least they provide common knowledge and a reference point to reduce knowledge gaps. They are therefore valuable just for that and often more.

In the scoping phase, establishing core guiding principles and/or north star metrics will be crucial. This is where an organization will define what is most important to them. This may also involve understanding the day-to-day operations of the end users, and understanding exactly how a specific technology will benefit them – ultimately, the end users and business would like to understand WIFM (What's In It For Me?) and the KPIs (Key Performance Indicators).

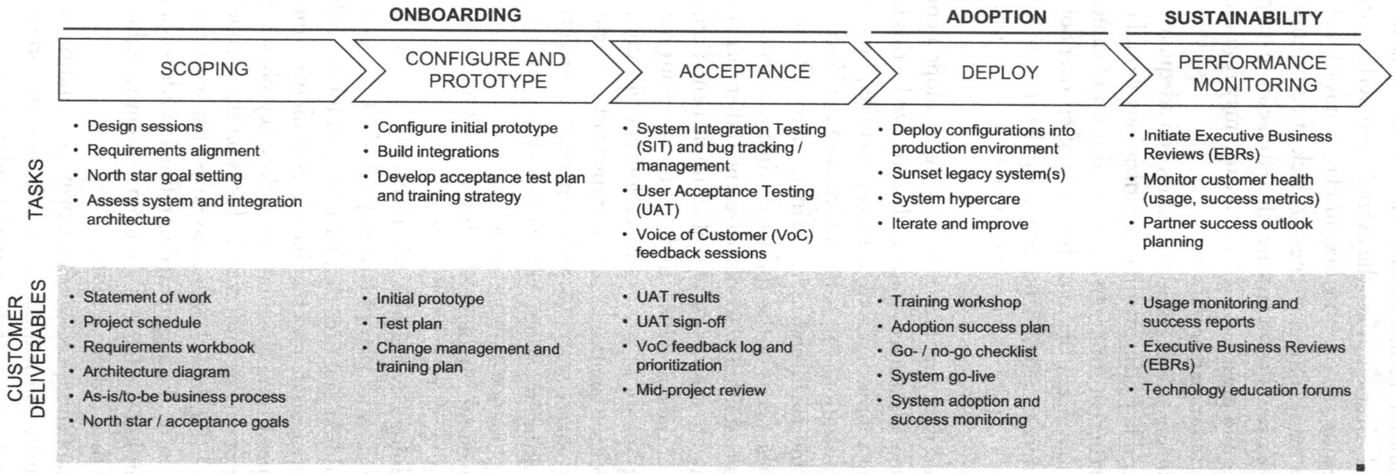

	ONBOARDING		ADOPTION		SUSTAINABILITY
	SCOPING	CONFIGURE AND PROTOTYPE	ACCEPTANCE	DEPLOY	PERFORMANCE MONITORING
TASKS	• Design sessions • Requirements alignment • North star goal setting • Assess system and integration architecture	• Configure initial prototype • Build integrations • Develop acceptance test plan and training strategy	• System Integration Testing (SIT) and bug tracking / management • User Acceptance Testing (UAT) • Voice of Customer (VoC) feedback sessions	• Deploy configurations into production environment • Sunset legacy system(s) • System hypercare • Iterate and improve	• Initiate Executive Business Reviews (EBRs) • Monitor customer health (usage, success metrics) • Partner success outlook planning
CUSTOMER DELIVERABLES	• Statement of work • Project schedule • Requirements workbook • Architecture diagram • As-is/to-be business process • North star / acceptance goals	• Initial prototype • Test plan • Change management and training plan	• UAT results • UAT sign-off • VoC feedback log and prioritization • Mid-project review	• Training workshop • Adoption success plan • Go- / no-go checklist • System go-live • System adoption and success monitoring	• Usage monitoring and success reports • Executive Business Reviews (EBRs) • Technology education forums

Figure 2.1 Implementation framework. Source: Sinzig, OSARO (2022)

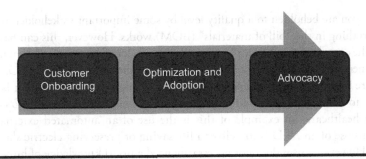

Figure 2.2 Path to advocacy framework. Source: Sinzig, OSARO (2022)

Involving end users at the beginning stages of a project will support both the end users and the integrator/partner. For the end user, this gives them the opportunity to express their desires and needs. For the integrator/partner, this gives them the opportunity to assess how a specific technology can be configured and tailored to meet the end users' needs – this may relate to understanding where a new technology might fit into an existing business process, or how a technology can be tailored to a specific use case. Ultimately, involving end users early on is essential for mitigating the potential for knowledge gaps.

Building quantifiable metrics will be important for the business to make key decisions – defining the acceptance criteria and KPI metrics that customers are looking to achieve will be essential. As part of a global Augmented Reality training implementation for a global medical device manufacturer, our ultimate business goal was to increase operator productivity, so we prioritized ramp rate to medical device yield production – this was the key to getting out of technology "Pilot Purgatory," and the technology rollout resulted in widespread value to the medical device manufacturer, globally (Figure 2.2). After establishing the core KPIs and proving that the system met the customers' objectives, our ultimate goal was to ensure that we optimized the solution with a core focus on adoption for the end users. This led to widespread end user advocacy for the system.

Measuring Knowledge Transfer Success: Measuring success is where the rubber meets the road. Knowledge transfer success should be monitored on a regular basis and covered with close executive alignment on a quarterly or semi-annual basis. Executive Business Reviews (EBRs) will help organizations realize the success of closing knowledge gaps – this is where key stakeholders have transparent discussions regarding goals against the key outcomes defined in the scoping phase. EBRs also provide the opportunity for the key stakeholders to align on what is, and what is not, working, and to be proactive about fixing what needs to be changed in order to meet the desired outcome. These quarterly/semi-annual health checks will now become one of the key drivers for long-term success.

Now that there is an understanding of the key north star KPIs that make a technology transformation successful, it's just as important that the end users understand how to utilize this technology.

Does the fact that I don't know how something works change the effectiveness of that thing? Ignorance is not bliss: There are several examples in industry where it has been seen as desirable to develop "vision terms" or marketing representation terms before the details are hammered out. While this is acceptable in consumer technology, it is less desirable in medical technology since all stakeholders in the supply chain would then be relying upon practical details or the absence of them to actually make things work. Therefore, if your outcome depends on the chain of elements that make up all the activities or technologies needed to drive the outcome, we would advise that you need to understand how things work. If you don't know how something works, you can't evaluate its effectiveness. If you are beholden to a quality level by some important stakeholder, then you better know how everything in the "bill of materials" (BOM) works. However, this can be harder than it sounds due to the complexity and multiple dependencies of a lot of systems in current use. That's why future systems need to be purpose built with an eye toward the ability to drive a BOM.

Could there be some temporal aspect about knowledge gap reduction? It is possible for a knowledge gap to be reduced or closed only temporarily while allowing a user to complete some useful task. In healthcare, an example of this is the use of an automated external defibrillator (AED). The purpose of an AED is to deliver a life saving or preserving electric shock to those in cardiac arrest. However, a user does not necessarily need a priori knowledge of human physiology to operate the device any more than a licensed driver needs knowledge of an internal combustion engine to operate a vehicle. The device (AED) itself acts as a supervisor of a human-in-the-loop, closed (cyber physical) system to determine and influence efficacy (e.g., cobots). As a corollary, we discuss, compare, contrast knowledge gaps with skills gaps.

For example, the university where one coauthor is engineering technology department chair and faculty promotes applied learning experiences (ALEs) which are "transformative learning experiences designed to engage students in active and reflective learning to further develop beyond-disciplinary skills and expand students' perspectives of self, community, and global environments" (https://www .tarleton.edu/ale/index.html). During the spring 2018 semester, faculty engaged with an instructor from the school of Kinesiology to establish an ALE. It combined students in Kinesiology studying therapeutic rehabilitation with engineering technology students studying machine design and/ or embedded systems to produce assistive technology designs. It brought together such a disparate group of students, allowing them to work in an area and with people they might not have known existed previously and to build empathy for one another. Moreover, it exposed students to the political, social, and economic issues, resulting from debilitating diseases and injuries such as equal access, affordable healthcare, and job retraining and retention. It is probably the most true-to-work experience I've been a part of. Finally, the students deserve a lot of credit for making the projects generally work even though we were unable to structure time for them as part of the classes. This DEI practice builds empathy for those who stand to gain from developments. It also builds allies with those who have to work together despite their disparate backgrounds.

How much do I need to know about something before I can declare the knowledge gap to be closed? Knowledge gaps should never be considered "closed." Because things are always changing, there can be unknown impacts on technical systems whether those changes are in tooling, standards, or regulations; expect change and expect there to be knowledge gaps. What this means in our minds is that you need to be diligent in cultivating multidisciplinary teams of teams. You need to have high-performing, highly functioning communities of practice where experts from different areas can learn from one another, constantly and on an ongoing basis. Our physician educator coauthor, Dr. Felix Ankel calls this a "knowmad" (Wang, Schlagwein, Cecez-Kecmanovic, and Cahalane 2020). While knowledge gaps can never be closed, if you are repeatedly asked by leaders in multiple segments to explain topics, you can comfortably imagine that you're considered a leader in that domain. Being a leader doesn't mean you've closed all knowledge gaps; being a leader means you are aware of them, including the presence of your own gaps, and that you have developed robust systems and networks to address them on an ongoing basis.

For example, the world is constantly changing and technology is changing faster than most things. Today's experts will be outdated in few years if they won't keep learning and updating their knowledge. It was true several years ago, but it is even more significant today, some 2.5 years into COVID. It seems that COVID expedited many areas, "squeezing" processes that would have taken 1–2 years. Companies that managed to evolve and adapt made it big time, while others ceased to exist. Specifically, the technology sector has made a huge leap, from technological to HR issues, such as remote work where as we examine reducing risks in the healthcare industry, we must remain adaptive and flexible in our desire to remove barriers and close knowledge gaps.

Reducing Risks in Healthcare

a. Just as there is nothing that is 100% secure can we assume that risks cannot be eliminated? That said, we should seek to identify risks, weigh or measure those risks, and prepare to mitigate those risks.
b. Considering that acceptable and unacceptable risks are ultimately determined either directly or indirectly by a human, the identification of risks is subjective. This might be characterized with being an individual who is risk-tolerant or risk-averse.
c. When considering a risk as acceptable, first decide if it is an acceptable reality, if that risk becomes realized.
d. Consider also that if knowledge gaps are reduced, the perception of risks will change.
e. Are people with less knowledge of something more likely to take risks or would the lack of knowledge make them more fearful of taking risks?

Reducing Knowledge Gaps and Encouraging Behaviors to Reduce Gaps: When discussing knowledge gaps, there are some important traits and behaviors that we need to recognize and embrace as it relates to establishing a foundation from which we can listen, empathize, and reflect. That said, listening skills are as equally important as communication skills (e.g., speaking and writing). Whereas, being able to establish empathy with a need as conveyed by a person or organization is the first step in building an understanding of needs. From there, we are able to establish a feedback loop of listening, asking questions, and reflecting on what has been conveyed and gathered from the stakeholder, thus ensuring that we don't start to work with limited information, incorrect assumptions, and without personal biases that negatively influence our perception of what is needed and what we are solving for.

Demystifying Technology/Relatable Tech: The proposal that technology can be demystified triggers some basic thoughts related to how we might go about explaining technology in relatable terms and to what extent we should seek to demystify technology. For example, as we have proposed in previous content, a person does not need to know about the inner workings of an automobile, in order to effectively operate a vehicle. Similarly, a patient doesn't need to know everything down to the cellular level of healthcare in order to lead a healthy and happy life. However, in healthcare, ignorance is not blissful, where the more a patient knows about their own healthcare, the more they can advocate for themselves.

For example, there are many technologies available to patients that can be used to determine their general health (i.e., blood pressure, blood oxygen levels, pulse rate, temperature, cardiac rhythm, etc.). But do patients understand how an alignment of the data from those technologies might paint a more detailed picture of their health state versus not aggregating all of the data, specifically at the point that the entire data set intersects? Probably not!! So, if we are the patient looking to augment our healthcare knowledge and our understanding of healthcare-related technology, where do we go?

If you can't explain it simply, you don't understand it well enough.

Albert Einstein

How do gaps in general knowledge influence/impact/expose risks in the medical industry?
Many individuals accept that there will be gaps in knowledge between a healthcare professional
and that of a patient. In some extreme cases, that acceptance has resulted in healthcare
professionals who will not directly interact with a patient. Thus, we widen the knowledge gap and
risk negatively impacting the quality of care that a patient is afforded. That said, we believe that
behaviors associated with empathy, speaking in relatable terms, and collaboration are key elements
that could increase or decrease knowledge gaps.

An actual example, associated with collaboration, was exposed during visits to both neuro
and cardiac surgical suites, where it was obvious that surgical prep methods and best practices
are not shared. Specifically, something as simple as intelligently preparing a patient's gowning/
draping (wrapping versus stapled to a patient) represents just one small element that may impact
a patient's overall care. Additionally, lack of empathy for a patient's needs (versus a doctor's perception of those needs) could mean that something has been missed/ignored. That "something"
also represents a gap in knowledge that may manifest as undesired results and ultimately reduce
a patient's quality of care.

Closing the gap between a doctor's knowledge and that of a patient represents a critical element
in demystifying healthcare. Specifically, learning to speak in terms with which a patient can relate
will help the patient make more informed decisions. Conversely, patients who do not know or
understand what is needed before and after surgery may make decisions that are fatal. With that
point in mind, patients can do their part in closing knowledge gaps by not fueling organizations
and individuals who do not possess certifiable skills in healthcare. In short, patients looking to
increase their knowledge of healthcare should ensure that the material they are reading is backed
up by evidence, such as found with empirical peer-reviewed research and scientific discovery.

Knowledge Transfer – Use Case/Example

Sometimes a point is not easy to relate to, so we practice a belief that it is better to show someone
something than to tell them about it. That said, even if you were not alive or directly involved
in the Challenger Shuttle Accident (Figure 2.3), we are confident that you have heard about the
event. So, when we speak about the accident, we are certain that most people know what happened and even how the accident changed many things at NASA (National Aeronautics and
Space Administration), but what might not be clear is how the event can be linked directly to the
concept of knowledge gaps.

Space Shuttle Challenger Accident – Results of Failed Knowledge Transfer

It is not always the case that gaps in knowledge can lead to disaster, but instead it is the inability
to transfer knowledge in a relatable form that can lead to disaster. Further, if knowledge is not
conveyed in usable form, then the resulting confusion/misinformation can widen knowledge gaps,
which can have disastrous results. Take for example the 1986 Challenger shuttle accident, where
those who possessed the knowledge and those who controlled the decision to launch were unable
to communicate or transfer knowledge in such a way as to avoid the accident. Specifically, though
it was perceived that the conditions surrounding the launch were within the margins of a no-go
scenario, the information related to how those conditions might impact the shuttle were not conveyed in a way that enabled leaders to make the right decision. Of course, without focusing blame
on any one group or individual and speaking from experience as a coauthor and person who was

Figure 2.3   Space shuttle clip art – free to use public domain. Source: Retrieved from https://www.clipartmax.com/max/m2i8H7m2H7b1b1N4/

working at Cape Canaveral in the 1980s, the Challenger accident does point to problems where it is not always the lack of knowledge that results in adverse effects. That said, we not only need to consider the lack of useful knowledge, but we also must examine the way in which we transfer knowledge (Abdalla, Renukappa, and Suresh 2022).

Using the Challenger accident as an example of how knowledge gaps form or can be widened, we must first seek to share knowledge and then find ways to convey that knowledge. Keeping in mind that we all want to be understood, it is not enough to convey knowledge as we perceive it, but to understand how others need to receive knowledge in such a way as it becomes relatable or useful. That point is easily illustrated when we observe the reaction of individuals with whom we are speaking (transferring knowledge). If they make gestures that the information went over their head, then the issue is not with them, but instead with us, the owner of specific knowledge. Thus, we must then look for other ways to convey knowledge through a process that involves developing empathy with the target of our knowledge transfer. For example, consider the background, experiences, education, etc. of the target and seek to use that information to reshape the knowledge into a form with which the target can relate. At the same time, learners have to take responsibility for their own learning agility, performance, and commitment. It's a two way street and the whole burden isn't on the person with the knowledge. All workforce members, in particular leaders, need to train and challenge themselves to be the highest level of learning athlete available to them intellectually and practically. This is both in rapidity and complexity of how multiple disciplines intersect related to essential medical technology details.

Swim Lanes (or Building Team Alignment and Synergy)

Healthcare organizations comprise numerous divisions, departments, and teams responsible for addressing the requirements for networking and IT functions, security, building operations, asset tracking, medical device security, and desktop and mobile device management, just to name but a few. While each has a specific mission, the mission of the overall healthcare organization can often be missed. The mission of individual groups may even overlap leading to duplicate effort and expense or be executed in a manner which is counterproductive to other teams.

It may be obvious that a more ideal approach is one where the organization has clear roles and objectives that reduce duplication and encourage collaboration between teams for optimal outcomes, but this synergistic ecosystem often requires top-down sponsorship to recognize and help define individual team goals and promote engagement of the different stakeholders on a regular basis. These periodic check-ins enable each team to share knowledge regarding their individual contributions and progress toward the overarching mission statement.

Such ideals can certainly apply to any organization, but healthcare uniquely struggles with these challenges due to the diverse and organic evolution of the network, connected devices, and applications, many that are introduced by numerous stakeholders, including medical staff. For example, a doctor may add a group of new clinical devices and applications not vetted by the IT, security, and server/application teams. In stark contrast, a typical financial institution consists of a much more homogeneous environment with a central team that mandates a common desktop, mobile, and application environment. IoT devices are more likely to be from a single or limited set of vendors and new devices are closely scrutinized and denied, by default.

Understanding the diverse makeup of your healthcare organization's people and devices can raise awareness into potential blockers and bottlenecks in achieving critical goals such as clinical device security to ensure reliable services without compromising patient safety or privacy. Multiple teams play pivotal roles in achieving this goal, from Healthcare Technology Management (HTM), sometimes referred to as Clinical Engineering or Biomedical Engineering, to Information Technology (IT), Security, Facilities, Desktop Management, Server/Application, Medical Device Manufacturers (MDMs), and others. With so many teams playing a role in clinical device security, it is easy to see how responsibilities can blur due to overlap or worse, be neglected.

Let's take a common scenario seen in different organizations. The HTM team is focused on ensuring all clinical IoT devices are tracked, assessed for bugs, vulnerabilities, and risk, and ensuring they are updated with the latest security software and patches. In some cases, they cannot be patched and therefore represent a higher risk and exposure to potential compromise. At the same time, the Physical Security, or possibly the Facilities team depending on your organization, may be tasked to track all cameras and their software, including those used in a medical setting such as patient or surgical rooms. The IT team is tasked with ensuring network connectivity and often manages the network access control (NAC) system for all connected devices – both medical and non-medical. The Security team sets policy and may also manage the organization's firewalls and other security devices. Still, many other teams are playing a key role in this scenario.

It is not uncommon for HTM, IT, and Security teams in this example to have unique objectives and expectations. Looking at your own organization, how often do your respective teams meet to discuss their roles, coverage, and key projects? Do they collaborate and know the main contacts responsible for critical services across the virtual divide? Or would the response from one group more likely be "We don't work with that team," or "That is not our job"? If the answer is the latter, then see this as an opportunity to unlock your organization's full potential. Even if teams are not outright dismissive of the other, passive aggressive behavior can disrupt and even halt progress for major initiatives. This is where top-down sponsorship can drive common vision and outcomes and promote collaboration to overcome obstacles.

Using the previous example, the HTM team's work to secure and patch vulnerable devices is certainly important, but it is also just as critical that the IT team secure network access to allow only trustworthy devices and limit their exposure to potential internal threats. Simultaneously,

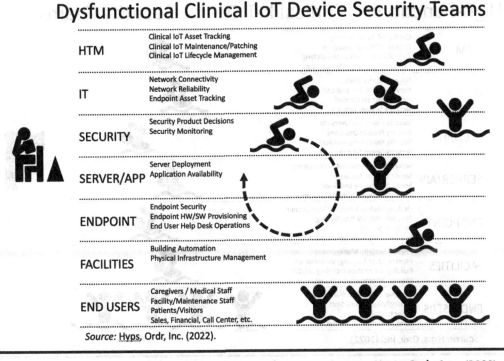

Figure 2.4   Dysfunctional clinical IoT device security teams. Source: Hyps, Ordr, Inc. (2022).

the Security team must implement the proper controls on the traffic to gateways, internal servers, and external systems (Figure 2.4). By securing the borders and internal communications, the risk to vulnerable medical systems is greatly reduced, and patching helps ensure systems are not compromised even if unauthorized traffic makes its way to the device. Recognize that the complementary efforts can help foster an environment of mutual respect and collaboration, and influence teams to offer support in areas that optimize results. Furthermore, it is important to recognize the ideas and the value each team and its members bring to the table. This recognition encourages new ideas and solutions to problems while increasing employee satisfaction and well-being (Figure 2.5).

Establishing a consistent set of terms and definitions is equally important when defining roles and responsibilities. While the taxonomy used across different organizations may differ, focus should be on ensuring all teams within your organization abide by the same set of terms. Align teams such that there is a common understanding as to which team is lead, or has primary responsibility for a topic, which team or teams have a secondary or supportive roles, and those which may perform a consultative role.

Security terms which merit clear definition and role assignment include trust, identity, authentication, authorization, access control, privacy, safety, protection, operations, audit, and interoperability. Roles can be use case–specific such that identity for connected medical devices may be handled by one team, identity for non-medical another, and identity for servers and medical applications yet another team. In general, security policy itself should be centralized as it sets governance for the entire organization. Similarly, privacy policy should be centralized. However, the execution of security and privacy policy can be led by individual teams responsible for a specific aspect of the security architecture.

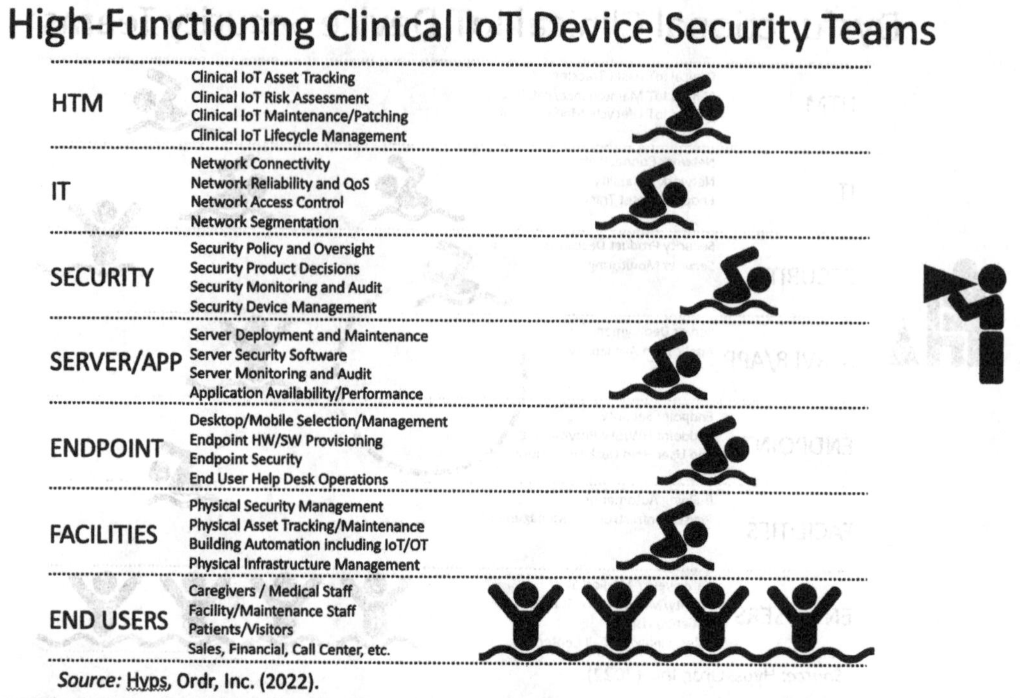

High-Functioning Clinical IoT Device Security Teams

HTM	Clinical IoT Asset Tracking Clinical IoT Risk Assessment Clinical IoT Maintenance/Patching Clinical IoT Lifecycle Management	
IT	Network Connectivity Network Reliability and QoS Network Access Control Network Segmentation	
SECURITY	Security Policy and Oversight Security Product Decisions Security Monitoring and Audit Security Device Management	
SERVER/APP	Server Deployment and Maintenance Server Security Software Server Monitoring and Audit Application Availability/Performance	
ENDPOINT	Desktop/Mobile Selection/Management Endpoint HW/SW Provisioning Endpoint Security End User Help Desk Operations	
FACILITIES	Physical Security Management Physical Asset Tracking/Maintenance Building Automation including IoT/OT Physical Infrastructure Management	
END USERS	Caregivers / Medical Staff Facility/Maintenance Staff Patients/Visitors Sales, Financial, Call Center, etc.	

Source: Hyps, Ordr, Inc. (2022).

Figure 2.5   High-functioning clinical IoT device security team. Source: Hyps, Ordr, Inc. (2022).

The following provides a list of related terms aligned by topic which must be commonly understood across teams as well as each team's primary/leading, secondary/supporting, or consultative/feedback role:

Security

- Security policy
- Integrity
- Availability
- Authenticity
- Privacy/confidentiality controls
- Security processes, practices, and principles

Privacy

- Privacy/confidentiality policy
- Data sharing
- Regulatory compliance
- Regional accommodations
- Privacy processes, practices, and principles

Trust

- Authorization
- Access controls

- Access rights and permissions
- Role-based access control (RBAC)
- Enforcement
- Segmentation
- Policy decisions

Identity

- Authentication
- Identity Stores
- Credentialing
- Certificates (X.509, PKI for identity, Trust Stores, etc.)
- Unique ID (UID)

Safety

- Hazards and harms
- Mitigation requirements
- Effectiveness
- Safety risk management
- Intended use and use environment

Interoperability

- Integration
- Testing
- Traceability
- Syntax
- Semantics
- Transformation

Again, a common agreement on terminology and team alignment can help ensure critical functions are properly covered, eliminate gaps and confusion, and optimize the contributions and outcome of the entire organization.

Large healthcare organizations are often hindered by separate, monolithic tools and processes. For example, within your organization, is there a standard process to onboard new clinical devices? Is the device properly tracked in a system visible to the HTM, Facilities, Security, and IT teams? Due to the convergence and interdependence of the different teams, holding data hostage to a given team is often counterproductive. Having separate tools and processes itself can be extremely costly in terms of licensing, deployment, and professional services, and make the system less useful if data cannot be shared between systems. Through organic growth, mergers, and acquisitions, many healthcare systems are composed of siloed applications with their own set of infrastructure, servers, and teams to manage. Here is yet another opportunity to streamline operations and foster interoperability and collaboration based on tool selection that serves the needs of the overall organization rather than individual teams. Of course, there are cases where the benefits derived from a special application merit its limited use to a specific team, but it is important to recognize opportunities to converge resources that will elicit greater benefit to the organization and its core mission. This can be particularly challenging as budgets tend to be granted to specific teams, and therefore may suppress altruistic behavior. This is where leadership can help execute on a common good.

To ensure proper coverage, the clear definition of roles and responsibilities can prevent gaps in clinical device security. As cited earlier, if there is a camera that is used in a clinical setting, there should be no confusion as to which team is responsible for its maintenance and security upkeep. Similarly, is a nurse call device managed as a medical device or media device? Is a workstation or mobile device that is used to transmit patient data managed by HTM or the Desktop/Mobile Management team? Ensuring clearly defined coverage and roles will help ensure teams are synchronized and potential gaps eliminated.

In summary, look inside your organization and objectively assess their level of collaboration and mutual understanding of roles, coverage, and unique capabilities to support other teams and the overall goals of the organization to deliver the highest possible outcomes. Like a swim team, maintaining swim lanes can help ensure there is no disruption to other teams or team members, but like a relay, it is the collective efforts that enable the organization to succeed (Hitchcock and Stavros 2017).

A Cognitive Approach to Understanding and Motivating Knowledge Sharing

Effective knowledge sharing as well as collaborative and innovative engagements are built on the foundation of autonomous motivation, which embraces an alignment of ideas from individuals who come from multidisciplinary and diverse backgrounds, cooperatively involved in creating technologically advanced solutions. Likewise, though motivation theory encompasses various concepts, including amotivation (i.e., lacking the ability to recognize positive outcomes associated with action), the focus is on the use of motivational concepts that are associated with self-determination theory (SDT). Accordingly, it is proposed that the motivational theories characterized within SDT are autonomous, controlled, extrinsic, and intrinsic motivation. Whereas, a discussion of autonomous motivation is more closely aligned with the prosocial behavioral characteristics that promote an individual's freedom to choose, while exhibiting moral and ethical attributes toward other individuals. With that in mind, it is postulated that individuals who display autonomous motivational behavior are able to effectively share knowledge, collaborate technologically advanced and innovative concepts, resulting in better outcomes and improved satisfaction, while realizing increased psychological well-being.

Conversely, though traditional collaborative theories focus on anticipating and resolving interorganizational conflicts, we examined the desired motivational concepts that reflect the constructive qualities of each collaborative participant. Further, with consideration for realizing effective collaborations and knowledge sharing, the positive attributes associated with psychoanalytic transference (i.e., improved collaborative bonds that support positive experiences) are embraced as foundational concepts that align the theories of autonomous motivation and collaboration: specifically, with a focus on the motivational characteristics of participant's independence, constructive results, and mutual accountability within collaborative groups that are oriented toward a common goal (Gray 1989). That said, let's examine a little more closely both motivation and collaboration theory, as we explore improved methods for knowledge sharing.

Motivation Theory: Motivation theory encompasses a number of theoretical concepts (e.g., autonomous, controlled, extrinsic, intrinsic), but when viewed through the lens of SDT, individual motivation is transformed into a cognitive process of self-actualization and psychological well-being (Deci and Ryan 2008a). That being considered, motivation theory is a concept and process found within sociocultural environments, where an individual's behavior is both externally and internally influenced, as exhibited through needs and desires.

Regardless of potential influences from an individual's cultural environment, autonomous motivation will produce the greatest personal satisfaction while maintaining a healthy mind and body. Accordingly, Chen and Jang (2010) infer that individuals who maintain constant awareness of motivational types while embracing a vacillating balance between intrinsic and extrinsic motivational influences will experience a greater sense of well-being, where they have chosen to be productive and socially responsible individuals.

Collaboration Theory: Garrison, Anderson, and Archer (1999) describe collaboration as concept that is not focused on simple exchanges, such as guided by the injection of specific declarative instructions, but as shared experiences between participants that are constructive and result in the creation of knowledge. Thus, collaboration is both a conscious and unconscious process that aggregates cognition with sociocultural development from the perspective of reaching a balance between psychological and sociological constructs.

Gajda's (2004) proposal that "collaboration is a journey, not a destination" (p. 69) supports the idea that collaboration theory represents an understanding of how individuals cooperatively engage in work to achieve a shared goal that embraces mutual accountability. That said, collaboration across interorganizational environments (i.e., business units, departments, and locations) empowers individuals to align on long-/short-term innovative solutions and goals that could not be realized through individual convergent thinking (e.g., working alone or independently).

Comparison of Collaboration and Motivation Theories: There is a causal relationship between motivation and collaboration, where the success of collaborative efforts is dependent on the motivational level of the participants who seek to move from focusing on individual work to tasks that focus on common shared goals, where making a difference is a driving factor. Specifically, effective collaboration cannot exist unless participants, problem statement owners, and leadership are properly motivated to trust collaborative processes, where job titles and egos should be set aside. From those statements, an examination of collaboration with respect to motivation exposes that effective collaboration cannot be realized unless participants are properly motivated (i.e., autonomous motivation). Correspondingly, the key or binding element that inspires successful collaboration and encourages desired motivation is skilled facilitation and mentoring, which serves to promote attributes of professional growth.

Aligning Collaboration and Motivation Theories: The theories of collaboration and motivation appear to be only vaguely aligned, where collaboration implies cooperative engagement and motivation is principally evaluated through the lens of the individual. However, it is posited that within technologically innovative environments, where sociocultural influences focus on developing prosocial behavior, individuals who are both extrinsically and intrinsically motivated (i.e., autonomous) are able to more easily align their goals within collaborative efforts. Similarly, individuals who seek to be collaborative from the perspective of increasing knowledge, where greater satisfaction is often derived from the journey toward their goal, possess aspirational alignment with the theory of autonomous motivation and intrinsic desires.

Controlled motivation approaches collaborative efforts from the perspective of extrinsically influencing individuals to engage in alliances, such that avoidance of punishment, guilt, and competition are the principal motivating factors. As is inferred by Kohn (1999), incentivizing (i.e., negative and positive) collaboration among employees, where individuals are manipulated into working together, may have worked during the industrial revelation, but it is an ineffective method such that it eventually leads to failure and harm to employees' morale. Thus, from the perspective of collaboration, "the more artificial inducements [that] are used to motivate people, the more they lose interest in what they are being bribed to do" (Kohn 1999, p. 73).

All things considered, successful knowledge sharing and innovative collaboration focused on integrating technology, with resultant long-term interorganizational alliances, can only be achieved when individuals possess the self-realized freedom to choose and act on those choices in directly connecting collaboration and goal-related successes with personal satisfaction. Thus, it is proposed that individuals who exhibit the characteristics of autonomous motivation, such as those characteristics that demonstrate the behavioral attributes of self-discovery and persistence, are the individuals who are more apt to effectively collaborate in creating innovative and novel technological solutions that are able to interoperably exchange and transform data.

Interoperability and Technology Integration – What Is the Purpose of Interoperability in Healthcare and Do Knowledge Gaps Affect It?

What's the current state and how can we drive toward an ideal state? So, when thinking in terms of interoperability and more specifically technology integration, are we considering the aggregation and the transformation of data through the lens of an attribute's "availability" versus "utility"? Additionally, when thinking of interoperability, shouldn't we be considering data security needs such as those represented by the IEEE/UL TIPPSS (trust, identity, privacy, protection, safety, and security) standard principles? Accordingly, there are critical attributes of interoperability that need to be considered (i.e., scalability, governance, conformity, compliance, integrity, quality, accuracy, reliability, reproducibility, and compatibility). For example, some of the use cases that we believe should be considered when discussing interoperability might be continuous glucose monitoring (CGM) (e.g., design, deployment, and use), and it would be good to consider the use case of the connected IoT home that can be classified as the "home healthcare." That point should emphasize that classic EMR's and EHR's band-aid approaches are from the perspective of an advanced technologist considered to be outdated as HL7 methods are for accomplishing ideal interoperability while exposing that interoperability should address the needs of patients, healthcare professionals, the payer, and the non-professional caregiver.

Considerations and Risks: Interoperability assumes that technology acceptance should evaluate technology and social aspects, to identify methods for technology integration, while also considering five factors that influence technology adoption:

- ■ Technology quality
- ■ Integration management
- ■ Solution ease-of-use and usability
- ■ Impact on institutional development
- ■ Sustainability

The combination of these five factors is appropriate at any phase of integration (e.g., assessment, decision-making, adoption, design, development, deployment, and sustainability), where the concept of solution interoperability embodies human, hardware, and software influencers. Respectively, though EHRs and EMRs originated as a billing solution, most organizations view

EHRs and EMRs as their interoperability solution. However, that point was missed during the implementation of EHRs and EMRs, where technology interoperability should have been examined in the exact same way as most organizations handle the integration of any new technological solution (i.e., with an eye toward working with existing systems). For example, though most healthcare organizations seek to adopt a technology solution that would accomplish the task of interoperability across all solutions, the examination of a technology's usefulness, versus interoperability, resulted in many technology adoption failures (Rosenbaum 2015). According to Rosenbaum, issues associated with successfully adopting EHR/EMR solutions emerged not because of any perceived *technophobic* behavior, but instead because the solution was not fully understood, nor had it been properly assessed related to its usability. From the inference that technology cannot be successfully adopted if it is not fully understood, it is proposed that theoretical models such as IASAM3 promote an understanding of more than just the advertised usability of a technology (Aizstrauta and Ginters 2017; Ginters, Mezitis, and Aizstrauta 2018) and more toward promoting an understanding of technology interoperability. With IASAM3 in mind, it is suggested that all future technological integration efforts be guided by the theoretical model of IASAM3.

Similarly, the negative consequences of a narrow focus on only a few phases of technology integration and the ability for a solution to function interoperably can be exposed if solution integrators looked beyond the ability for technology to solve a singular task and toward technological compatibility with existing solutions. From that point, which emphasizes increasing the scope of technology integration, it is suggested that EHR/EMR technology integrators would realize greater success if the technological experiences of all relevant stakeholders were included in the overall assessment of technology perspectives. That said, the healthcare industry lags other industries in its ability to integrate technology that functions interoperably with existing technology, which is exemplified by currently used methods, practices, electronic data, and healthcare equipment.

Fuel for Thought: Would you buy a vehicle (e.g., electric- or gas-powered) without knowing that you had a means to fuel it (i.e., the infrastructure associated with electrical power or combustible fuel)? Would you buy a new cell phone if it didn't work with your preferred mobile provider, run your desired apps, or protect your personal data? If you answered "yes" to those questions, it is proposed that you are accustomed to failure as well as wasting/losing money. That being considered and assuming that your answers were "no," why would a healthcare organization consider investing in a new system (e.g., inventory, diagnostics/sensing, health monitoring, or data storage/aggregation) without considering how that new system might integrate with their existing systems, securely transform data across dissimilar systems, or how that new system might be sustained?

The primary message that is embodied within the previous statements is that by using band-aid solutions as with (e.g., EHRs and EMRs) to overcome the issues associated with secure data transformation between poorly integrated systems, more burden will be added to an already overtaxed healthcare infrastructure (Christodoulakis, Asgarian, and Easterbrook 2016; Hosfield 2017). That burden has a direct impact on the cost of healthcare, physician/nurse burnout, an increase in human error, and a patient's quality of care (Owaid 2016; Shanafelt et al. 2016). Moreover, as was suggested by Christodoulakis, Asgarian, and Easterbrook (2016), Demarzo, Cebolla, and Garcia-Campayo (2015), Hosfield (2017), Mandl, Mandel, and Kohane (2015), and Venugopal (2015), systems characterized as EHRs or EMRs can be considered a poor attempt at molding a billing solution (Victores, Coggins, and Takashima 2015) into a method for interconnecting dissimilar systems and transforming data across those systems (e.g., collection and movement). Respectively, technology integrators must consider more than just the immediate needs of the healthcare organization's end users/solution sponsors and consider the ability for new technology to be interoperable compatible with existing technology (i.e., already present within healthcare organizations) and whether the new technology can be sustained for its expected lifecycle.

Use Case – Example: To help with understanding how gaps in knowledge can lead to failure, an interesting use case that we have reflected on is the issues that large corporations have experienced when attempting to apply a Euro/US model to their efforts to gain market share in India or China. For example, a US company that builds medical technology might first think to connect with a hospital or physician in India, thinking that those entities are the central point for connecting with potential patients. However, anyone who is even slightly familiar with the health structure in India knows that the key individual in India who manages the health of the community as well as the patient data is the pharmacist.

Taking the above example to the next level, we pondered how to remove the barriers between patients and healthcare professionals as well as how we could help patients advocate for themselves. That point triggers a reflection on a program that was initiated in India and China, which focused on bringing knowledge to rural villages.

Interoperability Summary

As discussed in the previous sections, there are many aspects of interoperability that must be understood, before diving head-first into building and deploying a solution. However, lack of consideration for those aspects represents major knowledge gaps across stakeholders associated with interoperable solutions. Obvious within that message is the point that if we don't first seek to reduce knowledge gaps in healthcare-targeted interoperable solutions, then we will continue to see increases in costs, an increase in human error, increases in user burnout, and ultimately the failure of solution to address the quality of care that patients deserve.

Takeaways toward supporting truth and tamping down on propaganda that perpetuates harmful knowledge gaps:

1. Become a master of information filtering. Reviewing the tech hype and misinformation chapter is a great resource for that.
2. Make a plan to practice writing with the goal to become a good writer which is an essential way of communicating with stakeholders.
3. Identify and mitigate the cause of exaggeration in self and your teams. Review and aim for the highest level, "scientist" in the "Hierarchy of Thinking" by Adam Grant (Grant, 2022)
4. To be credible in medical technology, you *must* paraphrase in your writing based on a solid comprehension of the material. This is to avoid plagiarism, you even must avoid plagiarizing yourself.

What Role Might AI Play in Reducing Knowledge Gaps? Artificial Intelligence (AI) in Reducing or Widening Knowledge Gaps

Without going too deep into defining what is AI or ML, it is important for us to establish a foundation for future work associated with AI. Specifically, what role does AI play in reducing knowledge gaps? The altruistic goal of AI/ML is to capture knowledge to improve some process. Such knowledge is generally provided in the form of data or as "expert" rules. The case we are certainly most interested in here is in healthcare outcomes. The former involves some type of supervised or unsupervised algorithm that attempts to associate cause and effect, while the latter attempts to emulate the knowledge of a subject matter expert. The usefulness of such systems was motivated in the discussion involving rural healthcare.

What Is Experiential Knowledge?

We don't often write material in academic terms, but we do know that many of our readers have solid academic backgrounds. We also recognize that many readers might be curious as to what is the foundation for our material. So, to satisfy those individuals who might be thinking "This is interesting, but from where does knowledge emerge?" we offer the following material that examines the concept of knowledge through the lens of human experience.

Knowledge and Human Experience: Instincts and intuition are personal tools for examining and explaining our understanding of the phenomena associated with human experience such that a refined hypothesis or idea can be formed that leads to the creation of knowledge through scientific reasoning. Moreover, and according to Dewey (1948), the exploration of the mind–body relationship with respect to human experience and prior knowledge is more than a study of observable stimuli and response such that experience is "a patchwork of disjoint parts, a mechanical conjunction, or unallied processes" (p. 356). Specifically, mind and body are connected through a seemingly disparate relationship between an individual's sensing and response, where the context of an experience influences its interpretation. With that in mind, it is proposed that the knowledge of human experiences and phenomenon cannot align singularly within the constraints of natural science and that the seed of knowledge is a system of orderly thoughts built off rational truth, which form the foundation for the discovery of further truths (Stumpf and Fieser 2015).

Subjective Analysis of Human Experience: Examining subjective analysis of the phenomena associated with human experience, from the perspective of empiricism, helps to determine its function, through isolation from objective analysis. That view of subject or interpreted assessment can yield fruitful data, but that data does not enable the individual to achieve results that can be considered as derived through scientific reasoning. With that in mind, insight gleaned from a few intellectuals may help to shed light on the concept of subjective analysis and enable an individual to reflect on its purpose.

Murphy, Alexander, and Muis's (2012) epistemological view of knowledge is best described in their statement that regardless of the theories of where knowledge comes from or where it resides, it cannot exist as unexamined beliefs. That single statement and its inference to the subjective assessment of phenomena further support the theory that knowledge is a cognitive construct (i.e., inside the mind) as well as a social construct (i.e., outside the body) (Murphy, Alexander, and Muis 2012).

Correspondingly, Dewey's (1948) explanation of subjective interpretation of human experience is defined as phenomena outside of the observable sensation and movement, such that the inclusion of "an anticipatory sensation" (p. 363) is introduced as part of the aggregation of parts that express the entire human experience. That said, human experience is the subjective study of phenomenon that are more than just the observable parts associated with stimuli and response (Dewey 1948).

Relating subjective interpretation to the method of inductive reasoning, it is posited that attempting to prove a hypothesis through inductive reasoning is flawed in that no matter how many theories or predictions can prove the correctness of a hypothesis (Wilkinson 2013). As is suggested by Watkins (1964), nature does not conform to a set of consistent rules, where theorizing truth without the consideration for the chaotic aspect of nature will result in disputable conclusions.

Objective Analysis of Human Experience: Following the examination of subjective analysis of the phenomena associated with human experience from the perspective of empiricism, it is then necessary to examine objective analysis from the perspective of rationalism. That examination will assist with the understanding that though both subjective and objective analysis methods can yield interesting results, applying the methods separately without the benefit of the other will yield results that either lack scientific methods or are too vague as to derive any conclusive results.

The theorist Edmund Husserl offers an interesting perspective on objective evidence where it is inferred that the observable connection between consciousness and an object is declared as objective evidence of phenomenon. Hussel's view may not have been the most popular at the time, but he built on solid system of beliefs in human consciousness where he proposed that as "an act of consciousness reaches its object, evidence will have occurred, and we will thereby have the experience of certainty" (Vernon, 2005, p. 290).

Husserl's view, which casts doubt on subjective interpretation, is best examined within his proposal that in transcendent consciousness there is a connection maintained with consciousness, where our vision resides outside of consciousness. Moreover, a modification of Husserl's transcendental consciousness theory might read as: a candle is lit and we see the light, where in seeing the light we see ourselves seeing, but we miss that the candle was lit, such that we could see.

Summary – Human Experience and Knowledge

Objective and subjective analysis can be used to form the foundation of instinctive/intuitive ideation that leads to the discovery of knowledge. Thus, the ontological view of interpreting phenomena and acquiring knowledge through inductive reasoning combined with intuition and instinctual assessments is supported through the proposed solution of scientific spectrum redefinition and subjective/objective unification. That concept is further supported by Dewey's (1948) inference to the role that ontology plays through the interpretation of human experience as well as Heidegger's proposal that ontology is the truest method for assessing human experiences such as pain, suffering, hunger, and procreation (Ciocan, 2008). Additionally, the data that supports this material's hypothesis is able to cast light on to the epistemological question of where knowledge comes from and where it resides. Specifically, knowledge is acquired through the formulation of a question and an idea, where it is composed of innate knowledge, instinctive/intuitive examination of phenomena, and interpretation of those phenomena (i.e., through induction). As an individual then uses the ontological results to deduce the truth of the results, knowledge will be exposed which will then be consumed, retained, and shared as it becomes an integral part of the mind and body experience.

Terms and Definitions

AI/ML – *Artificial Intelligence/Machine Learning*: Short for artificial intelligence (AI) and machine learning (ML), they represent an important evolution in computer science and data processing. Artificial intelligence generally refers to processes and algorithms that are able to simulate human intelligence, including mimicking cognitive functions such as perception, learning, and problem-solving. Machine learning and deep learning (DL) are subsets of AI.

BOM – *Bill of Material*: A list of raw materials, sub-assemblies, intermediate assemblies, sub-components, parts, and the quantities of each needed to manufacture an end product.

DEI – *Diversity, Equality, and Inclusion*

- **Diversity**: The presence of differences within a given setting. In the workplace, this generally refers to psychological, physical, and social differences that occur among any and all individuals. A diverse group, community, or organization is one in which a variety of social and cultural characteristics exist.

- **Equity**: Ensures everyone has access to the same treatment, opportunities, and advancement. Equity aims to identify and eliminate barriers that prevent the full participation of some groups. Specifically, barriers can come in many forms, but a prime example can be found in this study. In it, researchers asked faculty scientists to evaluate a candidate's application materials, which were randomly assigned either a male or a female name. Faculty scientists rated the male applicant as significantly more competent and hirable than the identical female applicant, and offered a higher starting salary and more career mentoring to the male applicant.

- **Inclusion**: Refers to how people with different identities feel as part of the larger group. Inclusion doesn't naturally result from diversity, and in reality, you can have a diverse team of talents, but that doesn't mean that everyone feels welcome or valued. Diversity, equity, and inclusion are mutually reinforcing principles within an organization. A focus on diversity alone is insufficient because an employee's sense of belonging (inclusion) and experience of fairness (equity) is critically important.

EMR/EHR – *Electronic Medical Records/Electronic Healthcare Records*: EHR and EMR are methods of healthcare record keeping that can exist in electronic or paper format, as might be found on a portable or desk-mounted piece of technology. According to Rosenbaum (2015), EHRs were created to improve methods for healthcare billing (versus improving the workflow of healthcare professionals). That point is supported by Zheng, Abraham, Novak, Reynolds, and Gettinger's (2016) material that proposed that EHRs were not created by the healthcare industry to solve the issues of transforming data between dissimilar solutions.

Formula 1 – *Formula 1* cars were given the name because they are the cars that are the fastest, most aerodynamic, power-efficient machines that have been designed on four wheels. The term "formula one" basically means the formula for the best.

IASAM3 – *IASAM3* is a new model that enables technology integrators to evaluate new technology through the lens of evaluating aspects of socioeconomical and technology characteristics that move beyond the assessment of pre-adoption phases and toward the examination of assessment, decision-making, integration, and sustainability of technological solutions (Aizstrauta and Ginters 2017; Ginters, Mezitis, and Aizstrauta 2018).

IEEE/UL TIPPSS – It is an IEEE/UL standard that establishes the framework with TIPPSS principles (Trust, Identity, Privacy, Protection, Safety, Security) for Clinical Internet of Things (IoT) data and device validation and interoperability. The standard includes wearable clinical IoT and interoperability with healthcare systems, including EHR, EMR, other clinical IoT devices, in hospital devices, and future devices and connected healthcare systems.

Interoperability – *Interoperability* is the ability for humans, hardware, and software technology to connect and communicate using a standard protocol, where data may be transformed across dissimilar technology (Wager, Lee, and Glaser 2017).

Knowmad – A *knowmad* is a nomadic knowledge worker who is creative, imaginative, innovative, and who can work with almost anybody, anytime, and anywhere. Knowmads are valued for their individual-level knowledge, and create new value by applying what they know, contextually, to solve problems or generate new opportunities.

VoC – *Voice-of-Customer* is the component of customer experience that focuses on customer needs, wants, expectations, and preferences. To determine the VoC, an organization analyzes indirect input or data that reflects customer behaviors as well as direct input or data that reflects what a customer says.

References

Abdalla, W., Renukappa, S., & Suresh, S. (2022). An evaluation of critical knowledge areas for managing the COVID-19 pandemic. *Journal of Knowledge Management*. https://doi.org/10.1108/JKM-01-2021 -0083

Aizstrauta, D., & Ginters, E. (2017). Using market data of technologies to build a dynamic integrated acceptance and sustainability assessment model. *Procedia Computer Science*, 104, 501–508. https:// doi.org/10.1016/j.procs.2017.01.165

Brach, C., & Harris, L. M. (2021). Healthy people 2030 health literacy definition tells organizations: Make information and services easy to find, understand, and use. *Journal of General Internal Medicine*, 36(4), 1084–1085.

Brugués de la Torre, A. (2016). Contributions to interoperability, scalability and formalization of personal health systems. Retrieved from https://upcommons.upc.edu/bitstream/handle/2117/96346/ TABT1de1.pdf

Cahyanti, D., & Permana, I. (2022). Comparison of book shopping before and during the COVID-19 pandemic using the FP-Growth algorithm at Zanafa bookstores. *Jurnal Teknik Informatika (Jutif)*, 3(2), 381–386. https://doi.org/10.1016/j.egyr.2022.02.125

Castro Rodrigues, D. de, Siqueira, V., Tavares, F., Lima, M., Oliveira, F., Osco, L., Junior, W., Costa, R., & Barbosa, R. (2022). Discovering associative patterns in healthcare data. In *Proceedings of Sixth International Congress on Information and Communication Technology* (pp. 371–379). Singapore: Springer.

Chen, K. C., & Jang, S. J. (2010). Motivation in online learning: Testing a model of self- determination theory. *Computers in Human Behavior*, 26(4), 741–752. https://doi.org/10.1016/j.chb.2010.01.011

Christodoulakis, C., Asgarian, A., & Easterbrook, S. (2016). *Barriers to Adoption of Information Technology in Healthcare*. https://doi.org/10.475/123_4

Ciocan, C. (2008). The question of the living body in Heidegger's analytic of Dasein. *Research in Phenomenology*, 38(1), 72–89. https://doi.org/10.1163/156916408X262811

Deci, E. L., & Ryan, R. M. (2008a). Facilitating optimal motivation and psychological well-being across life's domains. *Canadian Psychology*, 49(1), 14–23. https://doi.org/10.1037/0708-5591.49.1.14

Deci, E. L., & Ryan, R. M. (2008b). Self-determination theory: A macrotheory of human motivation, development, and health. *Canadian Psychological*, 49(3), 182–185. https://doi.org/10.1037/a0012801

Demarzo, M. M. P., Cebolla, A., & Garcia-Campayo, J. (2015). The implementation of mindfulness in healthcare systems: A theoretical analysis. *General Hospital Psychiatry*, 37(2), 166–171. https://doi.org /10.1016/j.genhosppsych.2014.11.013

Dewey, J. (1910). Systematic inference: Induction and deduction. In *How We Think* (pp. 79–100). Lexington, MA: D C Heath. https://doi.org/10.1037/10903-007

Dewey, J. (1948). The reflex arc concept in psychology, 1896. In W. Dennis (Ed.), *Readings in the History of Psychology* (pp. 355–365). East Norwalk, CT: Appleton- Century-Crofts. https://doi.org/10.1037 /11304-041

Dwyer, C. (2017). *Critical Thinking: Conceptual Perspectives and Practical Guidelines*. Cambridge: Cambridge University Press. doi:10.1017/9781316537411

Evans, S., Vladimirova, D., Holgado, M., Van Fossen, K., Yang, M., Silva, E. A., & Barlow, C. Y. (2017). Business model innovation for sustainability: Towards a unified perspective for creation of sustainable business models. *Business Strategy and the Environment*. https://doi.org/10.1002/bse.1939

Gajda, R. (2004). Utilizing collaboration theory to evaluate strategic alliances. *American Journal of Evaluation*, 25(1), 65–77. https://doi.org/10.1177/109821400402500105

Garrison, D. R., Anderson, T., & Archer, W. (1999). Critical inquiry in a text-based environment: Computer conferencing in higher education. *The Internet and Higher Education*, 2(2), 87–105. https://doi.org/10.1016/S1096-7516(00)00016-6

Ginters, E., Mezitis, M., & Aizstrauta, D. (2018). Sustainability simulation and assessment of bicycle network design and maintenance environment. In *2018 International Conference on Intelligent and Innovative Computing Applications (ICONIC)* (pp. 1–16). IEEE. https://doi.org/10.1109/ICONIC.2018.8601225

Gray, B. (1989). *Collaborating: Finding Common Ground for Multiparty Problems*. San Francisco, CA: Jossy-Bass.

Hitchcock, B. J. A., & Stavros, J. M. (2017). Organizational collective motivation. *OD Practitioner*, 49(4), 28–36.

Hosfield, J. (2017). Electronic medical records (EMR) adoption, patient mortality, and patient safety in US hospitals: A longitudinal study (2008–2014) (Doctoral dissertation, Capella University). Retrieved from http://search.proquest.com/docview/1877994521/previewPDF/6422078714DD4090PQ/1?accountid=7374

Janaswamy, S., & Kent, R. D. (2016, February). Semantic interoperability and data mapping in EHR systems. In *Advanced in Computing (IACC), 2016 IEEE 6th International Conference On* (pp. 117–122). IEEE. https://doi.org/10.1109/IACC.2016.31

Jardí, A., Webster, R., Petrenas, C., & Puigdellívol, I. (2022). Building successful partnerships between teaching assistants and teachers: Which interpersonal factors matter? *Teaching and Teacher Education*, 109, 103523. https://doi.org/10.1016/j.tate.2021.103523

Kobusinge, G., Pessi, K., Koutskouri, D., & Mugwanya, R. (2018). An implementation process of interoperability: A case-study of health information systems (HIS). Retrieved from https://aisel.aisnet.org/isd2014/proceedings2018/eHealth/1/

Kohn, A. (1999). Punished by rewards: The trouble with gold stars, incentive plans, A's, praise, and other bribes. Houghton Mifflin Harcourt. Retrieved from https://dspace.library.colostate.edu/bitstream/handle/10217/18176/JOUF_JSA1997.pdf?s equence=1&isAllowed=y#page=75

Mandl, K. D., Mandel, J. C., & Kohane, I. S. (2015, June 11). *Driving Innovation in Health Systems through an Apps-Based Information Economy*. Science Direct. https://doi.org/10.1016/j.cels.2015.05.001

Murphy, P. K., Alexander, P. A., & Muis, K. R. (2012). Knowledge and knowing: The journey from philosophy and psychology to human learning. In K. R. Harris, S. Graham, T. Urdan, C. B. McCormick, G. M. Sinatra, & J. Sweller (Eds.), *APA Educational Psychology Handbook, Vol 1: Theories, Constructs, and Critical Issues* (pp. 189–226). Washington, DC: American Psychological Association. https://doi.org/10.1037/13273-008

Owaid, O. (2016). The death of private practice: How the rising cost of healthcare is destroying physician autonomy. *Brooklyn Journal of Corporate, Financial & Commercial Law*, 11(6), 521. Retrieved from http://brooklynworks.brooklaw.edu/bjcfcl/vol11/iss2/9

Parker, M., & Partridge, B. (2021). Management and leadership distinctions required at stages of maturity in an EHR/EMR adoption model context. *Mobile Medicine: Overcoming People, Culture, and Governance*.

Rew, D. (2020). A surgeon among engineers. *The Bulletin of the Royal College of Surgeons of England*, 102(4), 154–157.

Rosenbaum, L. (2015). Transitional Chaos or Enduring Harm? The EHR and the Disruption of Medicine. *N Engl J Med*, 373(17), 1585–1588. doi: 10.1056/NEJMp1509961. PMID: 26488690.

Schwartz, R., & Stanovsky, G. (2022). On the limitations of dataset balancing: The lost battle against spurious correlations. arXiv Preprint ArXiv:2204.12708.

Shanafelt, T. D., Dyrbye, L. N., Sinsky, C., Hasan, O., Satele, D., Sloan, J., & West, C. P. (2016, July). Relationship between clerical burden and characteristics of the electronic environment with physician burnout and professional satisfaction. In *Mayo Clinic Proceedings* (Vol. 91, No. 7, pp. 836–848). Elsevier. https://doi.org/10.1016/j.mayocp.2016.05.007

Slight, S. P., Berner, E. S., Galanter, W., Huff, S., Lambert, B. L., Lannon, C., Lehmann, C. U., McCourt, B. J., McNamara, M., Menachemi, N., Payne, T. H., Spooner, S. A., Schiff, G. D., Wang, T. Y., Akincigil, A., Crystal, S., Fortmann, S. P., Vandermeer, M. L., & Bates, D. W. (2015). Meaningful use of electronic health records: Experiences from the field and future opportunities. *JMIR Medical Informatics*, 3(3). https://doi.org/10.2196/medinform.4457

Stumpf, S. E., & Fieser, J. (2015). Rationalism on the continent. In S. Greenberger (Ed.), *Theories of Inquiry*. New York: McGraw-Hill. Retrieved from http://www.gcumedia.com/digital-resources/mcgraw-hill/2015/theories-of-inquiry-bundle_ebook_1e.php

Venugopal, M. (2015). Application of SMAC technology. *Software Innovations in Clinical Drug Development and Safety*, 90. https://doi.org/10.4018/978-1-4666-8726-4.ch006

Vernon, R. F. (2005). Peering into the foundations of inquiry: An ontology of conscious experience along Husserlian lines. *Journal of Theoretical and Philosophical Psychology*, 25(2), 280–300. https://doi.org/10.1037/h0091263

Victores, A. J., Coggins, K., & Takashima, M. (2015). Electronic health records and resident workflow: A time-motion study of otolaryngology residents. *The Laryngoscope*, 125(3), 594–598. https://doi.org/10.1002/lary.24848

Wager, K. A., Lee, F. W., & Glaser, J. P. (2017). *Health Care Information Systems: A Practical Approach for Health Care Management*. Hoboken, New Jersey: John Wiley & Sons.

Wang, B., Schlagwein, D., Cecez-Kecmanovic, D., & Cahalane, M. C. (2020). Beyond the factory paradigm: Digital nomadism and the digital future (s) of knowledge work post-COVID-19. *Journal of the Association for Information Systems*, 21(6), 10.

Watkins, F. (1964). The Moral and Political Philosophy of David Hume. By John B. Stewart. (New York and London: Columbia University Press, 1963. Pp. xii, 422. $7.50.). *American Political Science Review*, 58(4), 990–991. doi:10.1017/S0003055400291430

Wilkinson, M. (2013). Testing the null hypothesis: The forgotten legacy of Karl Popper? *Journal of Sports Sciences*, 31(9), 919–920. https://doi.org/10.1080/02640414.2012.753636

Wineburg, S., & McGrew, S. (2016). Why Students Can't Google Their Way to the Truth. *Education Week*, 36, 22–28.

Zheng, K., Abraham, J., Novak, L. L., Reynolds, T. L., & Gettinger, A. (2016). A survey of the literature on unintended consequences associated with health information technology: 2014-2015. *Yearb Med Inform*, 10(1), 13–29. doi: 10.15265/IY-2016-036. PMID: 27830227; PMCID: PMC5171546.

Chapter 3

Tech Misinformation and Medical Misinformation as Evil Twins: The Misinformation Path to Destroying Trust in Medical Technology

Sherri Douville, Michael DeKort, and Kate Liebelt

Contents

Introduction

Technical and medical misinformation have much in common in that they both represent instances where the misinformation is used or even weaponized for an end to a means. They combine to create a perfect storm of circumstances that lead to mistrust and significant risk to stakeholders, the most important stakeholder being the patient. The practice of medicine relies on evidence

DOI: 10.4324/9781003348603-4

– and it is difficult to imagine a discipline where timely, objective, and accurate information and trusted means of communicating that information from a technology standpoint is *more critical to societal well-being than healthcare*. As the medical community wades through volumes of clinical and scientific data – from text messages exchanged between physicians to medical journal articles to magazine stories to patient testimonials on social media – they become highly susceptible to the risk of being attacked by stakeholders who believe in misinformation. Because technology is a relatively young field, there is a lack of awareness, knowledge, and discipline around the critical assessment of misinformation and the impact of misinformation on trust. It is an interesting coincidence that some of the most notable sources of medical misinformation are also known for the aggressive and often premature evangelizing of technologies. (which medical stakeholders would consider to be fraud) This chapter explores causes and impact of the technical and medical misinformation dynamic and offers perspective on countering this evil duo cratering trust for medical technology in medicine. Why should you care about misinformation? The Edelman Trust Barometer survey teaches us that trust with our workforce and stakeholders is urgent. It also points out that trust hinges on information hygiene, meaning information that is based upon facts that can be proven to be accurate (Edelman 2021).

Identifying Misinformation = Distrust

It's well understood that through research, students frequently struggle to evaluate the credibility of information online (Wineburg et al., 2016).

This chapter is written for:

■ Scientists
■ Physicians
■ Systems engineers
■ Bioengineers
■ Leaders who have to earn trust and maintain constructive working relationships based on trust and respect of scientists, physicians, systems engineers, and bioengineers

There are two layers to address for technology misinformation relevant to medicine:

1. Outright fraud and exaggeration
2. Ignorance about medical evidence; thus the inability to understand where one's technology stands with respect to medical evidence, the language that physicians speak

Unfortunately, it's not just those outside the medical community spreading misinformation. That's why at their recent annual meeting, the AMA voted for a report that characterizes any use of physician credentials as a time where they are bound by professional standards of conduct that would prohibit the spread of misinformation.

Evidence-based medicine (EBM) was originally defined as the conscientious, explicit, and judicious use of current best evidence in making decisions about the care of individual patients. The practice of evidence-based medicine means integrating individual clinical expertise with the best available external clinical evidence from systematic research (Sackett et al. 1996).

What does this mean when doctors are treating patients? If you're related to a doctor or multiple physicians, then you might get a chance to hear their conversations about patient care planning. What will strike you if you pay attention is how often they will bring up references to peer-reviewed papers in the majority of their sentences in many cases. They will often discuss their impressions of the quality of the papers such as number of participants, what the paper is analyzing, whether that is a single experiment, one at multiple centers, or a meta-analysis, and especially if the results are replicable. This progression roughly represents a positive progression of how they interpret the credibility of papers.

Because technology is a much younger field, there's not as much evidence. While certain segments of technology for high-reliability and high-performance industries such as computing chips have been generating evidence and publishing that evidence for some time, that hasn't been the norm in the consumer software industry. However, when there's less scholarship, the next best thing to evidence is sharing the basis, first principles facts, and reasoning by which something was decided. For example, physicians will occasionally use interventions that lack the highest levels of medical evidence taking into account first principles of science and a risk analysis to the patient's presentation. We think the latter needs to be a minimum bar for technology as well.

Examples of Gaps between Tech and Medicine

For example, let's take the firing of employees at firms recently for pointing out technical challenges. The purpose of this observation is to underscore how incompatible this censorship of research and data approach rooted in a philosophy of corporate secrecy is with evidence-based medicine, the scientific method and expectation for peer review, and the level of transparency that is expected in medical science. Prominent researchers and leaders in the field don't take kindly to scientific standards of conduct being violated which damages the reputation of firms in those research and related medical communities.

The point here isn't to shine light on specific poor behaving companies. The point is to underscore philosophical and expectations gaps between the industries of technology and medicine.

Note: There are examples of earlier stage purpose-built companies working to do the right things with software in healthcare in a way that's relevant to medical technology and literacy for clinical evidence according to our coauthor team. Those examples include but are not limited to Pear Therapeutics, Omada Health, Otsuka (coded pill), Akili Interactive, and Medigram.

"*Physicians are trained to review credibility, relevance, logical strength,* and *balance* clinical assessment which are productive principles for any situation; this helps us understand the quality, meaning the overall strength or weakness of any argument" (Dwyer, 2017).

An Excerpt by a Systems Engineer for Systems Engineers: A Call to Protect the Public

In this section, we explore the use of development and testing methods for autonomous vehicles that appear from a systems engineering perspective to be untenable from safety, time, and cost points of view. The safety issues include harming the public as test subjects. One of the main problems is a lack of domain and systems engineering expertise by the industry. Some industry stakeholders therefore lack awareness of appropriate methodologies for the risk level of the work, specifically aerospace and Department of Defense (DoD) simulation technology.

It must also be acknowledged that general and deep learning approaches are nowhere near where they need to be for this domain to do what "visionary evangelists" say they want it to. In the current technical development model, it is difficult to see how there will ever be a true driverless system. (Society of Automotive Engineers, SAE Level 4). The related gaps represent human cost to safety, lives, and potential damage to companies, shareholders, employees, and other stakeholders. Relevant and applicable aerospace and defense simulation technology and systems engineering approaches have existed for 25 years. If the right approaches are embraced, the holy grail can be reached. Only then can we drastically lower the roughly 38,000 traffic fatalities a year, provide aid for those who cannot drive, and lower the impacts of cars on global warming.

One challenging issue with technology, especially technology systems, is that the technology and healthcare, medical technology field, is very large and heterogeneous, analogous to medical specialty types; however, it is more sprawling and with the frequent absence of being bound by common values or purpose. We've seen a movie like this before, first with traditional automobile cars and auto safety until as our physician coauthor. Dr. Arthur Douville wisely points out the "Unsafe at Any Speed" Movement.

Today, it can be said that autonomous vehicles, to some degree, appear to sacrifice safety measures borne out of technical choices made by many vehicle makers. In a nutshell, autonomous vehicle engineers and executives may know what to do technically, but their culture can prevent them from doing what the best systems safety engineers know the right things are to prioritize human safety.

In medical technology, it's not typically or mostly a dark situation by intent or potential design. Though we're not out of the woods. The issue from our perspective is that many stakeholders don't know what to do and many can be overwhelmed by the technical, as well as deployment and implementation complexities inherent in medical technology. Our goal with this section is to shine a light on what is known to solve safety challenges in autonomous vehicles, so we can think about how this logic and how the challenges from autonomous vehicles can apply to designing the future of medical technology where arguably the stakes for safety are even higher than for cars (Figure 3.1).

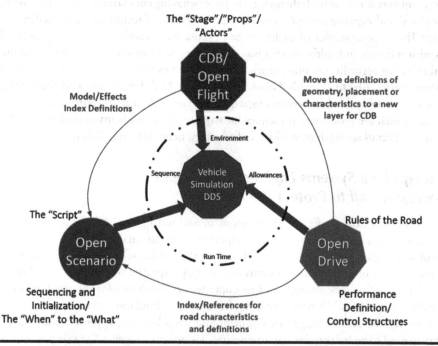

Figure 3.1   Ground Truth for Autonomous Vehicles.

For effective autonomous vehicles, ground truth is a comparison to the real world. In this domain, that is in both physics and visually. That includes position, the environment, topography, objects, etc. The only part up for interpretation is the required fidelity of each for any given scenario or use case. Intelligent systems engineers who are expert in the domain of autonomous vehicles widely recognize the solution that is to facilitate the concurrent use of the OpenFlight, OpenDrive, and OpenScenario standards, the latter for sequence. Specifically, the OpenDrive standard for simulation does not have the ability to normalize on positions because the standard doesn't force it. Though OpenFlight does. Essentially, it is known what the right thing is to do and without pursuing too many technical details here; the visualization above is a high-level summary. The important piece is that if it's known what to do, why isn't it the norm?

A significant example of challenges have involved the autonomous vehicle industry. Industry insiders characterize a very complex combination of over-exuberance and ignorance caused by IT and consumer web industry developers' acute lack of domain and systems engineering experience. Like many technologies, there has been hype, needless loss of life, and a public sentiment level growing which indicates a lack of trust with false confidence in the systems being touted. Here we have a unique circumstance where it can be said that the driver, passenger, pedestrian, or concurrent vehicle appear assumed to be the test both inside and outside the vehicles. The rationale is that the machine learning process used for development and testing of these systems can literally experience scenarios, over and over again to "learn" them. The critical issue is that most crash cases fall into this category.

This means that many people will literally have to go through crash scenarios directly for those experiences to create "learning." That might be tenable if this process eventually led to the creation of an autonomous vehicle and saved most of the roughly 38,000 people killed each year in automobile crashes as the stated goal. The problem is that this possibility doesn't appear technically feasible. The process requires so many "miles" to be driven repeatedly. What company can sustain the volume of activity needed? The solution is to transfer most of this real-world stumbling and re-stumbling on scenarios and object variations to simulation. The gap today is that many industry stakeholders rely on gaming-simulation technology. That technology, unlike the technology used by aerospace and defense, lacks the real-time and modeling fidelity required to handle complex scenarios and detailed sensor models. Thus, inhibiting its use to reach autonomy as well. And finally, there are efficiency and quality issues around the core machine learning processes being used. Those being general learning for scenarios and deep learning for objects. The current approaches for those are so nascent that billions of object and degraded object images and millions of scenarios and their variations must be experienced over and over for them to be memorized versus learned. This is because general learning is not capable of any significant levels of inference at this point. And deep learning gets confused over a small difference between object variations because it scans pixels inside out versus outside in as humans do. When pixels are discovered that have not been associated with an object detected, then the system gets "confused": like with shadows or dirt on a stop sign. This brings us to the most extreme examples on the market. One can go even farther astray in a situation where it relies on a camera for its only sensor system. Superior options would likely use LiDAR and radar in addition to cameras. There is an understandable ethical objection to having untrained customer test drivers or "safety drivers."

Could a 20-second or longer delay on a driver monitoring system be considered safe or responsible? Should a consumer technology company be able to bring unproven data to the public through making claims with the naming of its systems with words such as "Autopilot" and "Full Self-Driving" for examples? The question begs because neither of these names can be validated with technical performance evidence at this point. Each year for nine years they are predicted, only to be postponed again the following year. The Society of Automotive Engineering J3016 levels 1–5 are defined as follows: Level 1 is driver assist where the driver never cedes steering. L2 is driver assist where they can cede steering for very short periods but need to take over. This extends to development and testing for L3–L5. L3 is self-driving but only in certain operational design domains (ODD). L4 is self-driving in the public domain. L5 is self-driving anywhere the vehicle can manage to drive, including off road. To date, 13 people have died as the result of the development and "safety driving" process for one brand. (It should be noted here that to date three companies have said they have actually achieved a driverless state or L4 in micro-ODDs.)

Finally, there is the issue of scant regulations, standards, testing, and proof of capabilities. There are no "driver's tests" anywhere in the world. No one has had to prove their capabilities or demonstrate that they have reached a driverless state. Further, the only data being asked for involves real-world crashes and identification of disengagements. The shortfalls here include the fact that crashes in simulation are not included, crashes that would have occurred if a human did not disengage are not included, and there is no information on scenarios learned, especially crash scenarios. No information showing the models used in simulation, especially for sensors, has the right level of fidelity. One company even sued a government entity to avoid providing safety data citing IP concerns. In an ideal world, one would think that if a company actually created an autonomous vehicle, they would leap at the opportunity to prove it publicly, thereby increasing public trust and even differentiating themselves from their competitors. Instead, there is often hype in a drive to accelerate cash out via exit. In the end, how would any of these situations compare to some of the largest engineering debacles in history? We think this is an important discussion not just for any assessment of the self-driving industry; more importantly, it provides critical lessons and cautionary tales that in the context of medicine must be handled with more care and precision reflective of an even greater potential impact on human lives.

The plural of anecdote is misinformation.

– Dr. Jonathan Stea

I'm not sure if people adequately understand what is happening or recognize the consequences of a disinformation echo chamber which has merged conspiracy theories, normalized extremism, and glorified violence, but this is just the beginning. Be safe out there, my friends.

Taylor Nichols, Tweet Quote, December 5,

Here is this chapter's summary observation of medical and tech misinformation.

Hierarchy of Medical Misinformation

1. *Highest level*: disinformation for economic reasons
2. *Medium level*: siloed information sources without incorporating new disciplines or emerging facts
3. *Lowest level*: sheer ignorance of how medicine and science work

Hierarchy of Tech Hype Misinformation

1. *Highest level*: disinformation for economic reasons
2. *Medium level*: siloed information sources without incorporating new disciplines or emerging facts
3. *Lowest level*: sheer ignorance of how technologies in other parts of the tech stack work

Masks are aerosol physics, and physics is not political. It is either disingenuous or misinformed to overgeneralize various mask types, for example, comparing and lumping in cloth masks to regulated N95 respirators. It is more extreme than comparing apples and oranges. Making a conclusion about masks based on one type of mask like a paper mask instead of a fitted respirator is like comparing a pinto car's racing capabilities to a Maserati. This is inaccurate, unhelpful misinformation. But this is what happens when people point to studies or papers that are not apples for apples by talking about things they don't truly understand at a finely detailed level.

When Dogma and Physics Collide to Kill: Why Many People Don't Know That COVID Is Airborne

In an insightful WIRED article, the author had this to say about essential and unrecognized technical and engineering expertise essential to managing the pandemic:

> Marr was no stranger to being ignored by members of the medical establishment. Often seen as an epistemic trespasser, she was used to persevering through skepticism and outright rejection. This time, however, so much more than her ego was at stake. But she had an inkling that the verbal sparring was a symptom of a bigger problem—that outdated science was underpinning public health policy.

(Molteni 2021)

Resistance to outsiders is illustrated in Marr's challenge to bringing commonsense physics facts to light for risk reduction. This isn't the only barrier to bringing technology into medicine. Those trained in the scientific method are held by their professional colleagues and friends to the standard of communicating in an evidence-based way. This is why the way that technologies are marketed into medicine are often met with opposition. When describing one tech company's difficulties in medicine, industry insiders pointed to the resignation letters of physicians on their staff. The root cause of the misalignment between the physicians and the company is the barrier that secrecy presents in the obligation to the scientific method in that instance.

Conclusion

Scientists, physicians, engineers, and leaders (whose remit includes engagement with these stakeholders) can successfully create trust, influence, and make a significant impact by considering the risks of technical and medical misinformation.

Next Steps: Team Training to Bust Medical (and Medical Technology) Misinformation

1. Create and agree to a code of conduct that includes centering all conversations around ideally, peer-reviewed evidence levels.
2. Screen sources for failing fact checking and bias ratings. Aim for the highest facts ratings with no bias (/mediabiasfactcheck.com).
3. Onboard techies and outsiders to medicine for understanding by way of company policies. At Medigram, we use this tool for why evidence is pillar to what a medical stakeholder considers fake news or misinformation recommended by Medigram's Chief Medical Officer and practicing neurologist Dr. Art Douville: the GRADE EBM framework, hhttps://bestpractice.bmj.com/info/toolkit/learn-ebm/what-is-grade/

Studies show that people high in intellectual humility (IH) pay more attention to evidence and are interested in the reasons that other people disagree with them, rather than automatically squashing, diverging, or opposing ideas and people. This includes a reduction or absence of the use of what is referred to as ad hominem arguments. This is when someone engages in disagreement by attacking a person's character or unrelated attributes to the issue being discussed as a means of discrediting the person's ideas. We're pleased to refer you to the Medicine in Humility chapter 7 to learn more about the need for, current state of, and action plan for what you can do to foster necessary humility for overcoming misinformation in your organization.

References

Edelman. (2021). Edelman trust barometer. https://www.edelman.com/trust/2021-trust-barometer

Sackett, D. L., Rosenberg, W. M. C., Gray, J. A. M., Haynes, R. B., Richardson, W. S. (1996). Evidence based medicine: What it is and what it isn't. *BMJ*, 312, 71–72.

Molteni, M. (2021, May 13). The 60-year-old scientific screwup that helped covid kill. *WIRED*. https://www.wired.com/story/the-teeny-tiny-scientific-screwup-that-helped-covid-kill/

MANAGEMENT AND LEADERSHIP COMPETENCIES AND OBJECTIVES FOR DRIVING THE SCIENCE, MEDICINE, AND ENGINEERING OF ADVANCED MEDICAL TECHNOLOGY FORWARD

II

Chapter 4

Allyship in Reducing Medical Technology Risk: Why Partnerships Are Vital to Your Professional Success

William C. Harding, Mike Ng, Brittany Partridge, and Sherri Douville

Contents

DOI: 10.4324/9781003348603-6

Introduction

As industry leaders, it is essential that you set the highest standards associated with behaviors and results that keep your organization not just afloat but ahead of all others. Those behaviors might be your transformative leadership style, your decision-making skills, and/or your ability to surround yourself with talented individuals who contribute their own diverse perspectives. Specifically, as leaders, you should never work alone and never in a bubble surrounded by "yes" people. Instead, you must embrace the traits and behaviors associated with listening, empathy, communications, and transparency, which represent attributes that you must hone and continuously improve. Furthermore, you must seek to consider more than your own point of view and make informed decisions that are representative of a collective perspective. However, if you think that you have nothing to learn and are the master of your own domain, and then continue your practices of blaming others for your own failings and mistakes; then you risk driving your business into the ground. That said, this chapter is focused toward those individuals who have learned that they are not the only source of truth and that the best decisions represent the views of a collaborative organization.

Prelude

Humans are social animals who thrive when they collaborate toward mutually shared goals. Specifically, humans succeed when they work as allies. That same point is easily transferable to any discussion around innovation, ideation, or brainstorming. Where nearly every successfully deployed innovative life sciences solution achieved success because people worked together. Accordingly, there are some well-defined tools that help people work together on solutions (i.e., Design Thinking and Human-Centered Design). However, what might not be obvious is a critical concept that forms the foundation of collaborative tools such as Design Thinking and Human-Centered Design. That concept is diversity, equity, and inclusion (DEI), where the strength of the tools is in their ability to level-set all participants (e.g., no professional titles are acknowledged), to give voice to all participants, and to encourage different views as part of lived experiences (Pollock, Samuelson, and Silverthorn 2022). In other words, diversity, in its many forms (e.g., age, race, education, profession, gender, and ethnicity) enriches our experiences and helps us to evolve as social animals.

Thinking about DEI as being a foundational element associated with successfully deployed innovative solutions, it shouldn't be surprising to learn that DEI is strongly considered when potential employees are evaluating a prospective organization. Additionally, many individuals leave organizations because behaviors, actions, and practices do not reflect that an organization has truly embraced DEI as a recurring theme throughout all elements of the organization. Consequently, organizations will have a short life expectancy if they fail to recognize the need for creating an environment that embraces a DEI model. And companies who fail to align their corporate mission with a DEI model will find it difficult to retain employees (Brown and Pierce 2022).

Moving from our introductory discussion associated with allyships and how they are strengthened by the inclusion of diverse perspectives, we first focus on how alliances can reduce risks in the healthcare industry, by looking at what we consider to be either good or bad allies. Of course, the idea of a bad ally may not be entirely clear, so we have provided some use cases and examples that you can consider as you navigate your own environment, in search of "good" allies. That said, we don't suggest that you write off all "bad" allies, where instead we make proposals on how you as the "good" ally might assist other individuals in their journey to also becoming a "good" ally.

Aligning Allies with Reducing Risk and Increasing Success

The purpose of this first section answers the ever-important "why" allies are essential to reducing medical technology risk. You'll learn how to navigate and identify which allies best fit your team because even with the best intentions not all allies are good allies.

Pros and Cons of Allies

Good Allies: They think like a multiplier. They know that success for you means more success for them and reflects well on them. They don't fear more people becoming successful. They subscribe to a rising tide lifting all boats instead of having a mindset of scarcity.

Many worthwhile goals require taking risks, constantly and repeatedly (Bhide 1994). One of those is the act of sponsoring underrepresented talent. When a talent profile doesn't fit the status quo and there's not a lot of obvious pattern-matching available, sponsoring new kinds of talent is a risk that leaders choose to take or not.

One famous example of a successful person using their power to lift up a mentee is Marilyn Monroe and Ella Fitzgerald. Ella Fitzgerald was not allowed to perform in Hollywood's most popular nightclub, The Mocambo, because she wasn't considered glamorous enough. Marilyn Monroe, who was a big fan, called the owner and explained that if he booked Ella, she would be there every night, which guaranteed huge press coverage. He booked Ella, and Marilyn was there, front table, every single night as promised. Ella said, "After that, I never had to play a small jazz club again." She was an unusual woman, a little ahead of her time, and she didn't even know it (Krohn 2001).

We suggest that you make becoming a person of high self-esteem and identifying allies and colleagues with high self-esteem an explicit goal. So much value is lost in relationships and organizations when leaders or colleagues are insecure. Instead of authentically collaborating, egos present challenges that block success, and instead of people working to bring out the best in each other; they try to dominate each other when self-esteem is threatened. In "Why Bad Bosses Sabotage Their Teams," the author observes that insecure bosses sideline their best people and limit collaboration in order to protect their status and ego (Choo, Byington, Johnson, and Jagsi 2019).

Persons of high self-esteem (Figure 4.1) are not driven to make themselves superior to others; they do not seek to prove their value by measuring themselves against a comparative standard. Their joy is being who they are, not in being better than someone else – Nathaniel Branden (1994).

People exceptionally talented in the Self-Assurance theme feel confident in their ability to take risks and manage their own lives (Schwartz 1999). They have an inner compass that gives them certainty in their decisions. They will identify themselves as being masters of embracing conflict. They will challenge you with radical coaching and honesty, for example, if you make a recommendation request from them. Good allies recognize their success is tied to many others. While it's important that we drive and work hard, we can't attribute our success to ourselves alone.

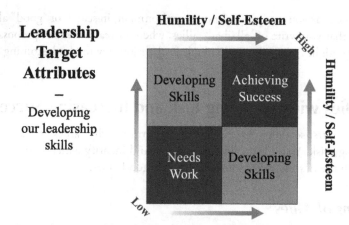

Figure 4.1   Leadership target attributes. Source: Harding, Medtronic, plc & Douville, Medigram, Inc. (2022).

Bad Allies: In Figure 4.1, we illustrate the target profile of a positive ally as someone with high self-esteem and high humility. This type of person can be a great role model or mentor, and be limited in the negativity they bring to the ally relationship. Much has been covered about the challenges to workplace environments, specifically goals related to governance and diversity in the *hubris* in technology chapter 6. In the section on impacts of malignant narcissism, if a former ally is threatened by newfound or rising success of a mentee, that can become dangerous (Levy, Katz, Wolf, Sillman, Handin, and Victor 2004).

The two keys are as follows:

1. Try to limit your time investment to the upper right quadrant when seeking out and prioritizing allies.
2. No one is completely free of narcissism as well as that human emotion of envy, even subconsciously. It is therefore very important to create abundance as you succeed. Look for the good things that others contributed. Give all the credit that's deserved. Stop and recognize everyone who helped with every success (Goldsmith and Mark 2007). Lift as you climb and you'll be more successful having long-term allies rather than those that turn on you.
   Note 1: There is also a reputational risk of misalignment always included between allies. Another suboptimal ally and one challenge is when people have power but don't recognize or know how to use it to make a difference.
   Note 2: Allies may underperform their ability to help and drive impact when they don't recognize their power (Hofer, Langmann, Burkart, and Neubauer 2022).

In the transformational leadership chapter, we'll explore another type of suboptimal ally, that is, the risk of the visionary ally who can't execute.

Evidence That Ally Relationships Work

Now that we have introduced you to the concept of "good" and "bad" allies, we hope that the power of allyships is starting to become clear. But why would you take our word for it? Honestly, we don't expect you to take our word for anything, but instead offer you evidence that supports our proposal that forming alliances are critical to your personal and professional successes. Additionally, with consideration for the evidence that we present, we want you to start thinking about how to recognize a "good" ally, how to encourage them to join your efforts, and to eventually learn how you might pay it forward, as you empower future allies.

Relationship Alliance – Evidence of Benefits: I can speak to the value of alliances and allyships through the lens of both personal and professional engagements. For example, though it is not often thought of as an alliance, the relationship with my spouse is the strongest and longest lasting alliance, which also exceeds the clear love and respect that is expressed between partners (Markman, Hawkins, Stanley, Halford, and Rhoades 2022). Further, alliance does not assume that anyone is in charge, more intelligent, stronger, or weaker. It basically assumes nothing, where there are no expectations or demands and only a mutually shared goal to experience life together. Thus, an alliance was formed that expresses capabilities that exceed what any one of us are able to express alone.

As for the professional side of recognizing the value of alliances, I like to reflect on the numerous patents and trade secrets that I have worked on. In my first efforts, I worked alone and in isolation, believing that it was the only way to be creative. However, as I shared my newly created IP with colleagues, I gathered so much rich data from their feedback that I realized that I could have made my initial patents exponentially better. From that revelation, I have never attempted to create a patent alone, where I have now tripled my IP output and the speed at which the IP has been accepted has nearly doubled.

Extra Evidence Associated with the Benefits of Alliances: The following small text captures and cited material represent further evidence toward the power of allyships. Each section is an excerpt from peer-reviewed material that provides clear and detailed evidence that allyships work.

- In our study, we inferred from a specific buyer–supplier setting to interorganizational relationships more generally. Clearly, replication of our findings in other inter-firm settings, such as joint ventures and strategic alliances, is needed to establish their external validity. Further, we see an interesting extension of the research in cross-national settings where cultural differences may alter the outcomes of trust in relational exchange (Zaheer, McEvily, and Perrone 1998).
- By utilizing trustworthy-based social capital, VHVs thus connect and ally various individual agents in the form of a social network to collaborate in protecting the community members from the pandemic effects by utilizing local resources and reaffirming the collaboration, alliance, and participation of individuals, families, and households to together implement the country's strategic healthcare policy in response to the disruption (Jiaviriyaboonya 2022).
- Striving for a trusting, cooperative, and open relationship includes establishing a stable work alliance, continuous cooperation, active listening, trying to understand the patients and their (family) life situation as well as an orientation toward his resources and strengths (Meier Magistretti and Reichlin 2022).
- Good alliances, like good transactions, require the unflagging focus of senior-level management. In the case of alliances, senior managers need to focus on them throughout the relationship or the alliance risks losing its intended value (Anslinger and Jenk 2004).

Recognizing, Recruiting, and Empowering Allies

It is not enough to assume that you are a leader or an effective ally, simply because you are in a position of leadership. Specifically, a title doesn't do anything, except maybe encourage others to have unrealistic expectations for you and from your position. However, you can change perception to reality by examining the attributes and behaviors of highly successful leaders and allies, such that you might learn from them and use their successful behaviors to guide your own professional growth. Specifically, there is nothing worse to hear than a newly appointed leader touting their title as justification for their dominating behavior. Of course, it is always reassuring to learn that someone has actual experience and the ability to use those experiences to collaborate with and mentor others. That said, it is suggested that regardless of your professional or academic rank, you practice humility and empathy with those around you as you seek out the ideal allies and develop yourself into an amazing leader.

Putting Myself in the Shoes of a Leader Looking to Learn How to Recognize a Solid Ally

Who are they? They have reputations as innovators who pioneer new business models, products, or services that have created billions of dollars in value. We recognize some of our sponsors as "superbosses." They are recognizable in the number of unusually successful mentees they have such as Wim Roelantds, who has mentored 34 CEOS across the United States, Europe, and Asia. They have had repeated and dramatic business success and profound personal fulfillment by repeatedly betting on people and spotting potential in people first, often before anyone else, even the talent themselves. They take a long-term, consistent investment approach to developing people and delighting in when their mentees achieve. We believe that tomorrow's superbosses will come in many new "packages" than what we've historically seen as exemplary, famous superbosses. That's what Chapter 5 is about: transformational leadership. It's about identifying the competencies and skills required to operationalize transformational leadership. One defining feature of being a truly transformational leader is that superboss capability of spotting and building/recruiting talent.

Methods for Recruiting Desired Allies (And Yes, Some Allies Might Not Be Desired)

When we think about recruiting allies, it might not be that we even recognize that an ally is needed. So, recruiting an ally might be the furthest thought from a person's mind. That said, the recruitment of an ally can be as simple as ensuring that both your door and your mind are open as well as receptive to working with others. From there, if there is a true desire to collaborate with others, you will have established the foundation for open conversations that are not impeded by egos or personal biases (Jeppesen, Andersen, Lauto, and Valentin 2014). Accordingly, the information that follows will help you reflect on your need for an allyship and prepare your mind for successfully working with others.

Broad Recruitment: A key to learning in career development is the ability to recognize when an ally is needed and in what ways they can help a career throughout all of its phases. As we make it a habit to note when an ally could play an impactful role in different situations, we realize how valuable they are. From there it follows that we begin to wonder how we go about recruiting allies.

The recruitment of allies can happen on a spectrum, from broad to targeted. The broad recruitment flows from a lifelong openness for connection and the organic relationships that grow from shared experiences and interests. Thinking about recruitment at a broad level, the first step is to do some inner work:

1. **Know yourself first: what are your strengths? Weaknesses? Career goals?**
   Mentors and allies can help to clarify these questions, but it is helpful to have self-awareness as you look for potential matches. Allyship is not a "command structure" or

dictatorship, it is also generally not an instructor–student situation where the mentor gives all the answers. Instead, it is a collaboration where thoughts are shared without bias, working toward a shared goal. Knowing your strengths will allow you to propose ways you can support your mentor or ally. Knowing your weaknesses allows you to reach out for guidance when particular situations arise (Kram and Higgins 2008). Knowing your career goals can drive the next steps around where to go to find allies.

2. **Use your knowledge of yourself and your career goals to show up and become a "familiar/known" face in your industry/system/areas.** For example, if your goal is to be a CTO, your probability of finding a mentor or an ally who is a leader with great technical acumen increases. Throughout one of the authors' careers thus far, there have been three areas that have driven ally connections and continued relationships:

   a. Hackathons: tight team circles, good opportunity to leverage passion and knowledge (Dr. Bob Wong).
   b. Conferences/Summits: volunteer to organize, volunteer to help with presentations (Patty Brennan, Dr. Paul Fu, Dr. Bill Hersh, Sherri, etc.).
   c. Volunteer at charities aligned with your passions and industries (Dr. Erlinger).

Once you have made the initial connections with potential allies/mentors, the most important part is to follow up. Use the connection point to reach out about further opportunities in areas of interest. For example, if you met at a Hackathon, drop them a note about another upcoming Hackathon and ask if they will be there.

Targeted Recruitment: Developing ally relationships broadly throughout your field can be vital to the continued evolution of your career. However, there is also an opportunity and necessity to develop those relationships internal to your organization in a more targeted manner.

First, identify leaders in roles that you aspire to, you should also want to identify roles that support those leaders and the leaders that serve laterally to the roles you aspire to (Kwok, Kwong, Wong, and Duan 2022). Looking at aspirational roles with a 360-degree view will allow you to find allies you might not have considered before and also identify ways to connect. Once you have identified these leaders, watch the way they interact with the organization, what things they get excited about, what types of projects they always gravitate toward. Through this observation you can learn their "love language," for some it might be a handwritten note, for others a public kudos on social media. This "love language" can be used to approach a potential ally with a suggestion on how you might help them.

Once an initial connection has been made with a potential mentor/ally, the most important step is follow-up and follow-through. Making yourself dependable and trustworthy will make the leader more willing to trust you with further opportunities. Take the opportunity to send the potential ally follow-ups and dot connections about things that you talked about in your initial meetings or projects they are working on. Do this consistently and at a regular cadence.

As the fledgling relationship unfolds, find ways to ask for small tangible things. This author has found that potential allies and mentors love to do small favors with low barriers. An example might be signing a letter of recommendation that you have written. Then when you are awarded the opportunity that the letter of recommendation is for, use it to thank the ally publicly (or in their love language). This allows the ally/mentor to share some of the win, and makes a connection for them around helping you, knowing that you will reflect back on them in a positive light. *Be a Better Ally* tasks mentors/allies to "nominate protégés on the basis of their potential, without expecting them to prove they can do a job in advance. This usually requires putting some social capital on the line-a risk sponsors need to get more comfortable with" (Melaku et al. 2020). As the mentee, creating continual low-risk interactions can help the mentor become more comfortable when those larger risks come along.

Recruiting Allies Summary: Allyships exist in many walks of life and in different forms to promote collaboration and goal alignment. For example, successful allyships are established in war, in the courtroom, in personal relationships, in business, and in politics (Marianno et al. 2022). Whereas, the need for alignment (in the form of allyships) might be represented by:

- Alignment on personal views and values (e.g., marriage, raising a family, maintaining a safe environment (i.e., neighborhoods)).
- Alignment on political agendas (e.g., to pass laws, to solicit votes, etc.).
- Alignment to protect mutual interests (e.g., countries/states, money, and property).
- Alignment spiritually (e.g., beliefs and religion).
- Alignment in the pursuit of scientific discovery (e.g., ISS, NSF, etc.).

Europe has a good model for promoting entrepreneurial and single-product businesses, where small businesses partner up and form social ties to provide a potential customer with a final high-quality solution. Taking that example even further, we look to the small businesses in India, where you might see an individual on a street corner who fixes bicycle tires, another individual on a corner who fixes bicycle seats, and more individuals who work on other unique elements associated with bicycles, which are the most popular mode of transportation in India. Specifically, the sort of allyship that is being seen in both India and Europe are proving to be extremely successful because individuals/organizations recognize their own limitations, are aware of what other organizations are doing, and where those entrepreneurs put aside their egos as they seek to form both short-term and long-term allyships (Pruthi and Tasavori 2022). With that all in mind and as we seek to recruit partners, allies, and collaborators, we might look to those entrepreneurial relationships that are found in India and Europe, to help guide our efforts.

Establishing a Model That Promotes and Strengthens Allies

Creating an ally model can feel like drinking from a firehose. There are so many ways to go about it and all may seem important. At the heart of establishing a model is creating a culture that values and prioritizes allyship. This chapter won't go into deep details about this but will point to other sources for further information. Check out Chapter 3, "I Can Love My Leaders (ICLML): Driving Innovation through Culture, Leadership, Management, and Learning" of the first *Mobile Medicine* book and *Culture Your Culture: Innovating Experiences at Work* (Jaw-Madson 2018) written by one of our coauthors. You'll find very helpful information and frameworks to apply.

In addition, there are two critical recommendations that will help you succeed (Epler 2022).

Offer Learning Opportunities: A great place to start is offering allyship learning opportunities. There are potential allies in each organization that have the desire to be allies but don't know exactly where to start. Using a professional development budget, an organization can invite DEI practitioners to lead interactive workshops.

Walk the Walk: Team members know when leadership is sincere or just giving lip service. As leadership, please communicate clearly and regularly company-wide why allyship is a priority at the organization and how you will demonstrate its priority. Then follow through. Will you revisit the gender pay gap? Create a plan to address microaggressions? Other potential allies will take notice and find the confidence to join you in their own areas. It just takes a spark.

Allyship Competency Checklist

With consideration for the previously discussed topics of diversity in alliances, "good"/"bad" allies, recognizing allies, becoming an ally, and recruiting allies, we wanted to provide you with some graphics and guides that you might take from this chapter. Specifically, the allyship competency checklist isn't so much a definitive list of what to do and what not to do, but instead some visual clues that you can tear out of this book (yes we are encouraging you to tear out pages and post them on your fridge or office wall). You can then glance at the visuals on those occasions when you are considering the attributes of a "good" ally and when thinking about the building blocks that you will want to use when forming strong alliances.

We Start Life in an Alliance: We are all born with an innate trust of our mothers and the strength as well as drive to survive. Building on those formative attributes, we then seek to understand our world and to communicate within it. Thus, as we continue to interact with others in our world, we recognize that if we collaborate, we have a greater chance for success. Whereas, building on our attributes, evolving skills, and our successes as well as failures, our knowledge of ourselves and our world increases. With those statements in mind, the traits and behaviors that make for a good ally can be learned and honed (Barclay 2016).

Of course, as we examine the traits and behaviors that are associated with each of the six principal ally qualities (Figure 4.2), we might say that they are desired in many aspects of life. However, if we examine those traits more closely and consider why they are associated with a particular quality, then we realize that the order and grouping form the unique persona that is desired in an ally. That said, possession of desired ally traits and behaviors does not imply that one ceases to grow. Instead, anyone who wishes to be an ally or who might be seeking an ally will need to continuously improve their skills and adjust their behaviors (Childs 2017).

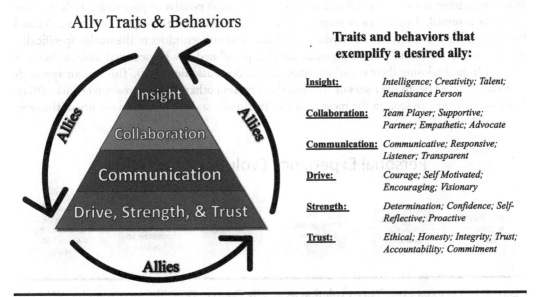

Figure 4.2   Ally traits and behaviors. Source: Harding, Medtronic, plc (2022).

Paying It Forward as an Ally, Mentor, and Role Model

The coauthors designed the below graphic (Figure 4.3) to depict what a journey pattern that they've recognized in themselves and others could follow. They start out thinking they're a good person, working to be a good person. Then at one point, they may learn from feedback or self-reflection that some of their behaviors did not necessarily contribute to being the most effective ally. They learn to manage their behaviors toward contributing to the goal of being an effective ally. The more wins they have, the more constructive ally behaviors they learn. As they become more effective allies with closer ally relationships and truly learn about lived experiences that are not their own, they begin to truly see objectively what inequality means. If they keep developing as authentic leaders, then they find courage to act against inequality. When they're ready, they scale this impact formally. One way would be to advise a formal program for directors of color and women such as the Santa Clara University (SCU) Black Corporate Board Readiness program. You can use this graphic to identify milestones and set goals. Use it to help you see what's next and reflect where you are now.

Ignorance Might Be Bliss

It might seem impossible that there are professionals out there, completely ignorant to the fact that workplaces have still not achieved equality. However, there must always be a starting point for any journey, and this state of unawareness is a place to begin. At this phase, in the progression to allyship, a person might think (or even say) things such as:

- "I don't see color."
- "My daughter is a professional woman, and she is excelling, there isn't a gap between males/females in the workplace any longer."
- "Don't ask, don't tell."

When these statements are made about minorities, they aren't positive or progressive as the speaker may have assumed. Turning away from differences means not seeing people's true identities and how those identities affect their everyday experiences and opportunities in the world. Specifically, we can look to the military as an organization that prohibited any homosexual male, lesbian, or bisexual from disclosing their sexual orientation (i.e., don't ask, don't tell). This was an approach to turning a blind eye to the issues of any sexual orientation other than heterosexual (Burks 2011). Most recently, we can look to the most recent ruling in one state that banded the use of the term

Personal Experience Evolution as an Ally

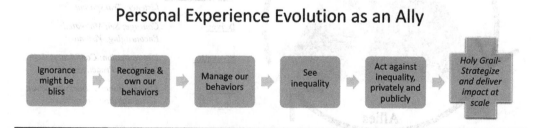

Figure 4.3 Personal experience evolution as an ally. Source: Douville, Medigram Inc. (2022).

"gay" (i.e., don't say gay). Thus, placing a paper bag over the public who might be uncomfortable with the concept that "love" can and does exist outside of non-heterosexual relationships.

Those that are in a state of unawareness might be well meaning, yet come from such a place of privilege that they don't want to acknowledge everyone wasn't given the same opportunities as them (Gonzalez, Riggle, and Rostosky 2015). They may be making incorrect assumptions about reality, based on the fact that the current system is comfortable for them. They might not want to look any harder to find the disparities as that means seeing the need for change.

An interesting exercise to highlight if you are living in this state of unawareness is called the Veil of Ignorance Thought Experiment. The experiment, originally outlined by John Rawls, has thinkers imagine themselves behind a veil where they know nothing about their own circumstances, natural abilities, age, race, or gender (Tremmel 2013). The thinkers must then design a society structure not knowing how those rules and institutions will affect them. At the end of the experiment, fictional factors about the thinker are revealed and they can assess if the rules they set up treated them fairly. This thinking transfers over well into the workplace. Take yourself through this experiment and design policies and structures at your place of work. Then remove the veil to find you are a very different race/gender/ability than your current state, do the policies and systems you have put into place still support you fairly?

Recognize and Own Our Behaviors

Just as we don't start technical projects without documenting the current state, we must identify our default behaviors around allyship before we can cultivate a plan of action to improve on and manage our behaviors. Through the Veil experiment above we looked at our perception level, now it is time to cultivate more understanding through the examination of ourselves. This process of recognizing our own actions and shortcomings can be uncomfortable. On Jennifer Brown's awareness continuum, this realization that we have work to do falls under the awareness category: "Knowing basic concepts but not yet activated."

We can begin to recognize our own behaviors by starting with our biases and our thoughts. One exercise that can highlight where our biases might list is the Circle of Trust Exercise (Stazicker and Woods 2022). After working through some of these bias identification exercises, we can begin to note when they might be impacting our actions and behaviors. Begin to ask yourself:

1. Why did I choose this person for my project? Is it because they are like me?
2. Why did I dismiss giving that analyst the new product? Is there something about their "otherness" that made me uncomfortable?

At this step in the ever-evolving road to allyship, we just want to note what biases are driving our decisions. This allows us to own our identified opportunities for improvement and create a roadmap for improving them. Thus, after identifying our biases, we also want to look at positive examples of ally behaviors and note if we are acting in accordance. There are likely some areas where we have stronger ally behaviors than others. Reflect on the identified behaviors and feel where it feels challenging, where does it feel easy? Some examples might be:

■ In meetings are you confident in using your voice to lend credibility and space to a minority voice in the room but struggle to think about and include others that aren't yet in the room?

- Do you happily share opportunities at another organization with minorities in your division but balk when it comes time to promote someone different than you on your own team?
- Are you extremely comfortable recognizing and apologizing for your own bias, but struggle to call out others of privilege like you that perform acts of microaggression?
- Use these examples to continuously recognize your own behavior, note them and own them, so that you can move forward in the management of your own behaviors.

Manage Our Behaviors

Once we have learned to recognize and identify our behaviors, we can take the next step and begin to manage them. An initial step in the direction of management might be taking the list of questions in the previous section and identifying action items from each that would allow us to better manage our behaviors. For example, if we wrote this note while identifying our behaviors, "Are you extremely comfortable recognizing and apologizing for your own bias, but struggle to call out others of privilege like you that perform acts of microaggression?" Then we might begin to manage our behaviors by writing out the following action plan:

- Next time I hear an act of microaggression, I will let the aggressor know that that type of language isn't tolerated.
  - The first three times I am privy to microaggression, I will communicate with the aggressor in the way I feel most comfortable (be it email, direct messaging, or taking the person aside after the meeting).
  - After three times of 1:1 communication practice, gaining confidence in my new behavior management, I will continue verbally addressing microaggressions out loud in real time as an example to those around me.

As we continue down the path to managing our own behaviors, it is imperative that we seek feedback from those that we are stepping up for. To be a good ally, you need to establish strong feedback loops with marginalized groups (Melaku et al. 2020). These relationships should be open enough to give you feedback not just on your behavior, but how well you respond to another member of the organization when they are undermining minorities.

The journey to managing our own behaviors can be rough, many even when they become competent at recognizing behaviors and see inequality may want to adopt a "not my business" attitude. However, becoming an ally is a skill that you build over time and you have to be willing to make mistakes (Obatomi, 2021). We add this reminder to encourage the readers who are just starting to manage their own behaviors, as it might not go well or perfectly the first time; however, it is your duty to continue down the path. One of our authors has seen this journey in their own organization, where the first time a leader called out the team for using a word recently added to the do not use list (master/slave in relation to servers), he received scathing looks. However, through repeated, calm persistence in subsequent meetings, the entire team has begun to step in and call each other out when any of the block listed words are used.

See Inequality

Our Worldviews and Inequality: Depending on your own story, seeing inequality might be harder for some while easier for others. This all comes back to our own worldviews (Gray 2011) we use to assess everyday situations. Take two people dining at a restaurant. Jessica was a former server while Stephanie has not. When the bill comes, studies show that Jessica because of her past work experience tips 5% more than Stephanie (Fisher 2014). Our worldviews began in childhood, took hold during future life events, and influenced our future actions and opinions. Objectively, our worldviews shape our outlook on inequality today.

Different Starting Lines: There is a group exercise where individuals walk up to a starting line (Thomas and Coleman 2015). This represents the strongly embedded view our society tells us, "we all start from the same place." The facilitator reads statements or questions and participants step forward or back if it applies to them. For example, "If your parents/guardian worked nights and weekends to support your family, take one step back." After going through the list, some participants are standing well ahead of the starting line, while others are further back. The facilitator stops and tells everyone that this represents your actual starting line. It is a very eye-opening and deeply personal experience.

This exercise is a very impactful visual of how our upbringing and circumstances outside our control have influenced our life opportunities. Regardless of our worldviews, life doesn't place us at the same starting point (Bowling 1991). Take two new college graduates, one graduated with student loans, while the other's parents paid for everything. Even if both graduates earned the same starting salary, they'll have different lifestyles and have drastically financial trajectories for the rest of their lives and for future generations.

Yes, we will always have inspiring individual rags to riches stories from marginalized groups that become the poster child. But we don't hear the thousands of other stories from the same circumstance not experiencing the same success. To see inequalities, it requires us to acknowledge each individual and group we interact with have different starting points.

Take the coauthors of this chapter. We are made up of a white man, a male person of color (POC), a female POC, and one white woman. Life has placed us in different circumstances and each of us had to navigate different situations in our careers. In certain larger groups, one or some of us may fit in easier than the others. Others might have a tougher barrier to entry. It all depends on the situation.

Seeing Inequality: Why has the proverb, "Birds of the same feather flock together," stood the test of time? Because it is still true today. We naturally gravitate toward others who are similar to us (McPherson, Smith-Lovin, and Cook 2001). We often do this subconsciously and even without malice. Seeing inequality requires a mindset to see our surroundings differently.

Look for Differences with Those around You: Many times inequality is tied to differences between individuals and/or groups. Some differences are self-evident like race and gender, while others aren't as obvious like socioeconomic or unseen disabilities. We often overlook these differences as we're going about our busy days. The first step to seeing inequality is identifying differences.

Have you thought about buying a new specific car model and then suddenly saw that model wherever you went? The same goes for identifying differences with others. Once you notice a difference others have from you, you'll suddenly see even more having these same differences.

I once worked on a team where many didn't have their own car. I witnessed extra challenges these individuals faced arriving to work on time or hardship working overtime since their ride couldn't wait. Up until this point, I was always surrounded by colleagues that had their own transportation and was unaware there's a large population that has extra steps to get to work. It was an eye-opening experience for me.

Listen to Those Different Than You: Connecting and cultivating relationships with others that are different will help expose inequalities others aren't aware of (Donaldson 2006). As I've gotten to know the other female coauthors in this chapter, I've learned more firsthand about the situations they face because of their gender. Because I'm hearing personal stories from individuals I deeply respect, it becomes more real and carries weight. These are things I don't have to consider and I was not aware of.

Workplace inequality statistics are great and helpful, but they're just data and numbers. In the allyship context, these statistics become more real and personal when one has stories to complement it. Allyship can be tough and often feel like an uphill battle. These personal stories are a source of fuel and energy to keep persevering in challenging situations.

Put Yourself in Their Situation: Recently, I had some foot issues and was unable to walk without assistance. I faced everyday challenges in my recovery as I hobbled around. What used to be easy to navigate like stairs became herculean tasks. I favored ramps when possible because it was so much easier. I had a small taste of what others face on a regular basis and made me more aware of their situation.

The same goes for seeing inequality. It's easier to see and accept inequality when we can empathize with others. Sometimes I take a step back and tell myself that I didn't ask for or work for the circumstances I was born into and neither did anyone else, especially those facing more challenges. This mindset accelerates one's ability to see inequality.

Act Against Inequality, Privately, and Publicly: Often as I reflect on the types of people whom I like to work with, I realize that there is no "type" of person that comes to mind. Instead, I arrive at the conclusion that I like to work with all types of people. Likewise, I realize that when I engage a diverse group of individuals, I gain richer insights into the world around me as well as a greater understanding of the needs of others. That point represents the framework from which I start any innovative effort, where I learned long ago that I will never succeed alone with the same magnitude as I might when working with others.

Thinking more deeply about working with diverse individuals, I believe that I take for granted the ease by which I as an older white male am able to work with anyone. For example, I used to wonder why others would not work with different varieties of individuals. But then I realized that it wasn't that they "would not" work with different people, it was that they could not work outside of their society-imposed boundaries. Moreover, I realized that others did not have the privilege that I was afforded, which was still a societal privilege even though I grew up poor. Thus, it was from that bit of enlightenment and the desire to enrich my world that I sought to remove the barriers that society had placed in front of the people with whom I wanted to work. Furthermore, I decided to not remain silent about those barriers that stood in the way of people, but instead, to embrace two of my favorite quotes as stated by Dr. Martin Luther King Jr. Specifically, the quotes that motivate me to openly fight to remove barriers are *"Our lives begin to end the day we become silent about the things that matter"* and *"The time is always right to do what is right."*

Of course, not everyone is as comfortable with breaking down barriers, but maybe if they thought in terms of their business going away or losing money, then they might realize how diverse and fresh perspectives increase their chances for success. Thus, by helping others, we help ourselves both with respect to immediate needs and as it relates to long-term strategic goals. That point exemplifies the key strengths of a leader, a colleague, and an ally. It also emphasizes that when we look for those who could be the best allies, we should not look for people like ourselves, but instead at those individuals who think differently. Likewise, when we open our minds to greater possibilities, through the lens of diversity, we make a conscious effort to create equality across all our spheres of influence (Archer, DeWitt, and Wong 2014).

Holy Grail – Strategizing Your Impact and Scale: Assessing Potential Partner Service Organizations

There's the old true adage: together we'll go far. I believe this to often be right; however, there are exceptions. Those exceptions of what could be a more universal truth seem to be due to competency gaps from my perspective. With 20 years of experience as a volunteer and board member across higher education, the arts, corporate, and healthcare, you can learn from my pain and wins. I truly do believe that partnering with the best organizations allows us to "go far." Though I caution folks to examine the team and capabilities present very carefully in light of this explosion of modern complexity and associated demands on all organizations' capability to succeed in today's environment. The "Philanthropy and volunteer system" many current executives grew up with was built in a different time when life was different. In particular, women need to be really careful with how the impacts of volunteering can negatively affect their working relationships, job application prospects, and careers in the corporate world in my view (Einolf 2011). In the best circumstance, the organization walks their talk and practically supports volunteering that is compatible with employee development. In an acceptable scenario, the organization permits volunteer activity. However, one needs to know when the demographics or culture of the employee base are not compatible with volunteering due to employees' lack of earlier life experience with volunteering. Further, if you're married, we advise you to design a joint set of goals with your spouse for how you will decide to volunteer.

In a recent Silicon Valley magazine interview, the editor of this book spoke to solving for the paradox of negative career impact and desire to make a difference by aligning volunteer efforts tightly around career ambitions. This is a function of where you are in your career, if you're in the *build* phase as opposed to a retirement phase, for example. As a younger person in their 20s, one can afford a wider range of activities which we encourage young people to do. As an aside, we are very fortunate to have found volunteer opportunities with IEEE, CHIME, and Santa Clara University that have been professionally productive and personally rewarding.

The holy grail for this chapter's topic, allyship for a good volunteer experience including as a board member would be if we could volunteer for an organization whose mission directly drives allyship. Ideally, it would be something rewarding that energized and didn't drain you. More possibilities could be realized if it contributed to your professional, intellectual, and emotional well-being while driving massive impact – all in a super-efficient way.

The reality of many NonProfits today is that they can be thinly staffed and/or have staff with less management and leadership experience (Guo, Brown, Ashcraft, Yoshioka, and Dong 2011). You, therefore, need to evaluate if the organization is sufficiently mature and well run to be a good partner on your common objectives. Or you need to determine if you have the time and energy to close the competency gaps.

Some things I look for as essentials to running effective organizations specifically considering board roles are the following 11 items:

1. Administrative or organizational competencies such as meeting planning and meeting management, conversational leadership, scheduling, calendaring, staff to take notes, and other office "housework" are housed and executed well under staff roles and responsibilities.
2. Project management work or leadership.
3. Program portfolio management or leadership: utilizing tracking, objectives, and key results, KPIs.

4. Event planning capabilities that the agency leads and executes where they are not asking volunteers to take responsibility for event management or planning or business operations.
5. Sales capability for driving fundraising operations.
6. Marketing both digital and classic communications capabilities.
7. Technology/data/IT capabilities. For example, can all the directors and staff use various collaboration tools? Is there a volunteer coordinator who uses CRM and other business operations and performance capabilities to manage volunteer work and its orchestration?
8. Operations and accounting/finance function.
9. Overall management and leadership assessment, scored objectively.
10. Dedicated legal staff/function.
11. Rating how compelling, effective, and the results of the core services the agency offers the community, i.e., education, healthcare, arts, homeless services, food, etc.

Nearly all organizations, both "for-profit" and nonprofit organizations, are extremely challenged today by the exponential forces and enormous requirements for driving success especially of "stakeholder capitalism" facing all organizations (community-building, collaboration, complexity, orchestration) (Hemphill, Kelley, and Cullari 2021). Some of the best volunteer experiences we've had have been with SCU, including through the Santa Clara University Leavey School of Business Silicon Valley Executive Center corporate board readiness programs. Organizations that succeed in building great boards and talent ecosystems will manage all the required operations, administration, events, orchestration, and coordination work whose requirements are exploding – all efficiently in my view.

While the editor has explored independent opportunities outside of work, chapter contributor, William C. Harding has experience in a work-related and sponsored venue to promote allyship.

Conclusion

In this chapter, we discussed the concept of an ally and more specifically "allyships." We characterized what is an ally, and how to recognize an ally, both within yourself and within others. We then proposed methods for recruiting allies and we even covered when an ally is not needed or inappropriate. But if we were to summarize the most important aspects of allyships, it was the need to be self-reflective, to condition ourselves such that we are receptive to forming effective allyships, and to influence the diversity of discipline within our realms. Additionally, we made a point of characterizing that the richest results can be achieved in ally relationships if we embrace the perspective that the inclusion of diverse perspectives will yield the greatest rewards. That said, the concept of an ally, as presented in this chapter, is someone who might be a mentor, a sponsor, or even a partner, where allies share our need to achieve mutually aligned goals.

References

Archer, Louise, Jennifer DeWitt, and Billy Wong. "Spheres of influence: What shapes young people's aspirations at age 12/13 and what are the implications for education policy?" *Journal of Education Policy* 29, no. 1 (2014): 58–85.
Anslinger, Patricia, and Justin Jenk. "Creating successful alliances." *Journal of Business Strategy* (2004).
Barclay, Pat. "Biological markets and the effects of partner choice on cooperation and friendship." *Current Opinion in Psychology* 7 (2016): 33–38.

Bhide, Amar. "How entrepreneurs craft strategies that work." *Harvard Business Review* 72, no. 2 (1994): 150–161.

Bowling, Ann. *Measuring health: A review of quality of life measurement scales.* Milton Keynes: Open University Press, 1991.

Branden, N. *The six pillars of self-esteem. Bantam Books, Inc.* New York, New York, 1994.

Brown, Elizabeth, and Jeannette E. Pierce. "2020 ACRL academic library trends and statistics survey: Highlights and key EDI findings." *College & Research Libraries News* 83, no. 4 (2022): 145.

Burks, Derek J. "Lesbian, gay, and bisexual victimization in the military: An unintended consequence of "don't ask, don't tell"?" *American Psychologist* 66, no. 7 (2011): 604.

Childs, Henry. "Strategies that logistics leaders use for achieving successful process improvement." PhD diss., Walden University, 2017.

Choo, Esther K., Carrie L. Byington, Niva-Lubin Johnson, and Reshma Jagsi. "From# MeToo to# TimesUp in health care: Can a culture of accountability end inequity and harassment?" *The Lancet* 393, no. 10171 (2019): 499–502.

Donaldson, Gordon. *Cultivating leadership in schools: Connecting people, purpose, and practice.* Teachers College Press, 2006.

Einolf, Christopher J. "Gender differences in the correlates of volunteering and charitable giving." *Nonprofit and Voluntary Sector Quarterly* 40, no. 6 (2011): 1092–1112.

Epler, Melinda Briana. "Managers, here's how to be a better Ally in the remote workplace." *Harvard Business Review* (January 27, 2022). https://hbr.org/2022/01/managers-heres-how-to-be-a-better-ally-in-the-remote-workplace.

Fisher, Katherine S. "How much should I tip? Restaurant tipping behavior as a result of prior foodservice experience." *Inquiries Journal* 6, no. 10 (2014).

Glas, Kathryn E., and Abimbola Faloye. "Unconscious (implicit) bias." *Journal of Cardiothoracic and Vascular Anesthesia* 35, no. 4 (2021): 991–992.

Goldsmith, Marshall and Mark. Reiter. *What Got You Here Won't Get You There: How Successful People Become Even More Successful.* New York, NY, Hyperion, 2007.

Gonzalez, Kirsten A., Ellen D. B. Riggle, and Sharon S. Rostosky. "Cultivating positive feelings and attitudes: A path to prejudice reduction and ally behavior." *Translational Issues in Psychological Science* 1, no. 4 (2015): 372.

Gray, Alison J. "Worldviews." *Int Psychiatry*, 8, no. 3 (August 2011): 58–60. https://www.ncbi.nlm.nih.gov/pmc/articles/PMC6735033/

Guo, Chao, William A. Brown, Robert F. Ashcraft, Carlton F. Yoshioka, and Hsiang-Kai Dennis Dong. "Strategic human resources management in nonprofit organizations." *Review of Public Personnel Administration* 31, no. 3 (2011): 248–269.

Hemphill, Thomas A., Keith E. Kelley, and Francine Cullari. "The ascendancy of stakeholder capitalism: What is its meaning for corporate governance?" *Journal of General Management* 46, no. 4 (2021): 262–273.

Hofer, Gabriela, Laura Langmann, Roman Burkart, and Aljoscha C. Neubauer. "Who knows what we are good at? Unique insights of the self, knowledgeable informants, and strangers into a person's abilities." *Journal of Research in Personality* (2022): 104226.

Jaw-Madson, Karen. *Culture your culture: Innovating experiences@ work.* Emerald Group Publishing, 2018.

Jeppesen, Jacob, Kristina Vaarst Andersen, Giancarlo Lauto, and Finn Valentin. "Big egos in big science: Unlocking peer and status effects in the evolution of collaborative networks." (2014).

Jiaviriyaboonya, Poonnatree. "Anthropological study of village health volunteers'(VHVs') socio-political network in minimizing risk and managing the crisis during COVID-19." *Heliyon* 8, no. 1 (2022): e08654.

Kram, Kathy E., and Monica C. Higgins. "A new approach to mentoring." *The Wall Street Journal* 22 (2008): 2008.

Krohn, Katherine E. *Ella Fitzgerald: First lady of song.* Twenty-First Century Books, 2001.

Kwok, Man Lung Jonathan, Raymond Kwong, Macy Wong, and Jinyun Duan. "Great leaders do everything: A moderated mediation model of transformational leadership, trust in leader, helping behaviour, and idiosyncratic deals." *Asian Business & Management* (2022): 1–21.

Levy, Bruce D., Joel T. Katz, Marshall A. Wolf, Jane S. Sillman, Robert I. Handin, and Victor J. Dzau. "An initiative in mentoring to promote residents' and faculty members' careers." *Academic Medicine* 79, no. 9 (2004): 845–850.

Marianno, Bradley D., Annie A. Hemphill, Ana Paula S. Loures-Elias, Libna Garcia, Deanna Cooper, and Emily Coombes. "Power in a pandemic: Teachers' unions and their responses to school reopening." *AERA Open* 8 (2022): 23328584221074337.

Markman, Howard J., Alan J. Hawkins, Scott M. Stanley, W. Kim Halford, and Galena Rhoades. "Helping couples achieve relationship success: A decade of progress in couple relationship education research and practice, 2010–2019." *Journal of Marital and Family Therapy* (2022).

McPherson, Miller, Lynn Smith-Lovin, and James M. Cook. "Birds of a feather: Homophily in social networks." *Annual Review of Sociology* 27, no. 1 (2001): 415–444.

Meier Magistretti, Claudia, and Beat Reichlin. "Salutogenesis and the sense of coherence in young adults not in education, employment, or training (NEET)." In *The Handbook of Salutogenesis*, pp. 151–165. Cham: Springer, 2022.

Melaku et al. Be a Better Ally, Harvard Business Review December 2020.

Obatomi, Maureen. "5 powerful steps to becoming a better ally." *Betterup*. May 25, 2021. https://www.betterup.com/blog/5-powerful-steps-to-becoming-a-better-ally.

Pruthi, Sarika, and Misagh Tasavori. "Staying in or stepping out? Growth strategies of second-generation immigrant entrepreneurs." *International Business Review* (2022): 101997.

Pollock, Jennifer S., Linda Samuelson, and Dee Silverthorn. "Innovating and building momentum for physiology's future." *Physiology* 37, no. 1 (2022): 2–3.

Schwartz, Victor. "Freud's practice of psychoanalysis." *American Journal of Psychiatry* 156, no. 6 (1999): 978–979.

Stazicker, Anne, and Nancy Woods. *Teaching international foundation year: A Practical guide for EAP practitioners in higher and further education*. Routledge, 2022.

Thomas, Marquita and Coleman, Chris. "What is privilege?" *Buzzfeed*. July 4, 2015. Video. https://www.youtube.com/watch?v=hD5f8GuNuGQ.

Tremmel, Joerg Chet. "The convention of representatives of all generations under the 'veil of ignorance.'" *Constellations* 20, no. 3 (2013): 483–502.

Zaheer, Akbar, Bill McEvily, and Vincenzo Perrone. "Does trust matter? Exploring the effects of interorganizational and interpersonal trust on performance." *Organization Science* 9, no. 2 (1998): 141–159.

Chapter 5

Extinction Is Eminent without Effective Transformational Leadership: Closing Leadership Gaps That Prevent Execution

William C. Harding, Michael Ng,
Brittany Partridge, and Sherri Douville

Contents

DOI: 10.4324/9781003348603-7

Introduction

Transformational leadership is a term that has been associated with many successful leaders, but that perception might be flawed. Specifically, because the leadership style of successful leaders cannot be defined in terms of a single style, attribute, or behavior. For example, public speaking and charisma have long been lauded as admirable leadership traits, and while they remain critical, they are woefully inadequate as any near complete definition of a true modern leader. Accordingly, in this chapter, we will examine what behaviors and attributes are embodied within transformational leaders and how successful leaders embrace more than one leadership style. We will also discuss how you might develop your own leadership style and how you can learn to adjust your behavior, such that your style can more closely align with that of a transformational leader. From there and building on developing your style, we will provide information that will help you identify various leadership styles, and we will reflect on the leadership styles that successful leaders have embraced. Further, we propose that the behaviors and attributes of the most successful leaders might not be the same ones that make for a great ally or mentor. With that all in mind, we believe that as you dig deeper into the theme of this book, you will further recognize that true transformational leaders are needed in this world of mobile medicine, which is clearly a gateway to a technologically advanced ecosystem. That point is further emphasized in the first mobile medicine book, where we described the promise and peril of complexity associated with mobile technology in the highly regulated industry of medicine (Douville 2021). We then went on to include the criticality of overcoming culture, people, and governance challenges to solve for the complexity of mobile computing, with a strong focus on compliance and security. Accordingly, within this book and specifically this chapter, we recognize the criticality of transformational leadership for scaling and growing competencies that reduce barriers often encountered when developing and implementing mobile medical and advanced health technology solutions.

What Is Transformational Leadership and Why You Should Care?

Leadership Styles

Successful leaders exhibit a unique combination of behaviors (mannerisms and characteristics), which may be considered disorders and possibly syndromes, though it can also be determined through an evaluation of individual leaders, the establishment of a personalized and unique balance between normal and abnormal behavior results in singular success. Moreover, successful leaders express behaviors that can be grouped into the categories of desirable personality traits, distinct leadership styles, and self-identity cognizance, where those behaviors may not always be desirable, but are critical in creating some of the most successful past and present business and technology leaders.

In this chapter, we will discuss different leadership styles and offer pros and cons associated with one specific leadership style. Specifically, in this chapter, we will discuss transformational leadership, where we will first synthesize peer-reviewed material to identify shared themes that emphasize important behaviors of successful leaders. One article evaluated by Lilienfeld et al.

(2012) examined potential positive aspects of leaders with suspected successful psychopathy personalities. Another article evaluated by Odom, Boyd, and Williams (2012), supports the concept of continuous learning through experiential learning activities and considers the experiences that help students to develop as leaders. Finally, another article evaluated by van Eeden, Cilliers, and van Deventer (2008) framed desirable and undesirable leadership styles defined as laissez-faire behavior, transactional leadership, and transformational leadership.

Further we will discuss the shared themes of desirable personality traits, distinct leadership styles, and self-identity cognizance, for the purpose of explaining the importance of each theme such that the aggregation of the evaluated articles conclusively illustrates the behaviors of successful leaders. That said, we will discuss the various positive behavioral attributes of successful leaders as well as the desired attributes that future leaders should seek to attain through increased personal awareness, strengthening of personality traits, and reinforcement of effective leadership styles.

Desirable Personality Traits

Every individual exhibits innate personality traits that can be generally characterized as habits and behaviors that are consistently displayed by an individual over long time periods (Fleeson 2004). Successful leaders are able to recognize and use personality traits that might even be considered psychopathic (Lilienfeld et al. 2012) and focus traits (such as agreeableness and daring/taking risks) into desirable strengths that enable those individuals to assume roles of authority.

With consideration of the personality traits of agreeableness and daring, Lilienfeld et al. (2012) suggest that "low levels of agreeableness, are modestly correlated with independently rated job performance" (p. 491), while van Eeden, Cilliers, and van Deventer (2008) associated agreeableness with the positive traits of transformational leaders, where successful leaders exhibited the behavior of a team player as perceived through their agreeable demeanor. Furthermore, the personality trait of taking risks or daring was a trait that all of the authors linked to self-confidence/enablement and identified with past and present successful leaders as well as a trait that aspiring future leaders sought to embrace (Odom, Boyd, and Williams 2012).

Distinct Leadership Styles

Leadership was the common shared topic discussed through numerous articles. However, the behavior associated with a distinct leader style was an attribute connected to successful leaders, which is characterized separate from desirable personality traits or self-identity cognizance and was identified as a unique theme within each of the studies. Leadership styles are a specific behavior that individuals exhibit, which might be confused with specific personality traits, but are an amalgamation of multiple personality traits and influential external elements associated with environment and circumstance (Kippenberger 2002). Unlike personality traits associated with successful leaders, an individual's style can change over short periods of time, such as when a style adjustment is required to adapt to current conditions and events.

According to van Eeden, Cilliers, and van Deventer (2008), three leadership styles that are associated with individuals in roles of authority are laissez-faire, transformational, and transactional; whereas transformational leadership is the style that is most desired by the global business community. Conversely, the leadership style of laissez-faire was in essence the "avoidance or absence of leadership" (van Eeden, Cilliers, and van Deventer 2008, p. 253) and was found to be most effective for groups and teams that are self-driven and capable of self-management.

With consideration for styles of successful leaders, transformational and transactional leadership styles were identified as the most desired styles either directly stated or implied within the reviewed articles. Accordingly, a transformational leader would have positive style attributes associated with the ability to communicate effectively, exhibit high intelligence, and be able to inspire others. By extension, Lilienfeld et al. (2012) also associated successful leaders with intelligence and being a team player, while Odom, Boyd, and Williams (2012) emphasized the style of a transformational leader as an individual with effective "public speaking skills, delegating, motivating, team-building" (p. 53).

Likewise, the style associated with transactional leadership was also a desirable style, where transactional leaders are more controlling, ambitious, result-driven, and exemplify leaders from the business community, where the style's negative impact results in subordinates that have little ability to inject personal creativity. Transactional leadership styles are often associated with dominant individuals who are egocentric and desire to control their subordinates (Lilienfeld et al. 2012; Odom, Boyd, and Williams 2012). For example, in a surgical arena, it might be desired that the principal surgeon embrace more of a transactional leadership style. Though for knowledge-intensive work related to data and technology, we will need to focus on the transformational style specifically.

Building on what is known of the three leadership styles, there are roles where it is needed that leaders exhibit greater control and where it may be undesirable to make decisions as a group. Thus, each of the three presented leadership styles may be needed, depending on the circumstances, but we suggest that a strong leader is someone who can pull aspects of each style to create a balanced multidimensional leader. Moreover, we have determined that an individual who more closely aligns with a transformative style is best suited for healthcare professionals who work with patients and other professionals.

Superbosses: What Are They?

Moving from an examination of the three principal leadership styles of transactional, transformative, and laissez-faire, we take a leap (and maybe a bound) toward a newly emerging leadership persona. That persona is embodied within the term "Superboss," where we believe that attributes and behaviors associated with transformative leadership coalesce to form a top leader. And with that vision of a top leader in mind, we now introduce you to the leadership persona of a "Superboss." A persona that characterizes those traits in a leader who we seek to learn from and emulate. A person, for example, who recognized years ago the need for and value of a truly digital transformation of life sciences/healthcare (Trubetskaya, Manto, and McDermott 2022).

Things that distinguish these business icons are as follows:

1. They spot, train, and develop future leaders. These individuals belong to a category beyond superstars: call them superbosses. A superboss is a leader who helps other people accomplish more than they ever thought possible.
2. The writer Finkelstein found that the top people in a number of industries, by pattern, nearly half of them, once worked for the same great leader. In professional football, 20 of the NFL's 32 head coaches trained under Bill Walsh of the San Francisco 49ers or under someone in his coaching tree.
3. In hedge funds, dozens of protégés of Julian Robertson, the founder of the investment firm Tiger Management, have become top fund managers.

In our world, Medigram board chair, Wim Roelandts, has deliberately sought out scale and worldwide impact by coaching 35 CEOs throughout the United States, Europe, and Asia. Even though our own ambitions or assessment of desirable colleagues and work partners as transformational leaders' roles may differ in purpose and scale, they will have certain patterns that appear similar.

Through Sydney Finkelstein's interviews with over 200 superbosses, they demonstrated key personality traits. They were extremely confident and brought what one would call "swagger" to their presence. They were once athletic and remain competitive in nature, and possess a high level of creativity or imagination. They have strong character and act with integrity. They aren't afraid to be authentic (Finkelstein 2019).

How do they execute on their superboss potential? Through a pattern of practices such as:

1. Their protégés are often unusually gifted people – individuals who are capable not merely of driving a business forward but of rewriting the target definition of success for a company or industry.
2. They hire for intelligence, creativity, and flexibility. Their hires were able to approach problems from new angles, handle surprises, learn quickly, and excel in any position.
3. They don't follow prescribed formulas from the past based solely on credentials. A superboss is likely to reject preconceived notions of what talent should look like. As a consequence, superbosses often show greater openness toward women and minorities.

They put the unique gifts and talents of their people over the structure of the organization, even changing the organization to enable the unique gifts.

1. They take the long road with talent which provides access to talent networks and future opportunities.
2. They put them on a steep learning curve.
3. "Superbosses give mentees much hands-on experience but also monitor their progress and offer instruction and intense feedback; they will jump in to work with them side by side when necessary."

How Transformational Leadership Helps Mobile Medicine and Other Advanced Technologies Succeed

Healthcare as an industry is complex with many barriers to innovation. Add mission-critical technology on top of an already fragmented system and it becomes clear that transformational leadership is vital to lasting change and success.

As described above, there are multiple leadership styles necessary for an organization, project, or technology to succeed. Looking through the lens of transformational leadership, there are a few traits that stand out in support of the needs of mobile medicine. First, transformational leaders guide and motivate common goals (Garcia-Morales 2008). In an industry where shiny object syndrome is consistently seeking to derail aligned movement, the ability to keep teams driven on a common path is vital. The ability of transformational leaders to encourage good communication networks and trust, thus allowing the transmission and sharing of knowledge, is also important in mobile medicine. As one of our authors has experienced the ability to learn from mistakes of others during go-lives as well as leverage knowledge from a vast array of teams can mean the difference between success and failure. These leaders not only encourage the sharing of knowledge but also support the creation of knowledge slack and absorption.

Knowledge slack is defined as a pool of knowledge resources in a firm in excess of the minimum necessary to produce a given level of organizational output (Garcia-Morales 2008). As a concept, it is imperative for organizations that wish to move fast with quickly changing technology and standards. Leaders that support this growth of knowledge and curiosity in their team members will position them well for rapid growth and implementation of mobile medicine and advanced health technology solutions. Finally, "Transformational leadership is also preventative, not reactive. The vision and actions of transformational leaders should also engage staff, partners and the community in addressing the root causes of health needs and inequities" (Texas A&M 2021). With technology that impacts human lives, there is nothing more important than being proactive, and transformational leaders applying this to mobile medicine and advanced health technologies is imperative.

Examining some of the barriers to technology-based innovation in healthcare and mobile medicine can help us to understand why certain leadership traits are needed. The first barrier, "Accountability/Compliance/Policy," is a gigantic area to tackle and some of the mitigation opportunities can be found in other chapters of this book. "It is important for innovators to understand the extensive network of regulations that may affect a particular innovation and how and by whom those rules are enacted, modified, and applied" (Herzlinger 2006). Transformation leaders are poised to empower their teams to gain this knowledge set through their focus in increasing knowledge slack and information sharing. Accountability also aligns with safety culture, which is necessary to consider when defining technology workflows that directly impact patient lives. A study done in 2013 by Hoffmeister et al. "used relative weights analysis to investigate the influence of individual facets of leader behavior on safety. Idealized attributes and behaviors emerged as consistently important predictors of multiple safety outcomes." Those leadership traits can be tied to transformational leaders. However, a subsequent study completed in 2022 by Wu et al. found that while "transformational leadership positively affects safety compliance through employees' felt obligation toward their leader. However, transformational leadership also can negatively impact safety compliance through safety risk tolerance."

Another barrier to innovation in mobile medicine is finances, both the funding for the initial development and implementation and the ongoing payment around services provided. When up against these barriers, the ability of transformational leaders to inspire and motivate others needs to be leveraged so that the team doesn't give up before they succeed. A 2020 study found that transformational leadership has a greater influence on financial performance, and knowledge sharing, a proponent of multiple leadership styles, is more significantly associated with operational performance (Son, 2020).

Technology infrastructure itself can be a barrier to innovation in the implementation of mobile medicine and advanced health technology solutions. From competing solutions to the lack of infrastructure in the implemention part of the organization, from technical debt in the hospital to lack of integration options, the technical and process (or lack of mature efficient processes) barriers can be vast. Transformational leadership qualities such as motivation toward a single goal and knowledge sharing work to bridge these barriers. An area in technology where transformational leaders really shine is post-implementation engagement for optimization of technology: "When IT managers display transformational leadership qualities, this sustained interest results in a greater intent by followers to contribute to system enhancement" (Yurov 2006).

Creating and maintaining secure technological infrastructure is imperative in all sectors; however, it is even more vital in healthcare with ePHI and sensitive information created and stored everywhere. Transformational leadership can be leveraged to overcome cybersecurity barriers in healthcare, where one of the biggest threats is the human element. Relationship-building, motivation, and optimism to go above and beyond, all traits of transformational leaders, are best for a human-centered security strategy where high reliability and individual integrity are paramount (Corpolongo, 2016). The 2016 article by Corpolongo goes on to explain that secure leadership and the sharing of information, a tenant of transformational leadership, avoids the "trust no-one" mentality and creates more secure organizations.

As this discussion has shown, there are many types of leadership with differing traits that can support the success of technology in healthcare. Transformational leadership has many tenants that are particularly suited to push expansion and implementation success in mobile medicine and advanced health technology undertakings. However, those that seek to lead health information technology systems, companies, and projects must not only become transformational leaders, but they must also incorporate further leadership qualities to truly be successful.

Key Transformational Leadership Traits You Can Implement

We have described transformational leadership and why it is essential for leaders to use in mobile medicine and advanced health technologies. This next section defines some practical and tactical traits you can develop if you haven't already. This is not an inclusive list but merely the essential traits from our experience.

Build People and Teams

Competent leaders are world-class builders of teams and the people and culture that make them perform and grow leveraging their communication skills. The best leaders are people that other people want to work with and learn from. They can build the best teams of not just the best credentialed individuals or stars, but of a range of diverse skills and personalities to meet the needs of a broad set of stakeholders and challenges. They have a clear vision for their programs, teams, and organization. The best leaders understand, collaborate to create a shared vision, and communicate what's possible for that vision (Harne-Britner and Hader 2022).

Lean into Humility

You must have humility to be capable of building the esteem and skills of team members. Learn more about humility in the Humility in Medicine (Chapter 7). They can be humble because they are secure in their own abilities and potential. Many insecure leaders demonstrate narcissism (Lang, Zhang, Liu, and Zhang 2021). Research shows that narcissism blocks effective teamwork (Simmons 2020). It also destroys the esteem of team members. The best leaders do not act superior to people, they instead build people, both their talents and esteem, instead of destroying the latter.

Know What Makes an Effective Executive and Be Honest about Your Gaps

We all have them. Respect the need to and constantly work toward the balance and juggling required to really lead which includes an emphasis on the foundations of character, management, execution, leadership, and relationship skills. Work to be honest with yourself and to solicit honest feedback from trusted advisors to have a clear understanding of areas of strengths and weaknesses, including the importance of not over-relying on strengths via teaming. Be obsessed with character. In terms of character, courage stands out as a requirement of character. Without courage, one cannot lead. As Winston Churchill famously said, "courage is rightly esteemed the first of human qualities because it has been said, it is the quality which guarantees all others" (Craggs 2022).

Take Responsibility for Tackling Conflict and Using It as a Tool for Improvement

Transformational leaders accept and seek out constructive conflict as necessary to the process of building and are skilled at conflict management and providing as well as receiving and acting on feedback that helps professionals and teams grow. The process, according to international executive coach, Lucy Georgiades, includes stating only facts, how it makes you feel, telling them how you feel about the facts, tying the behavior to global impact, and then asking gentle reflection questions to drive collaboration to solve.

Make Sure You Understand What Delegation Really Is

Transformational leaders understand how to and work on effective delegation for results. Delegation is not "set and forget it," like many people imagine. It involves identifying the clear task, objective, providing context for the why, identifying how to measure realistic completion, defining the priorities, getting buy-in from staff, giving authority, making a clear path, and making that clear to all, then checking in and following up ("Micromanagement is toxic").

Close Gaps between Visions and Execution

Many visionary leaders have the ability to see game-changing ideas that others can't. That's what makes them special. However, not enough of these leaders deliver these visions or value the execution process to make it a reality (Reddy, Hoskisson, and Smith 2022). Often, they wave their "magic wands," quickly turn away, and expect their teams or organizations to deliver without realistically understanding or appreciating the details, constraints, and timelines. We can all recall classic examples of leaders firmly saying, "just do it!" The tone doesn't allow for conversation or collaboration. This situation leads to frustration for everyone and often failure and burnout. Conversely, transformational leaders increase their success by valuing execution. These individuals may or may not be strong in operations but value and augment this skillset to the team. When the team asks clarifying questions or brings up potential risks, effective transformational leaders don't instinctively feel attacked or annoyed leading to a defensive posture. Instead, they can see others are helping bring these visionary ideas to life and in turn, welcome and engage in these conversations.

Case Study: Transformational Leadership in Action

What greater place to see transformational leadership in action than the four coauthors who wrote this chapter and Chapter 4. At face value, we're very different from each other (see Figure 5.1). We're at different places in our careers and have different backgrounds, strengths, and blind spots. However, we share some key values and perspectives that unify us. Each of us came to the table wanting to work together and was motivated to deliver fresh and game-changing ideas to you, our readers.

Figure 5.1   Allies building allies.

Building the Team

Our book editor aims to be a superboss and handpicked us because she saw the powerful potential and chemistry we would exhibit together. Rather than bring only senior leaders together, she strategically recognized the chapter would be more enriching bringing different stories and perspectives from individuals in different places in their careers. She continued this throughout other chapters bringing together individuals that needed to work together in the real world but often had differing views and perspectives.

Aligning Team Values and Establishing Culture

We were very fortunate that when we began meeting, we quickly created a trusting and safe environment. What did that practically mean? Despite our heavy individual personal and professional demands, we committed to meeting together regularly. We naturally created a space where we actively listened to each other. Regardless of seniority, everyone's voice was equally heard. We validated each other's ideas and contributions. We were fortunate, no individual had a personal agenda conflicting with the group. There were times we respectfully challenged each other to ensure our content met our chapter goals.

Thankfully, our team culture came together naturally but if it didn't, there are other ways we could have collectively defined the culture. Early on, we could have listed our team's values and include behaviors we would not allow. Then throughout the writing process, we could keep these values top of mind by reviewing them together at a set interval. Just carving out a few minutes once a month at a meeting to ask the group, "how are we doing in this area?" can really keep team values grounded in everyone. Having a set time for check-ins and team meetings no matter the cadence really helped with not only alignment but also accountability and keeping everyone on pace to finish the chapters. Willingness to be flexible around sched-ules, time zones, and work barriers was a necessity as the most impactful meetings occurred when everyone was in attendance, sometimes resulting in 0630 meetings before work or 1900 meetings after clinical shifts.

On a broader scale, we vigorously and intentionally created a coauthor book culture build-ing on the foundation of the first *Mobile Medicine* book's 27 coauthors' gifts. We are tak-ing best practices to elevate and shape our culture. As new industry leaders join our team, they can expect a unique co-creation and publishing experience which accounts for shared ownership of success both of the book itself and team members. Collaboration is one of our cornerstone values and that includes fostering an expectation that individuals bring industry-leading credibility, participation, and added value. We're thrilled with the successes that our authors have achieved through their participation in our book teams. One got tenure and was promoted to engineering department chair, another defined the agenda for a near company-wide conference, another got a promotion, title increase, and pay raise, and many are making their workplaces more effective.

Team Goal Setting

There were a variety of ways we could have taken these two chapter topics. As one of our coauthors appropriately pointed out, "we can't boil the ocean." Instead, we identified our target audience in detail and developed specific outcomes we desired you, our readers, to act upon after reading each chapter. These reader outcomes became our NorthStar. We reverse-engineered the outcomes and created a strong infrastructure and network to support them.

Team Delivery

We were fortunate that the majority of the contributors for this risk book also participated in the first *Mobile Medicine* book and recognized that a key component to successful chapter delivery was chapter management. Because we had multiple authors, more coordination was required to organize and operationalize the group from initial idea conception to final chapter delivery.

Our chapter group identified the individual who is more systems-oriented to not just coauthor but lead and drive the chapter operations from beginning to end. The chapter manager road-mapped the overall process and developed more granular weekly activities, and built a system to organize and monitor progress.

This individual further built agendas, identified team goals for each meeting, and led discussions to meet the goals. This ensured our limited time was well spent and engaged the coauthors. The chapter manager sent follow-up emails detailing each coauthor's action items within 24 hours after each meeting and sent a reminder email with the upcoming agenda 24 hours before the next meeting. The contributors found these emails very helpful allowing them to stay laser-focused in their area and not worry if we were on track.

Although one individual led the operational chapter delivery, it was a team effort. There were times the chapter manager sought the team's input and buy-in. The book editor also coached the chapter manager to continue his development and ensure overall alignment. The operational cadence allowed the coauthors to benefit from the structure to fulfill their individual responsibilities and deliver their action items on time. Without chapter management, chapter delivery would have been more frustrating and taken much longer.

Team Outcomes and Experience

Yes, we delivered the chapter content and achieved our goal, but the priority wasn't just the destination. The journey of working together was equally important. Our relationships with each other are stronger than when we started. We can look back at what we accomplished together. Without a doubt that each coauthor wouldn't have any hesitation reaching out to one another in the future. We also bring this positive transformational leadership experience to future groups we work with and have the confidence we can lead and contribute to this type of game-changing leadership style.

Conclusion

In this chapter, we discussed how a single leadership style, attribute, or behavior cannot define success in a mobile medicine or in an advanced healthcare technology context. We identified transformational and transactional leadership styles as the most successful styles. We then characterized those styles and identified the ways that their major traits impacted success. Next, we discussed how transformational leadership characteristics can support mobile medicine and advanced healthcare technology innovation initiatives. Through analysis, we identified that transformational leadership is not all encompassing. While transformational leadership is critical for scaling and growing competencies that reduce barriers to innovation and implementation of mobile medicine and advanced health technology solutions, it is a starting point. Further evolution of leadership traits beyond transformation is needed to continue to drive results to patients and clinicians through innovation in the mobile medicine and advanced health technology space.

References

Craggs, Andy. *The change mindset: The psychology of leading and thriving in an uncertain world.* London: Kogan Page Publishers, 2022.

Corpolongo, Kristen, Contributor. "Cybersecurity needs transformational leadership from women." *CIO*, January 12, 2016. https://www.cio.com/article/243018/cybersecurity-needs-transformational-leadership-from-women.html.

Delegation is the Cure (6 Simple Steps). "Micromanagement is toxic" YouTube video, 8:31. July 21, 2020. https://www.youtube.com/watch?v=6XUZrWzFmIE.

Douville, Sherri, ed. *Mobile medicine: Overcoming people, culture, and governance.* New York: CRC Press, 2021.

Fleeson, William. "Moving personality beyond the person-situation debate: The challenge and the opportunity of within-person variability." *Current Directions in Psychological Science* 13, no. 2 (2004): 83–87.

Finkelstein, Sydney. *Superbosses: How exceptional leaders master the flow of talent.* City of Westminster, London: Penguin, 2019.

García-Morales, V. J., Lloréns-Montes, F. J. and Verdú-Jover, A. J. "The effects of transformational leadership on organizational performance through knowledge and innovation." *British Journal of Management* 19, (2008): 299–319. https://doi.org/10.1111/j.1467-8551.2007.00547

Harne-Britner, Sarah. "Richard Hader visionary leader award 2021 visionary leader runner-up carol grove." *Nursing Management* 53, no. 4 (2022): 38–40.

Herzlinger, R. Why Innovation in Health Care Is So Hard. Harvard Business Review 2006. http://hbr.org/web/extras/insight-center/health-care/why-innovation-in-health-care-is-so-hard

Kippenberger, Tony. *Leadership Styles.* Oxford: Capstone Publishing, 2002.

Lang, Y, H. Zhang, J. Liu, and X. Zhang. "Narcissistic enough to challenge: The effect of narcissism on change-oriented organizational citizenship behavior". *Front Psychol* 12 (2022): 792818.

Lilienfeld, Scott O., Irwin D. Waldman, Kristin Landfield, Ashley L. Watts, Steven Rubenzer, and Thomas R. Faschingbauer. "Fearless dominance and the US presidency: Implications of psychopathic personality traits for successful and unsuccessful political leadership." *Journal of personality and social psychology* 103, no. 3 (2012): 489.

Odom, Summer F., Barry L. Boyd, and Jennifer Williams. "Impact of personal growth projects on leadership identity development." *Journal of Leadership Education* 11, no. 1 (2012): 50–60

Reddy, Arthur Hoskisson, and Jones Mifflin Smith. "Corporate leadership style and strategic plan implementation to local government in the United Kingdom." *Journal of Strategic Management* 6, no. 3 (2022): 12–20.

Simmons, Lee. "How narcissistic leaders destroy from within." Insights by Stanford Business https://www.gsb.stanford.edu/insights/how-narcissistic-leaders-destroy-within April 30 (2020).

Son, Than Thanh, Le Ba Phong, and Bùi Thị Loan. "Transformational leadership and knowledge sharing: Determinants of firm's operational and financial performance." *SAGE Open* 10, no. 2 (2020): 215824402092742. https://doi.org/10.1177/2158244020927426.

Tang et al. "Does Founder CEO Status Affect Firm Risk Taking?" *Journal of Leadership & Organizational Studies* 23, no. 3 (2016): 322.

Texas A&M. "Transformational Leadership in Healthcare", September 14, 2021. https://online.tamiu.edu/articles/mba/transformational-leadership-in-healthcare.aspx

Trubetskaya, Anna, Declan Manto, and Olivia McDermott. "A review of lean adoption in the Irish Medtech Industry." *Processes* 10, no. 2 (2022): 391.

Van Eeden, Rene, Frans Cilliers, and Vasi Van Deventer. "Leadership styles and associated personality traits: Support for the conceptualisation of transactional and transformational leadership." *South African Journal of Psychology* 38, no. 2 (2008): 253–267.

Wu, Ting, Yi Wang, Rebecca Ruan, and Jianzhuang Zheng. "Divergent effects of transformational leadership on safety compliance: A dual-path moderated mediation model." *PLOS ONE* 17, no. 1 (2022). https://doi.org/10.1371/journal.pone.0262394.

Yurov, Kirill and Richard Potter. Transformational Leadership in Technology Post-Adoption Period: A Motivational Factor for Acquiring Technology Enhancement Information. (2006)

Chapter 6

Ignore the Dangers of Hubris at Your Own Risk: Mitigating the Barrier to Trust and Respect for High-Performing Teams in Medical Technology

Karen Jaw-Madson and Sherri Douville

Contents

DOI: 10.4324/9781003348603-8

Introduction

"Well, sir, at the risk of sounding immodest, you see before you a young man without a flaw … Unless, of course, you find me immodest, in which case, I have one flaw."

– Alex P. Keaton, *Family Ties*, S2E12

Advancing science with technology for better healthcare outcomes has so much potential, but they come with combined risk and higher stakes. It's critical to know where these risks exist and how to mitigate and manage them. Perfection is impossible. Preciseness, on the other hand, is more attainable. With careful measurement and calibration, technologists can iterate their way through hardware and software alike. They can decide the what and where and calculate the upsides versus the downsides. Many are very good at this. The same concept applies to healthcare providers. "Do no harm," falsely attributed to the Hippocratic oath,

> *is* a reminder that we need high-quality research to help us better understand the balance of risk and benefit for the tests and treatments we recommend. Ultimately, it is also a reminder that doctors should neither overestimate their capacity to heal, nor underestimate their capacity to cause harm.

(Shmerling, 2020)

The very best leaders across healthcare and technology understand they must extend this practice of balancing and calculating risk beyond science and technology to the human sciences – that which includes the dynamics within their teams and organizations and, by extension, the "social side of health" (Yong, 2021). The thing is many don't do this, or not enough do. Even fewer realize this requires the same degree of curiosity and clinical discipline to explore any other scientific or technological challenge.

To neglect the human dynamics in science and technology takes hubris, what Merriam-Webster defines as "exaggerated pride or self-confidence," regardless of whether it's done by willful choice or ignorance. There is no better way to explain the consequences of this than by exploring hubris itself. This chapter will take on the challenge of doing so as a clinician would, limiting and deferring judgment that typically accompanies the connotations of the word "hubris." This discussion will describe its meaning and its implications on leadership and organizational dynamics as it relates to risk in digital healthcare. The focus will be given to what practically is and can be done instead of what "should be."

This chapter's goals are to:

■ Increase individual and organizational self-awareness of hubris – where it comes from, what it looks like, what drives it, and how to mitigate its associated risks.
■ Provide healthcare executives and their technology partners a way to explain suboptimal encounters, increasing appreciation for the importance of people skills as they do technical skills.
■ Inform choices and decisions made in medical technology design, development, and deployment with intention based upon calculated risk.

In a field that deals with life and death on a daily basis, every leader and every organization must decide if they are willing to take on more risk due to hubris. The hope is that this chapter will better inform that decision.

Conceptualizing Hubris

Human nature and socialization feeds the tendency to categorize whatever presents itself as good or bad. Suspending judgment, even on what is typically perceived as negative (like hubris), maintains the aforementioned clinical approach and leaves room for more meaningful learning.

Hubris is also often understood as a comparison to its perceived opposite – namely, humility. However, a closer look suggests that while related to one another in some respects, they are not necessarily polarities. Most definitions of hubris include an extreme degree of arrogance and/or confidence. As articulated in Chapter 7, humility in two forms, intellectual and relationship humility, itself does not suggest a lack of confidence or arrogance. Eliminating the contrast, if only temporarily, allows for greater depth and nuance in the conceptualization of hubris.

This chapter began with the dictionary definition of hubris. By itself, it involves an especially high degree of pride and overconfidence. Studies have provided a deeper and more nuanced understanding of the concept. It has been described in a host of different ways, suggesting that there's either some lack of alignment among scholars and/or varying degrees of this affliction.

Johnson et al. (2010, 406) describe hubris as reflecting the results of arrogance. Owen (2006) adds contempt for others in addition to confidence and pride to their definition, while Silverman and his colleagues (2012, 22) write that the "false confidence" of hubris is "without contempt towards others." Extreme overconfidence is determined when a leader can't back it up with job performance and cognitive ability (Johnson et al., 2010, 419) and/or there's a disconnect with reality (Silverman et al., 2012, 22). To flip Mohammed Ali's famous quote, it is indeed bragging if you can't back it up. Berglas (2014) says hubris starts with an act of defiance, a "reactive disorder," which Asad and Sadler-Smith (2020, 40) interpret to be more of a state and not a trait. Hubris, according to the latter, can therefore be acquired with the presence of "significant power," "overestimations of one's ability," and "lack of constraints" (Asad & Sadler-Smith, 2020, 40). Continuing with this logic, hubris can recede along with these elements. By examining a century's worth of US presidents and UK prime ministers, Owen and Davidson (2009, 2) go so far as to define this dynamic in medical terms as a sickness they call hubris syndrome (HS), which develops "only after power has been held for a period of time." They propose the following clinical features of HS below. Three or more must be present, with at least one of the five components (denoted with an asterisk) identified as unique to hubris.

■ Sees the world as a place for self-glorification through the use of power.
■ Has a tendency to take action primarily to enhance personal image.
■ Shows disproportionate concern for image and presentation.
■ Exhibits messianic zeal and exaltation in speech.
■ Conflates self with nation or organization.*
■ Uses the royal "we" in conversation.*
■ Shows excessive self-confidence.
■ Manifestly has contempt for others.
■ Shows accountability only to a higher court (history or God).
■ Displays unshakeable belief that they will be vindicated in that court.*
■ Loses contact with reality.
■ Resorts to restlessness, recklessness and impulsive actions.*

■ Allows moral rectitude to obviate consideration of practicality, cost, or outcome.*
■ Displays incompetence with disregard for nuts and bolts of policy making.

Source: Owen and Davidson (2009, 4).

Though this list above may seem repetitive or duplicative, they were sourced from the American Psychiatric Association's *Diagnostic and Statistical Manual of Mental Disorders (DSM)* with critical nuances not fully appreciated by the laypersons though they were validated through research. In their interpretation, Asad and Sadler-Smith (2020, 44) accept hubris syndrome, but characterizes it as an "adjustment disorder" more than an "acquired personality disorder." Also, the debate here is whether hubris is a disorder of "leadership position rather than a disorder of the person" (Asad & Sadler-Smith, 2020, 44). We believe that the existence of non-hubristic servant and/or values-based leaders in similar circumstances of power suggests that the leaders themselves are not just the afflicted, but active players in their own self-sabotage.

Clinical and academic definitions aside, there are clear, practical, and observable behaviors to watch out for that may indicate hubris in the picture:

■ Annoyance with others' opinions not aligned with their own.
■ Individual discomfort with working as a team.
■ Taking credit for successes and passing blame to others.
■ Retaliation based on jealousy.
■ A need to make decisions independently.
■ Rarely or never admits when they make mistakes.
■ Over-reliance on own expertise; not open to new information or changing views.
■ Too comfortable in a job position, social settings, knowledge centers.
■ Subordinates rarely challenge supervisors' thoughts, decisions, and/or direction.
■ Lashing out or otherwise receiving feedback poorly.
■ Comparatively lower performance and outcomes.
■ Prone to exaggerations that do not come to fruition.

While we avoided the comparison between hubris and humility (Figure 6.1), it is worth examining hubris in relation to how confidence and narcissism are perceived, along the same continuum. In doing so, leaders and their organizations can better calibrate when the more desirable confidence enters the less desirable territories of arrogance, hubris, and, at the far extreme, narcissism.

According to Silverman et al. (2012, 23), "Confidence is simply a factual and reality-driven belief about ability or standing." Johnson et al. (2010, 406) also agree: "Someone who is confident knows who they are and their ideas about themselves are built on information that is authentic or reality driven." Confidence is the aim of most people and their organizations. Rudman (1998) determined that "self-confidence and assertiveness are highly valued and rewarded," as explained

Figure 6.1 Hubris on a perception scale. Source: Karen Jaw-Madson, Co.-Design of Work Experience.

by Anderson, Ames, and Gosling (2008). "Confidence is an essential ingredient of success in a wide range of domains ranging from job performance and mental health to sports business and combat" (Johnson & Fowler, 2011, 317). It's no wonder that many executives, thought leaders, and their coaches invest in addressing imposter syndrome and developing confidence.

Confidence, however, can be overused as a behavior. When it is, the perception changes to arrogance. Writes Johnson et al. (2010, 406), "Those who are arrogant are likely to take this confidence to a different level, as they overestimate who they are and what they can do, along with acting in ways that make those around them feel inferior." They set about developing the Workplace Arrogance Scale (WARS) as a way to measure arrogance in the workplace as a "useful predictor of performance" (Johnson et al., 2010, 424).

To continue with the sickness analogy, arrogance is where health begins to deteriorate with the onset of exaggeration and the departure from reality. It impacts people at their expense, including themselves. It turns out the arrogant aren't as smart, don't perform as well, and have low self-esteem (Johnson et al., 2010, 423).

Arrogance is also designated as a career staller by Lombardo and Eichinger (2004). According to them, it becomes a problem when these behaviors occur:

- Always thinks he/she has the right and only answer.
- Discounts or dismisses the input of others.
- Can be cold and aloof, makes others feel inferior.
- May detach him/herself from others unless on his/her own terms.
- Keeps distance between him/herself and others.

Source: Lombardo and Eichinger (2004, 481).

One may question whether hubris is the same as narcissism. Asad and Sadler-Smith (2020, 41) determined that while there are overlaps between hubristic and narcissistic leadership, they are "theoretically separable." The research of Johnson et al. (2010, 414) confirms that while "arrogance was related to some dimensions of narcissism," they (arrogance and its more extreme form hubris) are "unrelated to other dimensions ... which suggests that the two constructs differ." Scholars have offered various arguments for how they vary. Particularly compelling is how they associate with power, as interpreted by Asad and Sadler-Smith (2020, 48): "Narcissistic leaders reflect a preoccupation with fantasies of personal power to garner the approval and admiration of others and bolster and enhance ego. Hubrists exercise power to achieve overly ambitious goals, both personal and organizational" (Brown and Zeigler-Hill, 2004; McClelland and Burnham, 2008; Rosenthal and Pittinsky, 2006).

Narcissism goes far beyond the excessive arrogance of hubris, though they can also coexist (Asad & Sadler-Smith, 2020, 44). Consideration for others is discarded with little pretense in favor of entitlement combined with extreme self-centeredness and selfishness. "Narcissism is characterized by a grandiose sense of self-importance, a lack of empathy for others, a need for excessive admiration, and the belief that one is unique and deserving of special treatment" (Psychology Today). It too has an instrument for measurement: Raskin and Hall's (1979) Narcissistic Personality Inventory (NPI), which delineates different degrees of narcissism. Indeed, many successful leaders today demonstrate narcissistic behaviors, but "only when these traits are inflexible, maladaptive, and persisting and cause significant functional impairment or subjective distress do they constitute narcissistic personality disorder" (Psychdb.com), which only a qualified mental health professional can diagnose.

Chen and Reissman (2017) chose two types of narcissists (out of many) likely to be encountered at work. The classic agentic narcissists, as described by Randall Peterson, are about "how they look and how they're seen." The other type, the communal narcissist, is more stealthy. "They are self-appointed saints who have unrealistic views of their contributions to others" (Peterson & Wakeman, 2017). These are the leaders who practice what this book's editor and chapter coauthor, Sherri, calls "faux-mility," where there is a fake effort to seem humble when what they really seek is recognition.

It's easy to separate ourselves from arrogance, hubris, and narcissism with "that's not me." That belief in and of itself is in fact, arrogant, hubristic, and narcissistic. Everyone has the potential to demonstrate these behaviors and fall prey to these conditions, hence the importance of being self-aware and educated in these concepts.

For a cause as noble as medical technology, the field is not immune to our main topic of hubris. A number of defining elements drive it, including:

- Excessive power (and fear of losing it).
- Not enough challenge from others.
- A homogenous workforce at odds with a highly diverse healthcare ecosystem and lack of trust in anyone but themselves.
- Us versus them mentalities.
- Risk avoidance applied in the wrong places.
- Complacency.
- Resistance to change and other defense mechanisms in organization and teams rarely experiencing or learning from failure (thus being a victim of one's own success).

Hubris has consequences, and there are many. The sheer volume is a strong indication of why it matters.

Consequences of Hubris

Benefits

There is a tendency to consider consequences in a negative light when the true meaning of the word is quite neutral. Simply put, consequences are results. The consequences of hubris can be positive and beneficial. According to Johnson and Fowler (2011),

> Some authors have suggested that not just confidence, but overconfidence—believing you are better than you are in reality—is advantageous because it serves to increase ambition, morale, resolve, persistence, or the credibility of bluffing, generating a self-fulfilling prophecy in which exaggerated confidence actually increases the probability of success.

Their analysis finds that overconfidence

> encourages individuals to claim resources they could not otherwise win if it came to a conflict (stronger but cautious rivals will sometimes fail to make a claim), and it keeps them from walking away from conflicts they would surely win.

(2011, 319)

In another example, Tang, Li, and Yang found a positive relationship between executive hubris and firm innovation in China (2012, 11), as measured by total patent applications and citations and yearly ratio of new product sales to total annual revenue (2012, 8). "When certain environmental conditions (such as less munificent or less complex environment) allow executives to focus their attention on the area of most interest to them (firm innovation, in our context), the main relationship becomes stronger" (2012, 18). It should be noted that the sustainability of executive hubris driving innovation has not been established, and that the flip side of these results is also true: more funds and complexity (which are inevitable as organizations mature) can have a negative relationship with innovation. This suggests the positive effect might be potentially temporary, but it's still positive.

Medical technology has no shortage of entrepreneurs. Hayward et al. (2009) investigated how confidence, even overconfidence, can (but not always) contribute to the emotional, cognitive, social, and financial resilience of serial entrepreneurs. The study found that, among other things, "more confident entrepreneurs will have greater emotional and cognitive resilience" and will therefore be more willing to "start subsequent ventures" (2009, 4), with "a necessary condition" that there is the confidence that they will succeed, as found by other research (2009, 8). The power of positive thinking trope is also notable: "Higher confidence increases the odds of securing coveted outcomes, from creating wealth to saving jobs and lives ... with overconfidence promoting positive effects that help to persevere and prevail" (2009, 8).

Leadership

Many arrogant leaders may be considered successful. Silverman et al (2012, 26) argue that they "may have been even more effective sans the arrogant behavior." We agree. Much can be lost in the potential of what might have been. So while there may be upsides to hubris, there are far more watch-outs, pitfalls, and dangers, starting with the individual leader themselves. To begin with, hubris is classified as a cognitive bias, which is also known as a psychological bias (Johnson & Fowler, 2011). This impacts where a leader dedicates energy and attention, as the study of firm innovation above established. There's a risk where attention is *not* paid as well. Overconfidence can lead to ignoring red flags in information and social relationships, for example. In both cases, arrogance neglects the burden of knowledge (knowing something and not acting) and ignorance (not knowing something that should be known) because attention is directed elsewhere.

Those suffering from hubris are less likely to seek feedback (Silverman et al., 2012, 22) or receive it well when offered (Silverman et al., 2012, 26). They are poor learners: "A weak learning orientation also causes people to identify others to blame when setbacks or failures are experienced, instead of revising performance strategies or uncovering why problems occurred" (Silverman et al., 2012, 26). Bias (perceptions and assumptions) with lack of feedback and/or learning could lead to poor decision quality, which is determined by how they stand the test of time (Lombardo & Eichinger, 2004). As a result, performance suffers.

We would be remiss if we didn't raise emotional intelligence (EQ), which "accounts for 58% of performance in all types of jobs" (Bradberry & Greaves, 2009, 18–19). EQ requires self-awareness, self-management, social awareness, and relationship management skills, none of which are particularly strong in hubristic executives, by definition. Hubris without EQ makes the odds of success that much more difficult. Too much hubris can render a leader's position "no longer fully functional" (Owen & Davidson, 2009).

Hubris creates a certain dynamic with entrepreneurs in medical technology. There are those who arrogantly attempt to force-fit their solutions without understanding the industry or audience.

That may work in other fields, but it won't work in the high-stakes, complex, highly regulated environments in healthcare. A founder may have achieved success in one context, but for them to assume that it translates to another is also hubris. The "move fast and break things" mentality so common in Silicon Valley is antithetical to the practice of medicine. The joke at Medigram is to instead "Move thoughtfully with compliance." Technologists are outsiders in the medical field. This is due to the hierarchical nature of medicine and its knowledge intensiveness. The starting point is that respect and a place on the hierarchy aren't granted, they must be earned. Entering any interaction without knowing this feeds the perception of arrogance.

That arrogance forms what Martell (2015) called the Founder's Achilles' Heel:

> the mind-numbing complacency that comes from arrogance is the Achilles heel that can take down even the smartest entrepreneur. Investors, partners, and clients won't care how great your product is if you constantly give off the vibe that you know everything and have nothing left to learn. A know-it-all attitude basically announces to the world that your growth has peaked, and there is nowhere to go but down.

> **(Martell, 2015)**

Contrarian styles feeding into their own cult of personality (as entrepreneurs can be known to do) will not charm, but actually clash with a physician's tendency to think critically. The technologist must demonstrate literacy of the medical and clinical context, or risk irreconcilable conflict between themselves and their stakeholders, including but not limited to physicians and other care team members, health system executives, and regulators. These parties can be innovation blockers in various ways, from active resistance to apathy – all of which limit the ability to impact the healthcare system.

Collaborators, Partners, Teams, and Organizations

Leadership is not a solitary activity. It requires other parties, whether they are collaborators, partners, teams, or entire organizations. Hubris, in its attempt to overpower people, actually pushes them away, with negative consequences: "workplace arrogance could prove particularly costly if it ends up costing the organization customer loyalty or satisfaction, team morale, leader-member relationships, or commitment to a project or task" (Johnson et al., 2010, 423). The universal truth is that arrogance is viewed with contempt and makes leaders less likable. They aren't always aware of this, however, forming "an overly positive perception of their status in a group" (Anderson, Ames, & Gosling, 2008, 90) when in reality they are less likely to even be accepted (Anderson, Ames, & Gosling, 2008, 97). Instead, behaviors such as self-enhancement (a hallmark of narcissism), "does not lead groups to overvalue the individual, but instead to punish that individual" (Anderson, Ames, & Gosling, 2008, 95). So instead of leveraging social capital that leaders need to influence, there is a preventable and wasteful "social cost" (Anderson, Ames, & Gosling, 2008, 90).

It doesn't end there. Leaders, through the power granted by their job titles, have significant influence on their organizational dynamics. "Executive hubris has been shown to influence firm-level decisions and outcomes significantly" (Hayward & Hambrick, 1997; Li & Tang, 2010), write Tang, Li, and Yang (2012, 3). Overconfidence is contagious (Johnson & Fowler 2011, 319). The combination of power and overconfidence sets the conditions for spreading dysfunction in organizations. The bullwhip effect, where "small changes in demand can ripple across the supply chain" (Bank of America) is an apt analogy for how a leader's actions could have a large-scale impact. With its lack of learning (among other elements discussed here), hubris breeds incompetent leaders which can lead to toxic culture (Chamorro-Premuzic, 2020). Signals of that toxic culture caused by hubris in medical technology and healthcare are visible from the outside: lack of communication between departments, top doctors not knowing who the CIO is, technical staff complaining about crying at work, doctors quitting in disgust and writing about it, etc.

Culture is created by the patterns of behavior and the boundaries of what's acceptable, unacceptable, or condoned. When hubris is a defining characteristic of an organization's culture, the following are repeatedly demonstrated and reinforced in daily operations:

■ Authority is not questioned.
■ Absence of meaningful feedback.
■ Poor behavior.
■ Distrust.
■ Uninspired performance.
■ Bad decisions.
■ Increasing incompetence without learning.
■ Exclusion and lack of diverse thinking.
■ Shortcuts.
■ Unethical actions.

All of these create tremendous and unnecessary risk. What could happen if a nurse doesn't challenge a doctor's assumptions? Or a product designed by an unhealthy workplace proves detrimental to patients? What about dismissing processes that are necessary to meet regulatory requirements because arrogance grants permission to break the rules, and not in a good way? Johnson and Fowler concluded that "although overconfidence may have been adaptive in our past, and may still be adaptive in some cases today, it seems that we are likely to become overconfident in precisely the most dangerous of situations" (2011, 320).

Business, Industry, and Society

Hubris in leaders, on teams, and across the organization has a definitive impact on their business, and potentially beyond that on medical technology and healthcare industries, and society as a whole.

For a company, arrogance could come with a high financial cost when business ventures fail. Some technology companies who have ventured into the healthcare industry, particularly with consumer-based solutions, have learned this the hard way. Legal and regulatory risk resulting from hubris could lead to false claims, lawsuits, consent decrees, fines, facility closures by federal authorities, criminal charges, and even incarceration.

News headlines where the scientific or professional community publicly challenges claims made by tech companies are telltale signs of hubris, where there's a disconnect with reality. It doesn't leave room for the scientific method nor accuracy, reliability, or validation – all essential for acceptance and adoption as canon.

The proliferation of misinformation also demonstrates the impact of hubris in our society, from politics to disaster management to medical technology and healthcare. Leaders and so-called experts are often sources of misinformation, colored by their hubristic lenses. They may lack full understanding outside their domains, not be willing to admit errors or misjudgments, or practice deception and spread it. Their audiences may be just as guilty of hubris. Author and security expert Daniel Bagge warns, "Remember that ignorance, arrogance, and fear all complicate one's ability to detect false information" (Bagge, 2019, 20). The same gap created by hubris and disinformation drives a wedge between technology and medicine. That gap is evidence.

Disrupting Hubris

The good news is that arrogance is a "cluster of changeable behaviors, driven by relatively malleable beliefs" (Silverman et al., 2012, 26), and because hubris is not inherent, but a state (Asad & Sadler-Smith, 2020, 44), it can therefore be managed. Doing so would not just ameliorate the risks and problems caused by hubris, but also "provide a competitive advantage to organizations as well as encouraging positive behaviors" (Johnson et al., 2010, 423). There's a lot of upside to addressing hubris.

Person to Person

If hubris is a "syndrome," then it takes antibodies to cure and provide immunity. The first step is to recognize the symptoms as soon as they appear and prevent a full-blown infection. We have delineated not only the behaviors that indicate hubris but some of the outcomes as well. Individuals can catch these in themselves and others and take action. Giving and receiving feedback is critical for everyone since hubris does not allow for it. In alignment with our values, the team at Medigram takes time every week for critique. We make it a habit to seek opportunities for improvement. One way to do this is to offer feedback (which is often retrospective) in the form of feedforward (feedback for the future). It's a common practice in Appreciative Inquiry using a simple format: "What I like about …" and "What might make it even better is …" Even narcissists may be more receptive to such an approach.

Developing other mindsets, capabilities, and behaviors may also compensate and offset hubristic tendencies. We recommend starting with the following:

Empathy. This is the ability to understand perspectives from other points of view and demonstrate that understanding (Wiseman, 1996). Those with hubris have blinders on. Empathy widens the field of vision, increasing the likelihood that important social dynamics are not missed. The practice of putting the scope of one's work in relation to the entire organization (and how it impacts others) can also ensure a broader perspective as well as discourage siloes and us versus them mentalities.

EQ. "Emotional Intelligence (EQ) is your ability to recognize and understand emotions in yourself and others, and use this awareness to manage your behavior and relationships" (Bradberry & Greaves, 2009, 17). To achieve EQ, one needs observation, objectivity, and active listening to develop key skills of self-awareness, self-management, social awareness, and relationship management.

Learning and Learning Agility. Learning is the ability to incorporate and act upon new knowledge. Learning agility is "what you do when you don't know what to do" (CCL SOURCE). The Burke Learning Agility Inventory (BLAI) identifies nine dimensions: flexibility, speed, experimenting, performance risk-taking, interpersonal risk-taking, collaborating, information gathering, feedback-seeking, and reflecting (Hoff & Burke, 2022, 4). Scrum and Agile are frameworks that practice learning and learning agility.

Conscientiousness. This is one of the "Big 5" building blocks of personalities and focuses on self-regulation and impulse control (Psychology Today). Recall that one of the conditions for hubris is unchecked boundaries. Conscientiousness provides a built-in self-checking mechanism.

Humility. Though not the opposite of hubris, it is a powerful tool against it (see Chapter 7). It practices openness, awareness, flexibility, and self-command, keeping the ego and hubris at bay. Overselling is a common problem with hubris, especially in medical technology. In the absence of regulating sales and marketing for tech in healthcare (as they do in the medical device and pharmaceutical industries), humility closes the gap created by hubris between claims and reality. It compels the introduction of evidence and other forms of backup needed to move the medical technology industry forward.

Earlier we shared behaviors indicating arrogance as defined by Lombardo and Eichinger (2004). They also described behaviors where arrogance would not be a problem:

- Listens and responds to others.
- Is approachable and warm.
- Interested in others' views even if they counter his/hers.
- Includes and builds others up.
- Values the opinion of others.
- Treats others as equal partners.
- Shares credit with others.
- Seldom pulls rank or tries to overpower others.
- Gets close to some people and interacts with many more.

Source: Lombardo and Eichinger (2004, 481).

Owen and Davidson (2009, 9–10) identify humor and cynicism as

> qualities protective against disproportionate hubris … but nothing can replace the need for self-control, the preservation of humility while in power, the ability to be laughed at, and the ability to listen to those who are in a position to advise.

The American Psychological Association (APA) adopted expectations and guidelines around civility (Plante, 2017) in the hopes that these fundamentals may create healthier norms and dialogue so desperately needed in our society. The Civility Operational Definitions are worth sharing, if only to highlight how different they are from hubris:

1. Think carefully before speaking.
2. Differentiate and articulate facts from opinions.
3. Focus on the common good.
4. Disagree with others respectfully.
5. Be open to others without hostility.
6. Respect diverse views and groups.
7. Offer a spirit of collegiality.
8. Offer productive and corrective feedback to those who behave in demeaning, insulting, disrespectful, and discriminatory ways.

9. Create a welcoming environment for all.
10. Focus corrective feedback on one's best and most desirable behavior.

Practicing and role modeling such behaviors as described here will address and prevent hubris.

Leadership Selection, Performance Management, and Development

Given their influence, leaders must be discouraged from hubris. Along with narcissism, it is a pervasive and destructive problem where leaders need literacy and tools to deal with it, such as leadership goals and development plans around the behaviors above. A variety of self, team, and 360-assessment tools are available to provide data, frameworks, recommendations, and common language. They are a great starting point for building self-awareness. Executive coaching and the right mentors have been shown to be effective interventions as well.

In addition, it is the responsibility of each organization's board of directors to talent manage the C-suite, one that they should not ignore from a performance and succession standpoint. They are uniquely positioned to set boundaries and expectations that otherwise may not exist within the rest of the organization.

Even better would be to prevent those with hubristic tendencies from ascending to positions of power. Recognizing the number (20%) of toxic leaders identified by its ranks, the US Army set about reinventing its leader selection process (Spain, 2020). There is no guarantee of 100% success (after all, people change and not always for the better), but signaling the expectations of leaders and discouraging hubris go a long way toward making progress.

Culture, Community, and Organizational Citizenship

In the first book of this series, *Mobile Medicine: Overcoming People Culture, and Governance* (Douville, 2021), we outlined in Chapter 3 what it takes to drive innovation through culture, leadership, management, and learning. These conditions definitively immunize against hubris, but it's worth providing additional commentary here.

At the organizational level, culture change becomes an effective intervention for addressing and preventing hubris, creating conditions where it *can't* exist, or is less likely to emerge at scale. We mentioned hubris as a cultural trait earlier. Establishing values and adopting contrasting cultural traits with a matching employee experience builds trust where it's most needed. Design of Work Experience (DOWE) (Figure 6.2) provides a framework and step-by-step process for co-creating and aligning culture and employee experience (Jaw-Madson, 2018).

Culture replicates the competencies and behaviors we propose for individual development across an entire firm, creating a "herd immunity" from hubris. It encourages acceptable behaviors and discourages unacceptable behaviors.

Culture can also provide the norms around reframing the organization as a community (Mintzberg, 2009), ensuring hubris has no place in it. bell hooks wrote, "I am often struck by the dangerous narcissism fostered by spiritual rhetoric that pays so much attention to individual self-improvement and so little to the practice of love within the context of community" (hooks, 2001). While we promote the development of the self, we believe it should happen in connection with others and for the greater good. "Community means caring about our work, our colleagues, and our place in the world, geographic and otherwise, and in turn being inspired by this caring. Tellingly, some of the companies we admire most … typically have this strong sense of community" (Mintzberg, 2009). Community distributes power away from seemingly infallible,

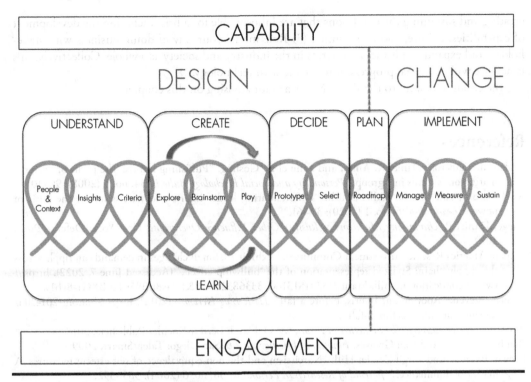

Figure 6.2   Detailed view of Design of Work Experience (DOWE) process. Source: Karen Jaw-Madson, Co.-Design of Work Experience

all-powerful figureheads into a shared model that still provides plenty of room for leadership. "Most sustainable improvements in community occur when citizens discover their own power to act … when citizens stop waiting for professionals or elected leadership to do something and decide they can reclaim what they have delegated to others" (Block, 2018). Instead of employees, managers, and executives, communities have citizens who are responsible for each other. There's nothing less characteristic of hubris than that.

Conclusion

The dangers of hubris are clear. If you are not taking proactive steps to stamp out the negative impacts of hubris across your organization and your job is to make medical technology work, you're taking huge risks and making a big mistake. Medical technology is the ultimate extreme team sport and anything with such an outsized impact must be considered a top priority. Much like narcissism, hubris appears to be a somewhat natural human failing that appears at times rewarded in popular (and arguably unhealthy) culture. Because of the intensity and breadth of teamwork required to be effective and successful in medical technology, we identified hubris as a collaboration blocker and show-stopping trait that every leader needs to be made aware of, have strong literacy for tools to counteract it in themselves as well as in their teams, and take proactive measures to address.

We leave you with a call to action: (1) Assess for hubris and other dysfunction in yourselves, your teams, and across your organization. (2) Address any immediate concerns and focus on creating and sustaining the conditions that are inhospitable to hubris, including the development of capabilities, culture, and community. (3) Role model your way of doing business with stakeholders and expand to advancing progress in the industry and society as a whole. Collectively, this is how we will make real progress in medicine with technology.

The co-authors want to thank Joe Mulhearn for advising on this chapter.

References

Anderson, Cameron, Daniel R. Ames, and Samuel D. Gosling. "Punishing hubris: The perils of overestimating one's status in a group." *Personality and Social Psychology Bulletin* 34, no. 1 (2008): 90–101.

Asad, Sarosh, and Eugene Sadler-Smith. "Differentiating leader hubris and narcissism on the basis of power." *Leadership* 16, no. 1 (2020): 39–61.

Bagge, Daniel P. *Unmasking Maskirovka: Russia's Cyber Influence Operations*. New York: Defense Press, 2019.

Bank of America Research Investment Committee. "Exhibit 10: Small changes in demand can ripple across the supply chain: Stylized representation of the 'bullwhip effect'." (Accessed June 7, 2022). https://twitter.com/carlquintanilla/status/1534181316073336832?s=20&t=l4s8nB95pTeoBXG1-8vb1w

Berglas, Steven. "Rooting out hubris, before a fall." *HBR Blog Network* (2014). https://hbr.org/2014/04/rooting-out-hubris-before-a-fall

Block, Peter. *Community: The Structure of Belonging*. Oakland: Berrett-Koehler Publishers, 2018.

Bradberry, Travis, and Jean Greaves. *Emotional Intelligence 2.0*. San Diego: TalentSmart, 2009.

Brown, Ryan P., and Virgil Zeigler-Hill. "Narcissism and the non-equivalence of self-esteem measures: A matter of dominance?." *Journal of Research in Personality* 38, no. 6 (2004): 585–592.

Chamorro-Premuzic, Tomas. "How to spot an incompetent leader." *Harvard Business Review* (2020). https://hbr.org/2020/03/how-to-spot-an-incompetent-leader

Chen, Daryl and Hailey Reissman. "The 2 types of narcissists you'll meet at work." *TED* (2017). https://ideas.ted.com/the-2-types-of-narcissists-youll-meet-at-work/

Douville, Sherri, ed. *Mobile Medicine: Overcoming People, Culture, and Governance*. New York: Routledge, 2021.

Hayward, Mathew L. A., and Donald C. Hambrick. "Explaining the premiums paid for large acquisitions: Evidence of CEO hubris." *Administrative Science Quarterly* 42, (1997): 103–127.

Hayward, Mathew, Forster, William, Sarasvathy, Saras, and Fredrickson, Barbara. "Beyond hubris: How highly confident entrepreneurs rebound to venture again." *Journal of Business Venturing* 25 (2009): 569–578. https://doi.org/10.1016/j.jbusvent.2009.03.002

Hoff, David F. and W. Warner Burke. *Developing Learning Agility*. Tulsa: Hogan Press, 2022.

hooks, bell. *All about love: New visions*. New York: Harper Perennial, 2001.

Jaw-Madson, Karen. *Culture Your Culture: Innovating Experiences@ Work*. Bingley, West Yorkshire: Emerald Group Publishing, 2018.

Johnson, Dominic D. P., and James H. Fowler. "The evolution of overconfidence." *Nature* 477, no. 7364 (2011): 317–320.

Johnson, Russell E., Stanley B. Silverman, Aarti Shyamsunder, Hsien-Yao Swee, O. Burcu Rodopman, Eunae Cho, and Jeremy Bauer. "Acting superior but actually inferior?: Correlates and consequences of workplace arrogance." *Human Performance* 23, no. 5 (2010): 403–427.

Li, Jiatao, and Y. I. Tang. "CEO hubris and firm risk taking in China: The moderating role of managerial discretion." *Academy of Management Journal* 53, no. 1 (2010): 45–68.

Lombardo, Michael M., and Robert W. Eichinger. *FYI For Your Improvement: A Guide for Development and Coaching, 4th Edition*. Minneapolis: Lominger Limited, 2004.

Martell, Dan. "Living life in permanent beta." *Startups.com* (2015). https://www.startups.com/library/expert-advice/living-life-permanent-beta.

Mintzberg, Henry. "Rebuilding companies as communities." *Harvard Business Review* 87, no. 7/8 (2009): 140–143.

Owen, Lord David. "Hubris and nemesis in heads of government." *Journal of the Royal Society of Medicine* 99, no. 11 (2006): 548–551.

Owen, David, and Jonathan Davidson. "Hubris syndrome: An acquired personality disorder? A study of US Presidents and UK Prime Ministers over the last 100 years." *Brain* 132, no. 5 (2009): 1396–1406.

Petersen, Randall S., and S. Wiley Wakeman. "The type of narcissist that can make a good leader." *Harvard Business Review* 3 (2017). https://hbr.org/2017/03/the-type-of-narcissist-that-can-make-a-good-leader.

Plante, Thomas G. "Even psychologists need help with civility guidelines." *Psychology Today* (2017). https://www.psychologytoday.com/us/blog/do-the-right-thing/201708/even-psychologists-need-help-civility-guidelines.

PsychDB.com. "Narcissistic personality disorder." (Accessed June 2022). https://www.psychdb.com/personality/narcissistic

Raskin, R. N., and C. S. Hall. A narcissistic personality inventory. *Psychological Reports* 45, no. 2, (1979): 590–590.

Rosenthal, Seth A., and Todd L. Pittinsky. "Narcissistic leadership." *The Leadership Quarterly* 17, no. 6 (2006): 617–633.

Rudman, Laurie A. "Self-promotion as a risk factor for women: The costs and benefits of counterstereotypical impression management." *Journal of Personality and Social Psychology* 74, no. 3 (1998): 629.

Shmerling, Robert H. "First, do no harm." *Harvard Health Publishing* (2020). https://www.health.harvard.edu/blog/first-do-no-harm-201510138421.

Silverman, Stanley B., Russell E. Johnson, Nicole McConnell, and Alison Carr. "Arrogance: A formula for leadership failure." *The Industrial-Organizational Psychologist* 50, no. 1 (2012): 21–28.

Spain, Everett. "Reinventing the leader selection process." (2020). https://hbr.org/2020/11/reinventing-the-leader-selection-process.

Tang, Yi, Li, Jiatao, and Yang, Hongyan. "What I See, What I Do: How Executive Hubris Affects Firm." *Innovation Journal of Management* 41, no. 6 (2015, first published 2012). https://doi.org/10.1177/0149206312441211

Wiseman, Theresa. "A concept analysis of empathy." *Journal of Advanced Nursing* 23, no. 6 (1996): 1162–1167.

Yong, Ed. "How public health took part in its own downfall." *The Atlantic* (October 23, 2021). https://www.theatlantic.com/health/archive/2021/10/how-public-health-took-part-its-own-downfall/620457/.

Chapter 7

Humility as a Core Value for the Adoption of Technology in Medicine: Building a Foundation for Communication and Collaboration

Brian D. McBeth, Brittany Partridge,
Arthur W. Douville, and Felix Ankel

Contents

DOI: 10.4324/9781003348603-9

Humility is a virtue which has been discussed and written about for millennia, among philosophers, religious scholars, and ethicists. From Socrates to Nietzsche, from Roman Catholic traditions to Buddhist writing, thinkers have struggled with defining humility in the human experience, while recognizing its centrality to understanding and relating to our world. This chapter will not attempt to summarize or cover the breadth and depth of this academic and philosophical exploration, but rather attempt to humbly explore the importance of humility as a core value for the adoption of technology in medicine.

Definitions of humility have challenged thinkers and researchers as well, and at times, humility is outlined in relation to a negative trait, sometimes dichotomized as its opposite: arrogance. Or even against psychopathology, such as narcissism. However, humility is more than a lack of arrogance and selfishness, and has at its roots an openness to perspective outside of the self and an acknowledgment of the pervasiveness of error in the human experience.

It is worthwhile to differentiate intellectual and relational humility, both of which will be important through the discussion in this chapter. Mark Leary, a social and personality psychologist at Duke University, defines intellectual humility as a recognition that the things you believe in might in fact be wrong – an awareness of one's cognitive fallibility. A humble intellectual approach acknowledges one's blind spots and shows active interest in learning from the perspectives of others (Leary et al 2017). The Templeton Foundation argues that intellectual humility "speaks to people's willingness to reconsider their views, to avoid defensiveness when challenged, and to moderate their own need to appear 'right.'" This suggests that this type of mindset encourages one to be "less influenced by our own motives and more oriented toward discovery of the truth" (Templeton Foundation 2022). Drs. Liz Mancuso and Stephen Rouse, two researchers from Pepperdine University, argue that intellectual humility ultimately will result in one's ability to be willing to revise one's most important viewpoints (Krumrei-Mancuso and Rouse 2015).

Relational humility is concerned less with ideas and thought, and rather one's intrapersonal (accurate view of self) and interpersonal (other-oriented rather than self-focused) perspectives which shape communication and human relationships. Psychologists Don Davis and Joshua Hook argue that this type of humility is essential for personal and communal social development and can be challenged in relationships where there is significant conflict – whether individual (like marriage), occupational, or societal/political (Davis et al 2011). Relational humility is also relationship-specific and includes an individual's ability to regulate expression of self-oriented emotions such as pride, guilt, or outward manifestation of ego while preferentially expressing positive social qualities such as compassion or empathy. Humility in a relational sense has the potential to promote higher levels of communication and conflict resolution in different contexts.

Both intellectual and relational humility will be important concepts as we consider the intersection of clinical medicine and health technology in our hospital systems. Table 7.1 outlines the working definitions for key terms referenced in this chapter.

Table 7.1 Definitions for Key Terms

Intellectual humility	Ability to revise one's viewpoint; awareness of one's cognitive fallibility
Relational humility	Ability to self-regulate emotion and outward manifestation of ego; use of humility to effectively shape human relationships
Narcissism	Entitlement and need for admiration; personality characterized by excessive self-interest, at times lacking empathy
Hubris	Excessive self-confidence; overbearing presumption of superiority of thought

Humility in Medical and Technical Education

If one assumes for a moment that humility is a desirable quality for care providers in a health delivery system, one might reasonably ask whether this is something that can be taught or if rather it is necessary to select students or learners who possess this quality already upon entrance to training. Certainly, when considering learners for admission to a medical school or residency, it would seem advantageous to select for applicants who already possess both relational and intellectual humility – this would indicate an open-mindedness to acquiring knowledge and building the toolkit that would allow a student eventually to provide informed, quality care. Intellectual humility from a learner would indicate that she could readily accept that her beliefs could be wrong, whereas research indicates that those lower in intellectual humility are more confident in their errors (Deffler, Leary, and Hoyle 2016). This could certainly present a challenge from an educator's perspective.

But do our educational systems and processes support a humble openness to knowledge acquisition with an acknowledgment of human fallibility? And if we strive to build these perspectives into our curricula and teaching approach, can this be modeled and ultimately transmitted to learners in a sustainable manner whereby the care is both informed and honest, transparently delivered to patients in a way that partners with them ultimately for their better health?

Much of the ethos of modern medical education has been influenced by the Flexner report (Cooke et al 2006). In the early 20th century, Abraham Flexner visited all medical schools in the United States and Canada to assess and report on the status of medical education for the Carnegie Foundation. His 1910 report was critical of the curricular content, educational methods, and structure of medical schools of this era. He recommended that medical schools should be aligned with universities and hospitals that engage in research and education, and this set the stage for the next century. Many small for-profit medical schools closed. Others transitioned into university-based medical schools associated with teaching hospitals. Students and residents were able to obtain preclinical and clinical training based on sound analytical thinking, robust educational curricula, rigorous assessment standards – led by faculty engaged in scholarship. It is only recently that educators have begun to rethink these foundational perspectives.

One of the unintended consequences of this system has been an emphasis on analytic over synthetic thinking, an emphasis on a strong and sometimes exclusionary professional identity, and an emphasis on hierarchical structures. Confidence is valued over humility and physicians often find themselves in intra- and inter-professional silos. Hierarchical structures that were built to assure accountability, decrease variability, and produce value now are viewed as limiting innovation and decreasing joy at work.

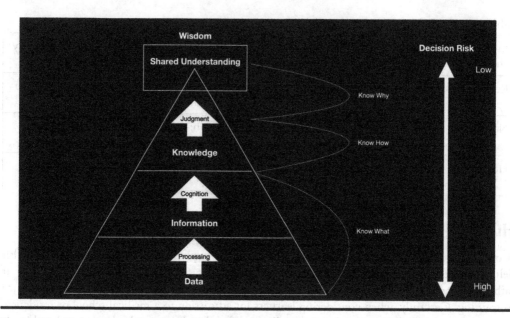

Figure 7.1   Data-Information-Knowledge-Wisdom (DIKW) Pyramid (adapted).

Felix, who has decades of experience educating students and residents, believes that three distinct relationships will affect the future of healthcare and will challenge the traditional physician relationship with knowledge, professional identity, and structure.

1. **The Physician's Relationship with Knowledge**

    There is a progression from data to information to knowledge to wisdom, as described in a model termed the Data-Information-Knowledge-Wisdom (DIKW) Pyramid credited to American organizational theorist Russell Ackoff – see Figure 7.1 (Ackoff 1999). Much of the value generated by clinicians in the last century has been in the space between information and knowledge: a patient would come to visit a physician with pieces of information about their health and well-being. The physician would analyze this information and create knowledge that would be a value to the patient. In the future, much of the data to information to knowledge progression will be automated. The value proposition for clinicians in the future will be in the knowledge to wisdom space. This space is a shared space: a space shared with patients, families, members of the healthcare team, and machines. To effectively thrive in this space, clinicians will need to move from a high-confidence to a high-humility framework. This shared space will also help physicians mitigate the Dunning–Kruger effect (DKE), a cognitive bias where high-confidence individuals overestimate their abilities because of lack of awareness (Dunning, Heath, and Suls 2004).

2. **The Physician's Relationship with Identity**

    Physicians have held an exalted position for the last century. Physicians were often viewed as leaders of the care team, the care structures, and the community. A big component of a physician's professional identity was the mindset of the physician as the vessel of medical knowledge. More recently, immediate access to knowledge resources and the advent of artificial intelligence are challenging this mindset. For physicians to thrive in the future, their analytic mindset will need to be complemented by a synthetic mindset. A synthetic mindset involves looking at disparate disciplines and seeing the connections between them. This involves humbly moving beyond silos and learning from others with diverse viewpoints. It is also clear that an integral but historically neglected viewpoint is that of the patient, and as physician identity evolves, relational humility is key to managing knowledge and engaging patients in their care.

3. **The Physician's Relationship with Structure**

Physicians traditionally have trained and worked in hierarchical structures. Hierarchical structures work well in manufacturing economies where there is a focus in reliably creating products of high value, having clear lines of accountability, and building infrastructures to support value production. Infrastructures such as finance, IT, communication, and planning support the work. In hierarchies, there can be a drift of original infrastructures becoming superstructures: rather than supporting the work, they drive the work. One of the unintended consequences of hierarchies is the challenge to rapidly innovate and create new services and products when structures where finance, IT, communication, and planning have become superstructures. Physicians of the future well need a humble approach to organizational structure. In some cases, hierarchies will support rapid value creation; in other cases, non-hierarchical structures such as networks will support rapid value creation – so-called ambidextrous organizations (O'Reilly and Tushman 2004).

Moving from Expertise to Master Adaptive Learning

Competency-based medical education is influenced by the Dreyfuss and Dreyfuss model of expertise (Dreyfus and Dreyfus 1980). Clinicians progress through competencies specific to their specialties through intermittent milestones that show a five-stage model of knowledge and skill acquisition (novice, advanced beginner, competent, proficient, expert). This expertise is achieved through repetition and feedback and leads to increased speed and accuracy in decision-making and procedural interventions. A parallel increase in confidence is often seen in these circumstances.

One of the unintended consequences of high-confidence individuals is the Dunning–Kruger effect (Dunning and Heath 2004). This effect, where unskilled individuals overestimate their performance, is present throughout multiple disciplines and challenges assumptions that physicians (and others) are effective in self-assessment. The DKE stands in opposition to intellectual humility and indicates a lack of awareness and insight that an educational intervention could be effective to address a gap in knowledge.

Another approach to professional development is to complement the focus of the physician as an expert with a focus on the physician as a master adaptive learner. The master adaptive learner model incorporates phases of planning, learning, assessing, and adjusting in a continuous loop (Cutrer et al 2017). Humility in the assessing phase allows for more accurate self-assessment and may help mitigate the DKE. There is now an increasing body of research showing that people who score higher on an intellectual humility scale are better at self-assessment (Krumrei-Mancuso et al 2020). Some experts may try to make points rather than make a difference and try to maximize a position rather than elevate a conversation. Master adaptive learners ask better questions, and one may suggest that physicians of the future need to ask better questions, rather than make better arguments.

As discussed above, intellectual humility serves an individual well, in terms of incorporation of new knowledge, revision of perspective, and collaboration. It would seem that this could have the potential to lead to improved innovation, and some educators in medicine have begun to incorporate structured reflection around cases for junior students, with the express intent to develop intellectual humility and critical self-awareness (Schei, Fuks, and Boudreau 2018).

There is a long-term need to develop skills that balance open-mindedness with critical scientific thinking, and attention to both intellectual humility and relational professionalism are of great importance in medical education (Spiegel 2012). There is also some recent research that suggests that intellectual humility could have a moderating effect on aging in the adoption of information and communications technologies, and Gloria Bernabe-Valero and her colleagues have argued that promotion of intellectual humility can have a "protective effect that cushions the effects of age on adaptability to new times" (Bernabe-Valero et al 2018).

If Bernabe-Valero and others are correct in asserting that intellectual humility could support adaptability as we age, this could potentially have profound implications for those interested in technology integration. Adoption of new technologies by physicians and other caregivers will be facilitated by a medical staff with a flexible and adaptable perspective to new techniques and ideas, as older physicians with limited technology skills are sometimes a challenging group to engage. A hospital and physician culture ideally embraces collaborative practice with technologists and acknowledges in a bidirectional way the work that must be done to support those who struggle to engage new platforms, while the physicians are encouraged to be open to innovative processes.

Humility in Technical Education and Culture

Technologists have typically been taught to attach self-worth to what they produce, what they earn, and what they know. The conversations are often around the valuation of a new product or service. Increasing output, productivity, and the who's who of thought leadership dominate blogs, podcasts, and feeds. In school when being trained to be a successful technologist, engineer, or scientist, they are often taught "to look and pay attention to hard, quantifiable, unambiguous, and repeatable data" – and that success was built on certainty (Cerri, 2016).

However, in healthcare as one moves from an individual contributor role to a leadership role, one must move from that certainty to being praised for having the "right" answer to defining success as "application of judgment when there is no right answer, but only answers that work, some better than others" (Cerri, 2016). Humility (and vulnerability) in health technology asks one to take a step back and examine if the products being created are impacting meaningful change in clinicians' and patients' lives. It requires admitting to oneself and others that an individual technologist does not have all the answers – one must get radically curious about the users and their barriers. It is coming to the realization and the core understanding that these solutions are not meant to be the next splashy headline, but rather the means by which a physician could get an hour more in the day to spend with the family or how a patient's quality of life might be improved by 10%.

The need for relational humility struck one of the coauthors when she was working with a team of developers that was creating a solution for a surgical team. As the lead Clinical Informaticist, Brittany was trying to lead the developers through workflows and outline the root problems and how they were affecting day-to-day OR operational performance. In each design meeting, the developers were excited about all the bells and whistles they were adding to the project; however, when she tried to refocus them on the surgeon's specific problems, they would look at her blankly and dismiss the concerns. Finally, she stopped the meeting and asked how many of the team had met a surgeon before, and only one-third of hands were raised. When asked how many had actually been in an operating room before, none of the hands remained in the air. No one saw this as a concern or barrier, and Brittany recognized immediately that a shift in mindset was required. Arrangements were made for the developers to scrub in with the surgeons, so they could watch and talk to them, and understand the challenges the physicians were experiencing. The follow-up meetings were markedly more focused and productive. The developers spoke of the site visit as a defining day of their careers, and they were now thinking about the user as a person. They were digging deep to address the real problems, while setting aside preconceived notions about how their tech would fix everything for the clinical staff.

> At the end of the day, developers, designers, and technologists in the healthcare space are not designing solutions for themselves. Rather, technology experts are working to bring solutions to clinical users who work in fast-paced, high-impact areas – areas where patients are dependent on the integrity and availability of data, and clinicians need to maintain a high level of trust in these systems. Technologists support clinical care delivery by coaching, outlining, and helping clinicians understand their systems, while the clinicians truly define the need and challenge of care delivery. Technologists, like physicians, need humility to allow them to be comfortable with saying "I don't know," to accept criticism of design, and to understand and empathize with the perspectives of the end users. When Edsger Dijkstra gave his Turing Lecture for the Association of Computing Machinery, he stated:

> We shall do a much better programming job, provided that we approach the task with a full appreciation of its tremendous difficulty ... provided that we respect the intrinsic limitations of the human mind and approach the task as Very Humble Programmers.

(Dijkstra 1972)

Those that seek to be "humble programmers" in healthcare might take that quote as a mission statement: to set down the fancy tricks, the vaporware, and the shiny objects, and come to the clinical space with humility and a willingness to accept when they are wrong and to learn from it.

Humility as a Foundation for Communication and Collaboration

Communication Models

As hospital culture has been slowly transitioning from hierarchical "top-down" administrative structures and physician-directed clinical management where "doctor knows best" to more team-based and holistic models, it is worth examining humility as a foundational construct in communication and collaborative practice. As far back as 1967, Paul Watzlawick was describing his axioms of communication theory, where models could range from simple transactional linear models where a source delivers a message to a receiver, to complex constructionist models that consider behavioral and societal influences on transmission of information (Littlejohn, Foss, and Oetzel 2017). Watzlawick's foundational communication theory is that not only the meaning (content level) of the message is important, but also the way that the communicator tries to be understood by others (relational level) (Watzlawick, Bavelas, and Jackson 1967). Although a deep dive into communication theory is beyond the scope of this chapter, humility could be considered as a foundational construct when looking at Watzlawick's fifth axiom of communication – symmetrical versus complementary communication. In short, symmetrical communication represents two participants on equal grounds with regard to power, whereas complementary communication represents a power differential between participants (Watzalwick, Bavelas, and Jackson 1967). When there is parity in symmetrical communication, there is a greater likelihood of transparency, though certainly no guarantee of optimal communication.

Complementary communication, with a power differential between the sender and the receiver, introduces a complexity to the interaction whereby humility can play a central role. Take first a simple example of administrative communication around a hospital decision to schedule an EHR software upgrade. Certainly, there is a power differential around this communication whereby the administration and IT leadership (sender) has the decision-making authority to delineate when necessary upgrades should occur, and this is subsequently communicated to staff (receiver). On one end of a spectrum of leadership humility, one approach would be to make a decision that's optimized for convenience of tech support staff hours or perhaps the revenue cycle of daily bed occupancy charges, without input from the clinical staff or consideration of individual patient needs – and then communicate that out as a directive by email 48 hours in advance. On the other hand, leadership could employ a process that incorporates clinical staff input into the timing of necessary software upgrades, selecting a low census time (3 a.m. to 5 a.m.) and attempting to minimize disruption to services. The complementary communication around this could be multimodal and bidirectional, communicating the need for the upgrade as well as the process which was utilized to solicit staff input and perspective with regard to timing. On larger scale administrative projects, incorporating dynamic and bidirectional communication strategies that approach planning and engagement of staff with humility and respect for the perspective of caregivers at the "sharp end" will increase odds of success (McBeth 2021).

Integrating Humility into Communication

With complementary communication, humility is critical for both the sender and receiver, and this is certainly evident in medical education. A teacher who is transparent about her limitations in knowledge will not risk losing credibility for the medical learners she is teaching at the bedside. Students and residents who approach patients with a humble appreciation and gratitude for the opportunity to grow their fund of knowledge and care experience will be more receptive to incorporating this learning regardless of its source. Opportunities to learn are abundant in clinical settings, regardless of one's level of training and often do not come from the professor or the experienced clinician. Patient advocacy groups have been pointing this out for some time, and we are reminded of William Osler's famous words: "He who studies medicine without books sails an uncharted sea, but he who studies medicine without patients does not go to sea at all" (Becker and Seeman 2018).

Another of Watzlawick's axioms of communication is that we cannot *not* communicate. "Activity or inactivity, words or silence all have message value: they influence others and these others, in turn, cannot not respond to these communications and are thus themselves communicating" (Watzlawick, Bavelas, and Jackson 1967). Nowhere is this more relevant than with regard to medical education and medical care in general. Certainly, words, but also action with regard to interaction with patients and other staff, are witnessed by impressionable learners and even established clinicians, and over time these everyday interactions establish and endow a hospital's culture and reputation. When a student witnesses respectful communication and a humble approach by her mentor, she is more likely to emulate this and academicians who study mentorship have argued that faculty facilitate an "identity transformation" in professional development, and that both teachers and institutions have a "moral responsibility" to promote professional virtues (caring, integrity, etc.) as well as competency and abilities (Johnson 2003). Conversely, exposure to dismissive or paternalistic interactions with others, or an approach lacking compassion and humility can certainly have the opposite effect – especially over time. At its worst, if a hospital and medical staff are known for disruptive behavior and lack of physician empathy and humility, it will represent infertile ground for development of empathetic learners and will also struggle to retain and recruit staff in general (McBeth and Douville 2019).

Yet communication is not always about collaboration, and one of the tasks of leadership is decision-making for an organization, and clarity in communication around these decisions. Conflict resolution classically (Thomas Kilmann model) employs different modes and strategies that balance assertiveness and cooperativeness (Thomas and Kilmann 1974). For example, leaders in medical technology and administration will find greater success taking a thoughtful approach to deployment of a new mobile medical product for clinicians that balances an accommodating and competitive perspective along with collaboration. In this context, leadership humility may represent both an acknowledgment of different perspectives and challenges for the various players affected by the technology, and also a pragmatic awareness of the role of the leader to effectively guide an organization and most effectively communicate its vision and strategy.

Likewise, physicians at times need to communicate difficult medical recommendations at the bedside which are not always well or easily received and understood. Certainly, there is a skill seen in an experienced clinician at the bedside, balancing appropriate confidence and also conveying clinical uncertainty for a suspected new cancer diagnosis, for example. Approaching such difficult communication with an intellectual humility and an ability to recognize the limitations of both self and system frames the interaction in a more transparent way that could improve patient understanding. More fundamentally though, this approach could reduce hierarchical barriers between patient and provider if a provider sees herself as a team member who has lined up alongside the patient to guide in an ongoing battle with uncertainty around diagnosis and treatment options. A patient may feel more support with such collaborative communication and facilitating this has been shown to improve patients' understanding and adherence to recommended treatment, with potential for other improved outcomes as well (Ha and Longnecker 2010).

Communication Styles that Challenge Humility

As biomedical information and technology inexorably expand, the application of knowledge in safe and effective patient care becomes increasingly dependent upon communication within healthcare systems. By communication, we mean a flow of information that is increasingly dependent upon information technologies. In the healthcare setting, communication styles as modulated by personality factors operate at individual, team, and organizational levels.

Some patient care events are essentially dyadic, playing out in the interaction between a healthcare "provider" or team and a patient. This might be a physician, nurse, or other healthcare team member interacting directly with a patient in a hospital room or a medical office setting, an operative team led by a physician doing a procedure, for example, an open-heart surgery – certainly a paradigm of complexity – or creating and applying a radiotherapy treatment plan, also involving information regarding the impact of a particular type of radiation therapy on a particular kind of cancer problem, as well as the application of a complex, often computer-driven technology over a treatment program lasting weeks. There is also the arc of care of an illness process and treatment requiring integration of inpatient, outpatient, and home care management. These processes may involve different organizations within the continuum of care, each with different managerial styles, practitioner skill sets, and financial constraints.

In this context, all healthcare systems face the challenges of the "human factor" with all its attendant opportunities for miscommunication, temperamental misjudgment, biased thinking, narcissistic departure from safe practice, pressures for financial and time-based efficiencies, and the added structural design flaws that reportedly yield many thousands of preventable medical care–related deaths and injuries every year (Institute of Medicine 2000, Makary 2016).

While the safety and effectiveness of patient care are paramount, other key performance indicators of healthcare systems are impacted by leadership styles and communication factors. These include profitability, broadly understood as the capacity to create a financial margin that allows for investment in new facilities and technology, ability to attract the kind of knowledge workers needed to operate and maintain them, and organizational processes that create competitive advantage in the largely market-based healthcare system found currently in the United States.

It Starts at the Top

The role of the chief executive officer of a healthcare organization cannot be overstated in that a paramount role of the CEO of any organization is to set the tone of interaction of almost everyone serving at almost any level. In the context of today's consolidated healthcare markets, regional and national structures share that challenge. Even in staff models emphasizing employment and vertical integration of services, "employment does not equal alignment" (Rastello 2012). Heavily unionized public healthcare organizations combine a spirit of public service with recognition of the limits of executive influence, which is often dispersed between hospital CEOs, regional program directors, hospital and system boards, and governmental organizations. Personal charisma is less a factor than balancing all these influences with a public face of service and humility. So long as public funding is generous, developments proceed calmly if not quickly.

In the private sector, whether "for profit" or not, the challenge of clinical transformation occurs in a setting of market competition demanding strong cost management disciplines and an aggressive approach to pricing. CEOs in this setting may be more likely to demonstrate personality traits that tend to permeate the entire organization, sometimes with a strongly narcissistic lack of humility. Narcissistic leaders with a bent for creative strategy and risk-taking present complex trade-offs between service versus profit-oriented approaches to organizational structure and function (Dinesh 2019). At worst, overweening and narcissistic leadership can lead to processes seen as fraudulent, with severe organizational consequences, for example, when a large and rapidly growing healthcare system was levied over $1.7B in fines and restitutions (Department of Justice 2003). More commonly, leadership under pressure to demonstrate profitability may tolerate individual physician behavior that is destructive of team morale as well as slowness in developing programs that emphasize team coordination. Sometimes, strong leadership can be associated with a "shame and blame" approach to management that tends to encourage similar behaviors up and down the chain of command, as well as a tendency to avoid admissions of error or delay addressing quality issues.

Assad suggests differentiation of hubris and narcissism by "analyzing their differences in relation to power and leadership." In his view:

> Hubris and narcissism overlap, and although extant research explores relationships between them in terms of characteristics, attributes, and behaviors, we take a different view by analyzing their differences in relation to power and leadership. Drawing on a psychology of power perspective, we argue that narcissistic and hubristic leaders relate to and are covetous of power for fundamentally different reasons … Unbridled hubris and narcissism (i.e. searching for and facilitated by unfettered power) have important ramifications for leadership research and practice. Leadership discourse, preoccupied with and predicated on positive aspects of leadership, should assess these two potent

aspects of leadership because misuse of power by hubristic and narcissistic leaders can create conditions for, or directly bring about, destructive, and sometimes catastrophic unintended outcomes for organizations and society.

(Asad and Sadler-Smith 2020)

The role of narcissistic leadership is complex, in that individuals with these traits may be well suited to innovation and leadership during periods of challenge to organizations (Maccoby 2007, Dinesh 2019). Other analysis suggests that the narcissistic personality style is most often toxic in its effect on organizational effectiveness and culture (O'Reilly, Chatman, and Doerr 2021). Systematic evaluation of the effect of narcissistic personality styles and organizational effectiveness emphasize the complexity of the problem and the need for further study (Liu et al 2021).

In a practical sense, narcissism and disruptive behavior can have a corrosive effect on organizations that can manifest in major incidents and blow-ups, as well as be insidious. Over time, such a situation can bring about the erosion of trust and a crisis in the professional culture of a healthcare organization. Two of the authors – Brian and Art – addressed a situation with an accumulation of unaddressed incidents in professional behavior by the medical staff of one hospital (prior to current ownership), and instituted a "Physicians' Citizenship Committee" that was self-governed and integrated into the medical staff governance structure. It was successful in empowering physicians to address their own internal behavioral challenges and restore trust within the organization (McBeth and Douville 2019). The reporting structure and workflow is shared here as one example of a structured mechanism to support professional discourse and promote organizational humility (see Figure 7.2).

Leadership Hubris and Situational Narcissism

Hubris can be described in terms of a narcissistic social style that can evolve in a setting of success as defined by the broader culture. Internally, the experience is of a sense of exceptionalism that by right allows the imposition of will on others. By necessity, it exists in a field of power that ultimately will find its test in contact with a similar player or events that either support or undermine the social illusions upon which it feeds. It is potentially sociopathic to the extent that it is amorally self-centered and self-aggrandizing, and thus is natural to narcissistic personality disorder, but may emerge in settings where the reinforcement of success as realized in social prestige, wealth, influence, and social opportunities also reinforce an individual sense of special value. Thus, a competent manager or successful entrepreneur elevated to a position of power and wealth can demonstrate the dynamic of hubris even if their original personal style was merely obsessive, flexible, introverted, extroverted, other-directed, inner directed, etc. The expression of hubris and the degree of narcissistic demonstration are predicated on "narcissistic supplies" bestowed by the larger culture, e.g., "situational." Success at this level can infect entire organizations to the point that the organization remains internally self-absorbed, convinced of its superiority and unable to learn or grow from contact with those outside its milieu. Individuals at all levels begin to express similar hubris through "narcissistic identification" with the "brand." In the end, leaders and followers alike may display what Maccoby notes as unproductive narcissistic traits, including "arrogance, grandiosity, not listening to others, paranoid sensitivity to threats, extreme competitiveness, and unbridled ambition and aggressiveness" (Maccoby 2007). The reader may recognize these traits in both individuals and organizations in the healthcare field and consider the impact on the success of this kind of enterprise in the broader context of the requirements of a successful healthcare system.

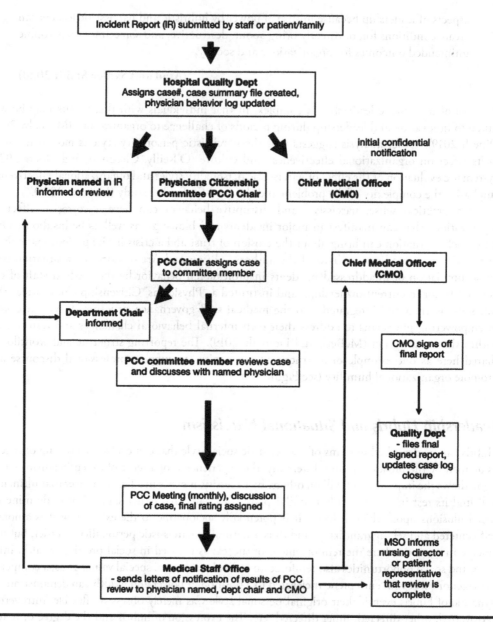

Figure 7.2   Physician Citizenship Committee to Address Disruptive Physician Behavior 2.

Table 7.2 highlights the productive and unproductive effects of narcissistic leadership with regard to workplace relationships and organizational culture.

Interface of Technologists and Care Providers – Why Humility Is Critical

From a knowledge perspective, physician and hospital leaders can and should apply the principles of intellectual humility to technology projects. First is an acknowledgment of one's own

Table 7.2 Narcissistic Leader and Organizational Impact

	Personal Reality	*Organizational Impact*
Productive narcissism	Takes risks to feed self-ego	Visionary and Strategic
	Self-projection activities	Elevates the Organization in Crisis
Unproductive narcissism	Situational narcissism of egocentric leadership	Organization becomes unable to grow and learn beyond the "brand"
	Arrogance, self-absorption, extreme ambition, hubris	Organization fails to recognize existential threats with increased costs, loss of competitive advantage, or outright failure

limitations as a leader and need to understand others perspectives. There is ample research in medicine and psychology that suggests confidence in one's understanding and skill does not correlate with objective measures of knowledge (Tracey et al 1997). Nobel Prize recipient Daniel Kahneman published extensively on the "illusion of validity" whereby even experts overestimate their ability to interpret and predict accurately outcomes, even when educated around potential bias and cognitive pitfalls in overconfidence (Kahneman, Slovic, and Tversky 1982). Also described earlier in the chapter was the Dunning–Kruger effect which can contribute to increased rates of error, misdiagnosis, and patient injury (Dunning et al 2003, Dunning, Heath, and Suls 2004). However, most physicians are not overconfident when it comes to technology, and embedded in the practice of clinical medicine is also the construct of consultation, which fundamentally represents an acknowledgment of the limitations of one's fund of knowledge and skill set. While there is other obvious professional "scaffolding" that prevents an internist – for example, from deciding to perform a cardiac bypass surgery (credentialing and privileging, hospital bylaws, etc.) – appropriate humility frames a decision as to when to engage another service or physician on clinical care. A physician steps back and acknowledges that best practice dictates that he or she is not the one to manage this challenge and additional help and expertise is needed.

Likewise, physicians who have successfully engaged technology systems within the hospital generally employ a humble approach, with regard to both the systems themselves and the experts who manage them. Contrast for a moment a hierarchical physician-centered perspective in which a physician sees himself at the top of hospital care delivery pyramid, where he dictates and is ultimately responsible for care delivery, and nurses and other ancillary staff answer to his orders and direction. The technology expert would fall under him as well in the support structure, with deliverables that should serve the physician first and foremost. The physician leader in this case may see technology in this model as integral to his success or may view it as a necessary evil that is required by a third party, which ultimately may detract from his performance.

On the other hand, a collaborative model – where humility is a core value and the care team includes the physician as a leader, but also values everyone's contribution – looks quite different. A physician who recognizes that patient care in a hospital cannot occur without the entirety of the care team will likely incorporate a higher level of appreciation and gratitude, for example, to the environmental service staff member who cleans the operating theater and patient rooms. Likewise, this recognition would be more likely to include the importance of expeditious access to patient health information, as facilitated by a robust EHR supported by technology experts who manage data security and access for the clinicians who need these data for clinical decisions. Gratitude is tied to humility as a core value, and there is psychology research that suggests that managers who incorporate communication with gratitude may see increased employee performance (Grant and Gino 2010). Paul Atkins et al. have published on what they term the "Prosocial" approach to collaborative management and strategy, in which evolutionary theory sets a framework for examination of group collaboration, decision-making, and conflict resolution. Although a detailed examination of their work is beyond the scope of this chapter, humility and respect within and between groups, as well as a recognition of autonomy and justice set the ground rules for a theory of behavioral relationships and also offer practical tools for leaders in a variety of management settings (Atkins, Wilson, and Hayes 2019, Eirdosh and Hanisch 2020).

Returning to a technology context, at the risk of seeming reductionist, one cannot overstate the importance of medical and technology specialists interacting in a respectful, humble, and collaborative approach. There are concrete organizational approaches that may facilitate the development of such a hospital culture. Standing meetings to bring clinicians together with technology experts can facilitate interdisciplinary discussions and provide opportunity for team-building and sharing perspectives. Defining clinical "champions" on technology project builds and rollouts helps define and develop leadership, potentially improving engagement in the clinical space, as well as prospectively avoiding conflict with anticipation of project challenges with implementation. Integrating leadership with technology expertise in the executive team, and educating physicians and other clinical leaders around innovation and developing technologies in the healthcare space will support collaborative interactions and a humble pragmatism with regard to project development.

How Can Technologists Bring Humility to Clinical Users?

Beyond the organizational approaches listed above, each technologist has the opportunity to show up with humility in their interactions with clinicians every day. Some small concrete ideas that can make an impact:

1. **Going to the *gemba* (Six Sigma terminology for the location of the work).** The goal here is to validate workflow and ask questions – the first step in ensuring that the design will fit into the users' workflow is to analyze what is happening currently. Define the problem, or the metric needing to be captured and put boots on the ground to see how it is being addressed currently. Follow multiple clinicians around: watch what they do, ask questions, look for inconsistencies, ask more questions. In essence, demonstrate radical curiosity to understand the perspective of the clinician. Then create a current state workflow map, and validate it with your users/clinicians.
2. **Ensure clinicians are included in EVERY step of the design process.** It is the role of the technologist to present design decisions and to suggest possibilities to meet the functionality or needs of the hospital or clinic. Then it must be the role of the end user – be it physicians, nurses, or ancillary staff – to make the final decisions. They must be involved in design discussions, workflow analysis, and final testing. Even the most state-of-the-art application will fail, if it doesn't fit into the workflow of the clinician or have clinician usability.
3. **Consistently follow up post-implementation for optimization.** In order to build trust, a consistent follow-up cycle is needed post-implementation of any new technology in the clinical setting. Teams are more likely to attempt to use technology if they believe there will be follow-through on requested changes post-implementation. Through checking in at

regular intervals, and consistently making promised changes, a strong relationship for future rollouts can be forged. However, if tech is regularly dropped into clinical workflow never to be touched again, resentment and frustration will form.

4. **Acknowledge frustration and follow up**. This can be a big learning curve and change in focus. As mentioned, technologists are trained that success looks like having the right answer, in a similar way to the way physician confidence is rewarded in medical training. Build trust and improve communication between clinicians and technologists. Early in Brittany's career, not having the solution would cause her to withdraw and cease communication because she thought she was failing her team. However, she learned over time to be transparent. For example, a software application with which she currently works doesn't have the capability to integrate with interpreters. Having first validated that this is frustrating, she then suggests another option to use interpreters, and communicates each time she escalates with the two vendors who need to integrate. Is this the one "right" way to handle this? No, but it is an improvement from cutting off communication.

Figure 7.3 integrates key components of these suggestions into a "Trust Loop."

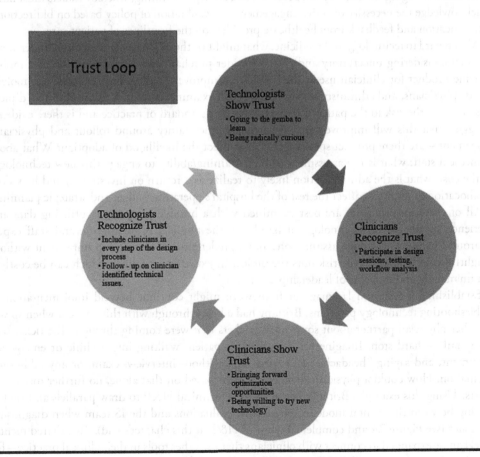

Figure 7.3   Trust loop. Source: Partridge, UC San Diego Health (2022).

Risk and Humility in Healthcare Systems

Consideration of risk in the clinical space has the potential to be an opportunity for common language between technologists and clinicians. Physicians are familiar with the prevalence of uncertainty in clinical medicine. In diagnosis, clinicians employ a Bayesian approach with application of a test or a physical exam finding, for instance, to refine a differential diagnosis and predict likelihood of a given condition (Gill, Sabin, and Schmid 2005). An experienced diagnostician is humble, understanding that as additional information becomes available, one must refine the diagnosis in a dynamic way and while it might be appropriate to be confident, absolute certainty often eludes. The Academy of Medicine has developed a dynamic model which builds uncertainty, and in a sense, humility into the diagnostic process (National Academies of Sciences, Engineering and Medicine 2015). Additionally, in treatment decisions, a surgeon or internist may try to convey odds of success of a given procedure or therapy and quantify risks of complications. The art of medicine incorporates the skills of being able to communicate risk and uncertainty to patients and yet inspire trust and confidence at the bedside.

Outside of clinical practice, Brian has published a model of administrative policy communication and staff engagement that also relies on dynamic modification of policy with outcomes and staff feedback (McBeth et 2021). It is an argument for humility with respect to organizational communication, acknowledging uncertainty in operations and strategic planning, whereby leaders must humbly acknowledge the necessity of active engagement and calibration of policy based on bidirectional communication and feedback from healthcare providers on the frontlines of patient care.

With regard to technology and medicine, what might be the common language and understanding when considering uncertainty and risk? Consider purchase and implementation of a mobile medicine product for clinician use at the bedside to improve the diagnostic process. Technology experts, physicians, and administrators would want to examine associated risk in multiple dimensions. What is the risk to the patients versus the current standard of practice and is there evidence to suggest that this will improve care? What is the uncertainty around rollout and physicians' engagement – are there product specifics that may affect the likelihood of adoption? What about the medical staff: what is the assessment of their training/ability to engage this new technology? For the cost, what is the administration likely to realize as a return on investment, and how does this allocation of resources affect the rest of the hospital's operating budget and strategic planning?

All of these uncertainties are best examined with a humble perspective, utilizing data and experience specific to the technology at hand and the physical infrastructure and staff expertise around its use. As leaders, assumptions and a predetermined decision to implement without thoughtful examination of the risk puts the project in jeopardy for failure, which can be costly – both financially and in terms of leadership capital.

Establishing a common language or framework might continue beyond implementation to troubleshooting technology problems. Brittany had a breakthrough with this concept when speaking to her physician partner about some of the tickets that were coming through. The ticket said "phone fail" – hard stop. Imagine for a moment, a patient walking into a clinic or emergency department, and saying "headache" as a hard stop, without interview, exam, or any additional information. How could a physician diagnose or treat based on that alone, no further discussion or tests? Using this example, Brittany began using a technical H&P to draw parallels and understanding between the mental model of physicians' evaluations and the IS team when diagnosing tech issues (see Figure 7.4 and complete Technical H&P at this chapter's end). This shared mental model can be expanded to connect with clinicians that use other tools in their clinical practice – for example, triage protocols and SBARs can easily be adopted for tech to allow for common language.

The Technical H&P

Chief Complaint (CC)

Difficulty hearing the patient when conducting a video visit

History of Present Incident

Device is a three year old iPhone 8, that is used to conduct video visits in the emergency room. About seven days ago it began to have intermittent sound while being used for video calls, patient can hear the physician, physician cannot hear the patient. There is no associated loss of visual both sides can see each-other clearly. Physician is able to access application and begin video call without any trouble. Moving the iPhone to different locations doesn't seem to relive the issue. Issue is not exacerbated by time of day of video visit. Force quitting the application and restarting the phone hasn't helped. There are no error messages that present when audio fails. Moving from Canto to Doximity allow the visit to complete with audio intact.

History

Negative for recent trauma (being dropped/thrown)

Negative for history of liquid immersion

Social/Family:

No other recent issue tickets submitted on device.

Other iPhones in the department are not experiencing this issue. iPhone 8s have experienced dropped audio on phone calls 4 months ago, however issue was resolved with latest update.

Review of Systems/Physical Exam:

Software (Application): Canto V. 12.6, Safari V. 13.2, no updates needed both running most recent version

Operating System: iOS 14.2, update available to iOS 14.5

Hardware: iPhone 8, doesn't look to have an defects. Ancillary devices used – Airpods via blue tooth.

Network: Health System Web, no recent changes to the network, not other symptoms of instability

Management: System managed device, ED- Physician Profile

Access/Security: User is in the ed_physician security group, no recent changes, is able to perform visits with necessary audio on other devices including other iPhone 8s.

Pertinent Tests:

Application Security: Mic is allowed in application settings, toggling it on and off didn't resolve the issue

Figure 7.4   The technical H&P. Source: Partridge, UC San Diego Health (2022).

Humility, Transparency, and the Acknowledgment of Error

Another lens through which one might assess the importance of humility in healthcare and technology is transparency and risk – as related to patient safety, the disclosure of medical error, and medical malpractice claims. Thomas Gallagher, an internist and researcher at the University of Washington, has published extensively on the importance of early disclosure and transparency with regard to medical error that leads to patient harm (Gallagher, Student, and Levinson 2007,

Gallagher, Mello, and Levinson 2013). His research and work with the Agency for Healthcare Research and Quality (AHRQ) has led to disclosure tools and programs, called Communication and Resolution Programs (CRPs). These programs are based on principles of humility and honesty and promote disclosure of error, transparency, and communication, as well as fair compensation for those injured by medical error (Gallagher et al 2018). A number of hospitals and health systems have moved in this direction – such as the University of Michigan. The "Michigan Model," as they term it, involves adoption of a transparent and more open approach, and has yielded fewer malpractice claims, quicker resolution of claims initiated, and a lower total cost of litigation for the health system (Kachalia, Kaufman, and Boothman 2010, Mello, Boothman, and McDonald 2014).

How could system transparency and acknowledgment of error facilitate collaboration and ultimately better integration of health technologies for patient care? First, it is clear that we are in an era when patient engagement, autonomy, and access to one's own personal health information is a focal point of care delivery. Although health literacy and understanding has not necessarily kept pace, more consumers are asking for access and control of their medical records and the Congressional Cures Act, which went into effect in 2021, supports this. With certain exceptions, providers and health systems must provide access to patients' healthcare records for "reasonable and necessary" activities and at a patient's request (21st Century Cures Act 2016). Concurrent with this increasing patient demand for access and control of their healthcare data is the greater national awareness of medical error over the last two decades, integration of patient safety goals into routine patient care, and an international patient safety movement incorporating transparency and disclosure.

Hierarchical leadership structures and culture within healthcare organizations have not necessarily kept up with these changes, though government legislation such as Cures establishes compliance expectations that leaders cannot dismiss. Other healthcare quality organizations direct hospital systems to structure their systems around important though less specifically defined objectives – such as the Institute for Healthcare Improvement's (IHI) Triple Aim: improving population health, enhancing patient outcomes and experience, and reducing cost (Berwick, Nolan, and Whittington 2008, Institute for Healthcare Improvement 2022). In the context of evolving consumer demands for personal data accessibility, rapidly evolving mobile medical technologies, and changing regulatory requirements around data, both physician and administrative leaders in healthcare need to recognize and embrace technology experts as partners in a transformative evolution that ultimately can elevate quality of patient care. Without a humble leadership perspective and valuing of information technology expertise, organizations may be left floundering in the tumultuous and complex sea of data access, security needs, risk and regulatory requirements, patients' demands, and physicians' dissatisfaction. This partnership in strategic and operational leadership is not a means to an end of care delivery, but rather a critical component of the ultimate goal: a more equitable healthcare delivery system that values contribution from all its constituents, recognizing the need for integration in information and collaboration of all team members to improve health and care. Ultimately this also aligns well with the IHI Triple Aim – the goals of safe and quality care, with a more just and equitable healthcare delivery system – built on a foundation of humility and collaborative practice.

Personality Factors and Medical Error

If the costs of organizational and leadership hubris and narcissism can be large, narcissistic traits at the individual practice level can also exact a price in terms of organizational cohesion and patient safety. Most healthcare administrators have encountered healthcare professionals who exhibit traits that one might consider the opposite of "humility." Sometimes this is a manifestation of "system arrogance" (Berger 2002) in which the patient becomes a faceless cog in the machinery of the healthcare organization, and the physician, depersonalized by time pressures and the eroding autonomy and professionalism of the role, exerts a form of psychological tyranny, typing away at the computer keyboard with hardly a glance at the patient during the allotted 15-minute clinic appointment. In one study of medical error and personality (Babaei et al 2018), physician and nursing staff without recorded medical errors during the study period "showed higher scores in conscientiousness, extraversion, and agreeableness, and lower scores in neuroticism than those with medical errors."

Other studies found it harder to demonstrate a relationship between interpersonal behaviors and healthcare outcomes. In one systematic review featuring an attempt to correlate quality of processes of care, healthcare outcomes, and interpersonal physician behaviors, most studies reported found little or no effect of clinician's personality traits or interpersonal behaviors on the quality of patient care (Boerebach et al. 2014). Nonetheless, extensive comments in internet searches indicate strong feelings among many patients about how their physicians' behaviors affect their willingness to participate in a treatment program and their belief that it might be effective. The five-factor personality inventory or model has been shown to be a good predictor of success and teaching programs involving medical residents and students, but it is not clear whether this can be extended to effectiveness in patient care. The five factors include conscientiousness, extraversion, emotional stability, agreeableness, and openness (Scheepers et al 2014). As these traits impact communication with patients, they might also have an impact on patient satisfaction, compliance with recommended treatment programs, and response to medical error whenever it occurs. It is well known that patients are less likely to develop a litigious attitude in relation to medical error that is confessed in the context of communication that has been consistently active, open, and compassionate.

Solutions

Elsewhere in this chapter, we have discussed the potential role of selection and education of future physicians in promoting a collaborative style that requires humility and willingness to work with other professionals in a team setting. Other areas of potential impact include organizational awareness in selecting leadership that fosters positive communication styles resistant to hubris and arrogance. Conscious avoidance of individuals demonstrating frankly narcissistic personality disorders or clues to outright sociopathic behavior can be advocated at system levels tasked with choosing leaders. At the same time, there needs to be recognition that narcissistic personality styles may also be associated with the visionary leadership sometimes necessary for organizational innovation and competitiveness.

Programmatic approaches include adoption and training in courses such as TeamSTEPPS™, SEIPS Model, Just Culture, or Servant Leadership. TeamSTEPPS™ emphasizes the role of anyone at any level of the healthcare team in calling attention to the potential for error or patient harm. Training involves "a willingness to cooperate, coordinate, and communicate" (King et al 2008). This approach is based on crew resource management training in the aviation industry and has found wide acceptance in healthcare.

The developed *SEIPS 3.0 model* goes beyond the original systems workflow and organizational analysis to include outcomes evaluations for caregivers, clinicians, and organizations themselves, including caregiver burden and stress, quality of working life, and organizational turnover, absenteeism, and presenteeism (Carayon et al 2020). By emphasizing a systems approach to healthcare design, including management of unnecessary treatment variation and care provider engagement, it might be expected that organization design itself will modify the impact of overly narcissistic character styles on the teams that deliver care and its outcomes.

Just culture adopts coherent strategies and tools to enhance performance and safety (Boysen 2013). By looking beyond patient care events into the system processes that may promote error and accepting the concept of latent errors inherent in complex systems, there is less room for the "blame and shame" management style often seen in organizations run by whims of narcissistic managers. At the same time, caregivers can expect accountability for actions that might place patients and their organizations at risk (Frankel, Leonard, and Denham 2006).

Servant leadership emphasizes personal and leadership communication styles that would seem ideally suited to a profession that claims to value service and commitment to the well-being of others (Trasek, Hamilton, and Niles 2014). On the leadership side, helping members of the healthcare team achieve a high degree of mutual support and trust through personal and team development requires qualities of listening, empathy, healing, awareness, persuasion, conceptualization, foresight, stewardship, commitment to the growth of people, and building community (Spears 2010).

Humility as a "Use Case" for Technology Implementation

It may be useful to consider humility from the perspective of the leader invested in the development and implementation of a health system technology project, to illustrate its importance as well as potential pitfalls, should it be neglected. Here we will examine humility from the vantage points of the physician leader, the technology expert, and the hospital executive.

Humility and Technology as a Physician Leader

With the implementation of hospital technology projects, it has long been established that the engagement of physician leaders is critical. Typically, larger organizations have a Chief Medical Information Officer (CMIO) responsible for oversight of an organization's strategic plan, who is often a physician with some degree of formal health informatics training. Smaller organizations may not have this as a formally designated role, but still will have a physician liaison to bridge the gap between clinicians and the technology support and staff. These physician leaders are of critical importance in their ability to interact meaningfully in both worlds of medicine and technology, to translate the language of clinical need and technological frameworks, and to support successful implementation of projects.

Although CMIOs may also work clinically in smaller organizations, this is usually not the case in larger health systems, as the administrative and organizational demands on their time are significant. Still, a clinical background affords an awareness of the challenges and pressures of care delivery at the bedside. A background of medical education affords one the ability to converse in the language of diagnosis and treatment and thus some communication with clinicians may be more readily received. As an example, humility and flexibility serve a leader well when he is considering timing of an upgrade and its potential disruption of clinical service – perhaps in the context of understanding the challenges of a hospitalist burdened with 15 complex patients or a surgeon pressured to high efficiency trying to turn fit her cases in a tight operating room schedule while attending to patient communication needs. Understanding the role and integration of the technology within the team's care delivery in a humble way – for both the individual patient and the system as a whole – will reduce friction and likelihood of alienation of the medical and hospital staff.

A successful CMIO also has a deep understanding of the technology professionals and experts supporting the day-to-day operations, the hardware, the software, and the data security in the hospital. She has respect with regard to their expertise and limitations, and employs humility with an awareness of her own knowledge and leadership role. A humble leader here will have savvy – knowing when to push and when to prioritize, as well as avoidance of professional cultural assumptions and bias that may originate in medical training.

There will often be an additional layer of physician leaders supporting a CMIO, who might be labeled in different ways: "early adopters" of technology, informatics specialists or enthusiasts, or physician "champions." These professionals may be engaged in formal or informal roles, but are critical to larger organizations and major tech project implementations and maintenance. Take for example the rollout or upgrade of the Electronic Health Record (EHR) of a hospital. Each department will need elbow-support for its cadre of nurses, physicians, and support staff, and physician leaders dedicated to this task will prevent clinical services from grinding to a halt during implementation but are also critical to supporting maintenance of such systems longitudinally. Humility by these professionals and an ability to communicate will facilitate problem-solving and support of the healthcare staff working directly with patients. While some of these labels such as "champion" may seem incongruous with humility (perhaps bringing to mind a Greek Olympic champion dominating athletic competition), one might also consider this construct as a leader or champion of problem-solving and communication. There is no small amount of patience and relational expertise required in these roles, in addition to the technical requirements.

To contrast for a moment the physician leader who approaches these projects from a perspective of overconfidence or even hubris, the likelihood of communication breakdown and unanticipated problems increases. Physicians and other care providers at the "sharp end" of care delivery may see these leaders as driving an agenda, separate from their needs to deliver care. Friction and challenges with technology may not be communicated or addressed and there is a risk of staff alienation and "work arounds" that circumvent technology safeguards. With high pressure and stressful clinical environments, the risk of alienation and failure of these systems increases. This is especially true in critical care areas like the ICU or emergency department, where Brian works clinically and has witnessed "splitting" at times between clinical staff and administration over technology projects, with potential for finger-pointing and mistrust.

Humility Approaching Medicine as a Technology Expert

The technology expert brings a skill set of tools and knowledge to the medical team from a perspective and background which is quite different from medical training. Engineering and technology in many ways are more predictable compared to the vagaries of human physiology, presentation of pathology, and response to treatment. A technologist who recognizes the inherent uncertainty in medicine and appreciates with humble perspective the challenges of clinical practice may interact with the medical team with a greater empathy to the stress of day-to-day patient care. As outlined above, the more the time spent with the clinical team, the better the understanding of their needs and challenges around the technology. Contrast a technology expert who first listens to clinicians, following them in their routine with patients and humbly trying to understand the pain points, with one arriving with a predetermined and inflexible agenda with regard to how technology needs to be used.

A humble technologist will share the clinician's perspective that, ultimately, the work of the hospital team is to support the best patient care and health possible. If this organizational vision is embraced by the leaders deciding strategy, the prioritization of technology projects can be more easily determined and transparent. Technologists will also likely have greater satisfaction in their roles on the team and see the fruits of their labor manifest in physicians and caregivers who are eager to support new projects which improve quality of care.

If a technologist lacks humility in discourse with the medical team, her projects will be tougher to implement. Without a willingness to compromise and understand the clinical perspective, communication will be difficult and ultimately trust is harder to build. Integration into the care team in a functional sense is challenged in these situations, and efficiency of work and success of technology implementation may suffer without deep engagement and understanding of clinician and patient care needs. Suggestions for bringing humility to the perspectives and operations of technologists are outlined above.

Humility in Technology Implementation by the Hospital Executive

Finally, hospital, nursing and physician executives are also critical leaders with regard to technology adoption in their organizations and their approach will often reflect a broader leadership perspective and organizational strategy. In general, hierarchical organizations with an emphasis on "top-down" leadership are at higher risk for challenges with engagement of staff and understanding the complexity of needs on the front lines of care delivery. Hierarchical leadership doesn't necessarily lack humility, but there are fewer "guardrails" to ensure that a leader is incorporating appropriate advisory perspective and understanding. When a hospital executive listens with humility to input from all levels of the organization, she will more likely have an understanding that anticipates pitfalls and plans for challenges. To the extent that these can be anticipated and communicated to her staff, this will help manifest an impression of transparency and engagement by the leadership. As mentioned earlier, bidirectional and dynamic communication strategies can significantly add to staff engagement and perceptions of leadership accountability (McBeth 2021, McBeth et al 2021).

Collaborative and humble approaches from the C-suite set the tone for organizational culture and conflict management. Much has been written about "Just Culture" in recent years, a concept based on the decriminalization of error and accountability of organizations for the systems they have designed (Dekker 2017). As already discussed, transparency around error disclosure acknowledges the reality that human error is part of medical care, while shining a critical light on an organization's processes to continually drive improvement. Humility by an executive leader will include acknowledgment of system flaws and personal error in a constructive way – promoting safety reporting and honest disclosure.

Technology has the ability to improve our efficiency, reduce error, and ultimately to drive improvement in safe care delivery. The executive who understands this and can delegate the granular detail of project development with this ultimate goal of improved care delivery in mind will approach these endeavors with humility. Like any good leader, she will acknowledge her own limitations in expertise and surround herself with a team that shares her collaborative vision and will support safety and communication, regardless of the project. When obstacles manifest with regard to technology performance, implementation, or staff engagement, she will share accountability and engage to support creative solutions that reflect the ultimate goal of safe and quality care delivery.

Conclusion

This chapter makes an argument that humility is a core value and critical for successful development, implementation, and sustainability of technology in hospitals and other healthcare settings. Its critical role in medical and technology education sets a foundation for learning and professional discourse, and its presence mitigates risk on multiple different levels – with humility being a common thread to effective communication among providers, support staff, and patients. Risk reduction around technology adoption and interface is demonstrated by high-functioning executives, physicians, and technology experts who approach problem-solving with intellectual humility that acknowledges personal limitations and embraces collaboration with incorporation of diverse perspectives. Relational humility mitigates conflict and tension around challenging clinical situations and technological interfaces. Transparency and acknowledgment of human error is a fundamental component of quality care delivery, and ultimately humility is foundational to drive improved patient engagement and health outcomes.

Bibliography

21st Century Cures Act, Pub. L. No. 114–255 (2016). https://www.govinfo.gov/content/pkg/PLAW-114publ255/pdf/PLAW-114publ255.pdf.

Ackoff, Russell. "Ackoff's Best: His Classic Writings on Management ," 170–72. New York: John Wiley & Sons, 1999.

Asad, Sarosh, and Eugene Sadler-Smith. "Differentiating Leader Hubris and Narcissism on the Basis of Power." *Leadership* 16, no. 1 (2020): 39–61. https://doi.org/10.1177/1742715019885763.

Atkins, P. W., D. S. Wilson, and S. C. Hayes. *Prosocial: Using Evolutionary Science to Build Productive, Equitable, and Collaborative Groups.* Oakland, CA: New Harbinger Publications, 2019.

Babaei, Mansour, Mohammad Mohammadian, Masoud Abdollahi, and Ali Hatami. "Relationship between Big Five Personality Factors, Problem Solving and Medical Errors." *Heliyon Open Access* 4, no. 9 (September 17, 2018): e00789. https://doi.org/10.1016/j.heliyon.2018.e00789.

Becker, R. E., and M. V. Seeman. "Patients Are Our Teachers." *Journal of Patient-Centered Research and Reviews* 5, no. 2 (2018): 183–86.

Berger, Allan S. "Arrogance among Physicians." *Academic Medicine: Journal of the Association of American Medical Colleges* 77, no. 2 (2002): 145–47. https://doi.org/10.1097/00001888-200202000-00010.

Bernabe-Valero, Gloria, Isabel Iborra-Marmolego, Maria J. Beneyto-Arrojo, and Nuria Senent-Capuz. "The Moderating Role of Intellectual Humility in the Adoption of ICT: A Study Across Life-Span." *Frontiers of Psychology* 9 (2018). https://doi.org/https://www.frontiersin.org/articles/10.3389/fpsyg.2018.02433/full.

Berwick, Donald M, Thomas W Nolan, and John Whittington. "The Triple Aim: Care, Health, and Cost." *Health Affairs (Milwood)* 27, no. 3 (2008): 759–69. https://doi.org/10.1377/hlthaff.27.3.759.

Boerebach, Benjamin C. M., Renee A. Scheepers, Renee M. van der Leeuw, Maas Jan Heineman, Onyebuchi A. Arah, and Kiki M. J. M. H. Lombarts. "The Impact of Clinicians' Personality and Their Interpersonal Behaviors on the Quality of Patient Care: A Systematic Review." *International Journal for Quality in Health Care* 26, no. 4 (August 2014): 426–81. https://doi.org/10.1093/intqhc/mzu055.

Boysen, Philip G. "Just Culture: A Foundation for Balanced Accountability and Patient Safety." *The Ochsner Journal* 13, no. 3 (2013): 400–406.

Carayon, Pascale, Abigail Wooldridge, Peter Hoonakker, Ann Schoofs Hundt, and Michelle M Kelly. "SEIPS 3.0: Human-Centered Design of the Patient Journey for Patient Safety." *Ergonomics* 84 (2020): 103033. https://doi.org/10.1016/j.apergo.2019.103033.

Cerri, Steven T. *The Fully Integrated Engineer: Combining Technical Ability and Leadership Prowess.* New York: IEEE Press, 2016.

Cooke, Molly, David M. Irby, William Sullivan, and Kenneth M. Ludmerer. "American Medical Education 100 Years after the Flexner Report." *New England Journal of Medicine* 355, no. 13 (September 28, 2006): 1339–44.

Cutrer, William B., Bonnie Miller, Martin V. Pusic, George Mejicano, Rajesh S. Mangrulkar, Larry D. Gruppen, Richard E. Hawkins, Susan E. Skochelak, and Donald E. Moore. "Fostering the Development of Master Adaptive Learners: A Conceptual Model to Guide Skill Acquisition in Medical Education." *Academic Medicine* 92 (2017): 70–75.

Davis, Don E., Joshua N. Hook, Everett L. Jr. Worthington, Daryl R. Van Tongeren, Aubrey L. Gartner, David J. II Jennings, and Robert A. Emmons. "Relational Humility: Conceptualizing and Measuring Humility as a Personality Judgment." *Journal of Personality Assessment* 93, no. 3 (2011): 225–34.

Deffler, Samantha A., Mark R. Leary, and Rick H. Hoyle. "Knowing What You Know: Intellectual Humility and Judgments of Recognition Memory." *Personality and Individual Differences* 96 (2016): 255–59. https://doi.org/10.1016/j.paid.2016.03.016.

Dekker, Sydney. *Just Culture: Restoring Trust and Accountability in Your Organization.* Third. New York: CRC Press, 2017.

Dijkstra, Edsger. "The Humble Programmer." Presented at the Turing Lecture for the Association of Computing Machinery, 1972. https://www.cs.utexas.edu/users/EWD/transcriptions/EWD03xx/EWD340.html.

Dinesh, Kumar V. "Multiple Faces of Narcissistic Leadership in Medical Education." *Journal of Advances in Medical Education & Professionalism* 7, no. 2 (2019): 103–105. https://doi.org/10.30476/JAMP.2019.44705.

Dreyfus, Stuart E, and Hubert L Dreyfus. "A Five-Stage Model of the Mental Activities Involved in Directed Skill Acquisition." Research Report. Operations Research Center, University of California at Berkeley, February 1980.

Dunning, David, Chip Heath, and Jerry M. Suls. "Flawed Self-Assessment." *Psychological Science in the Public Interest* 5, no. 3 (2004): 69–106.

Dunning, David, Kerri Johnson, Joyce Ehrlinger, and Justin Kruger. "Why People Fail to Recognize their Incompetence." *Current Directions in Psychological Science* 12, no. 3 (2003): 83–87. https://doi.org/doi:10.1111/1467-8721.01235.

Eirdosh, D., and S. Hanisch. "Can the Science of Prosocial be a Part of Evolution Education?" *Evolution: Education and Outreach* 13, no. 5 (2020). https://doi.org/10.1186/s12052-020-00119-7.

Frankel, Allan S., Michael W. Leonard, and Charles R. Denham. "Fair and Just Culture, Team Behavior, and Leadership Engagement: The Tools to Achieve High Reliability." *Health Services Research* 41, no. 4 Pt 2 (2006): 1690–709. https://doi.org/10.1111/j.1475-6773.2006.00572.x.

Gallagher, T. H., M. M. Mello, and W. Levinson. "Talking with Patients about Others Clinicians' Errors." *New England Journal of Medicine* 369 (2013): 1752–57. https://doi.org/doi:10.1056/NEJMsb1303119.

Gallagher, T. H., M. M. Mello, W. M. Sage, S. K. Bell, T. B. McDonald, and E. J. Thomas. "Can Communication-And-Resolution Programs Achieve their Potential? Five Key Questions." *Health Affairs* 37, no. 11 (2018): 1845–52. https://doi.org/doi:10.1377/hlthaff.2018.0727.

Gallagher, T. H., D. Student, and W. Levinson. "Disclosing Harmful Medical Errors to Patients." *New England Journal of Medicine* 356 (2007): 2713–19. https://doi.org/doi:10.1056/NEJMra070568.

Gill, C. J., L. Sabin, and C. H. Schmid. "Why Clinicians are Natural Bayesians." *BMJ* 330, no. 7499 (2005): 1080–83.

Grant, A. M., and F. Gino. "A Little Thanks Goes a Long Way: Explaining Why Gratitude Expressions Motivate Prosocial Behavior." *Journal of Personality and Social Psychology* 98, no. 6 (2010): 946–55.

Ha, Jennifer Fong, and Nancy Longnecker. "Doctor-Patient Communication: A Review." *The Ochsner Journal* 10, no. 1 (2010): 38–43.

"Improving Diagnosis in Health Care." The National Academies of Sciences, Engineering and Medicine (NASEM) – The National Academies Press, 2015.

Institute of Medicine – Committee on Quality of Health Care in America. *To Err Is Human: Building a Safer Health System.* Edited by Linda T. Kohn, Janet M. Corrigan, and Molla S. Donaldson. Washington, DC: National Academies Press, 2000. http://www.ncbi.nlm.nih.gov/books/NBK225182.

Johnson, W. Brad. "A Framework for Conceptualizing Competence to Mentor." *Ethics and Behavior* 13, no. 2 (2003): 127–51.

Kachalia, A., S. R. Kaufman, and R. C. Boothman. "Liability Claims and Costs Before and After Implementation of a Medical Error Disclosure Program." *Annals of Internal Medicine* 153, no. 4 (2010): 213–21. https://doi.org/doi:10.7326/0003-4819-153-4-201008170-00002.

Kahneman, D., P. Slovic, and A. Tversky. "Judgment Under Uncertainty: Heuristics and Biases." *Science* 185 (1982): 1124–31. https://doi.org/10.1126/science.185.4157.1124. ISBN 978-0521284141. PMID 17835457. S2CID 143452957.

King, Heidi B., James Battles, David P. Baker, Alexander Alonso, Eduardo Salas, John Webster, Lauren Toomey, and Mary Salisbury. "TeamSTEPPS™: Team Strategies and Tools to Enhance Performance and Patient Safety." In *Advances in Patient Safety: New Directions and Alternative Approaches, Vol. 3: Performance and Tools*. Agency for Healthcare Research and Quality, 2008.

Krumrei-Mancuso, Elizabeth J., Megan C. Haggard, Jordan P. LaBouff, and Wade C. Rowatt. "Links between Intellectual Humility and Acquiring Knowledge." *Journal of Positive Psychology* 15, no. 2 (2020): 155–70.

Krumrei-Mancuso, Elizabeth J., and Steven V. Rouse. "The Development and Validation of the Comprehensive Intellectual Humility Scale." *Journal of Personality Assessment* 98, no. 2 (2015): 209–21. https://doi.org/10.1080/00223891.2015.1068174.

"Largest Health Care Fraud Case in U.S. History Settled: HCA Investigation Nets Record Total of $1.7 Billion." Department of Justice, 2003. https://www.justice.gov/archive/opa/pr/2003/June/03_civ _386.htm.

Leary, Mark R., Kate J. Diebels, Erin K. Davisson, Katrina P. Jongman-Sereno, Jennifer C. Isherwood, Kaitlin T. Raimi, Samantha A. Deffler, and Rick H. Hoyle. "Cognitive and Interpersonal Features of Intellectual Humility." *Personality and Social Bulletin* 43, (2017): 1–21.

Littlejohn, S. W., K. A. Foss, and J. G. Oetzel. *Theories of Human Communication, 11th Edition.* Long Grove, IL: Waveband Press, Inc., 2017.

Liu, Dege, Ting Zhu, Xiaojun Huang, Mansi Wang, and Man Haung. "Narcissism and Entrepreneurship: A Systematic Review and an Agenda for Future Research." *Frontiers of Psychology* 12 (2021): 657681. https://doi.org/10.3389/fpsyg.2021.657681.

Maccoby, Michael. *Narcissistic Leaders: Who Succeeds and Who Fails.* New York: Harvard Business School Press, 2007.

Makary, Martin A., and Michael Daniel. "Medical Error – The Third Leading Cause of Death in the US." *BMJ* 353, no. i2139 (2016). https://doi.org/10.1136/bmj.i2139.

McBeth, B. D. "Facilitating Trust and Communication by Hospital Administrations during a Pandemic: Importance of a Multimodal Approach." *Health Systems and Policy Research* 8, no. 6 (2021): 1–3.

McBeth, B. D., and A. Douville. "A Brief Report of Implementation of a Physician Citizenship Committee in a Community Hospital to Address Disruptive Physician Behavior." *Journal of Hospital Management and Health Policy* 3, no. 31 (2019): 1–5.

McBeth, B. D., Y. Karanas, P. H. Nguyen, S. Kurani, and M. Bhimani. "Improving Communication between Hospital Administrations and Healthcare Providers During COVID-19: Experience from a Large Public Hospital System in Northern California." *Journal of Communication in Healthcare* 14, no. 4 (2021): 274–82.

Mello, M. M., R. C. Boothman, and T. McDonald. "Communication-And-Resolution Programs: The Challenges and Lessons Learned from Six Early Adopters." *Health Affairs* 33, no. 1 (2014). https://doi .org/10.1377/hlthaff.2013.0828.

Nance, John J. *Why Hospitals Should Fly: The Ultimate Flight Plan to Patient Safety and Quality Care.* Bozeman: Second River Healthcare Press, 2008.

O'Reilly, Charles A., Jennifer A. Chatman, and Bernadette Doerr. "When 'Me' Trumps 'We': Narcissistic Leaders and the Cultures They Create." *Academy of Management Discoveries* 7, no. 3 (2021): 419–50. https://doi.org/10.5465/amd.2019.0163.

O'Reilly, Charles A., and Michael L. Tushman. "The Ambidextrous Organization." *Harvard Business Review*, April 2004. https://hbr.org/2004/04/the-ambidextrous-organization.

Rastello, Peter. "Physician Employment Isn't Always Equal to Alignment." *The CFA Perspective* (blog), January 17, 2012. https://www.charlesfrancassociates.com/CFA-blog/bid/80665/Physician -Employment-Isn-t-Always-Equal-to-Alignment#:~:text=This%20is%20not%20necessarily%20so. %20As%20we.

Scheepers, Renee A., Kiki M. J. M. H. Lombars, Marcel A. G. van Aken, Maas Jan Heineman, and Onyebuchi A. Arah. "Personality Traits Affect Teaching Performance of Attending Physicians: Results of a Multi-Center Observational Study." *PLoS ONE* 9, no. 5 (2014). https://doi.org/10.1371/ journal.pone.0098107.

Schei, Edvin, Abraham Fuks, and J. Donald Boudreau. "Reflection in Medical Education: Intellectual Humility, Discovery, and Know-How." *Medicine, Health Care and Philosophy* 22 (2018): 167–78. https://doi.org/10.1007/s11019-018-9878-2.

Spears, Larry C. "Character and Servant Leadership: Ten Characteristics of Effective, Caring Leaders." *Journal of Virtues and Leadership* 1 (2010): 25–30.

Spiegel, James. "Open-Mindedness and Intellectual Humility." *Theory and Research in Education* 10, no. 1 (2012): 27–38. https://doi.org/10.1177/1477878512437472.

Templeton Foundation. "Intellectual Humility." Accessed April 16, 2022. https://www.templeton.org/ project/intellectual-humility.

Thomas, K. W., and R. H. Kilmann. "Thomas-Kilmann Conflict Mode Instrument." Xicom, Inc., 1974.

Tracey, J. M., B. Arrow, D. E. Richmond, and P. E. Barham. "The Validity of General Practitioners' Self Assessment of Knowledge: Cross Sectional Study." *BMJ* 315 (1997): 1426–28.

Trastek, Victor F., Neil W. Hamilton, and Emily E. Niles. "Leadership Models in Health Care – A Case for Servant Leadership." *Mayo Clinic Proceedings* 89, no. 3 (2014): 374–81. https://doi.org/10.1016/j .mayocp.2013.10.012.

"Triple Aim for Populations." Institute for Healthcare Improvement. Accessed January 23, 2022. http:// www.ihi.org/Topics/TripleAim/Pages/default.aspx.

Watzlawick, P., J. B. Bavelas, and D. D. Jackson. *Pragmatics of Human Communication: A Study of Interactional Patterns, Pathologies and Paradoxes, 1st Edition.* New York: WW Norton & Co, 1967.

Chapter 8

Burnout in Information Security: The Case of Healthcare

Mitch Parker

Contents

DOI: 10.4324/9781003348603-10

What Is Burnout?: A Basic Background

What Is Burnout?

According to the World Health Organization, in the 11th Revision of the International Classification of Diseases (ICD-11), burnout is defined as a syndrome conceptualized as resulting from chronic workplace stress that has not been managed. It has three dimensions (WHO 2019):

- Feelings of energy depletion or exhaustion
- Increased mental distance from one's job, or feelings of negativism or cynicism related to one's job
- Reduced professional efficacy

COVID-19 and its associated changes, including working from home, isolation, and caring for people that have had it, even ourselves, have exacerbated many existing situations. The significant change of taking a large portion of the workforce and moving them toward home has also contributed. The use of toxic positivity as a motivating tool, which stigmatizes and encourages people to bury away their anger and depression, also creates an undercurrent of stress and bad feelings that don't have an outlet and can fester inside, eating people away while showing a façade that everything is OK.

The Mayo Clinic has identified various factors, including (Mayo Clinic 2021):

- *Lack of Control:* An inability to influence decisions that affect your job, such as schedule, assignments, or workload. This can lead to job burnout.
- *Lack of Resources*: A lack of resources needed to do work.
- *Unclear Job Expectations*: If you're unclear about the degree of authority you have or what is expected of you by your supervisor or others, you're not going to be comfortable at work.
- *Dysfunctional Workplace Dynamics/Toxic Workplace*: If you deal with an office bully, feel undermined by colleagues, or work for a micromanaging boss, this can contribute to stress.
- *Activity Extremes*: When a job is either monotonous or chaotic, you need constant energy to keep focused. This can lead to fatigue and job burnout.
- *Lack of Social Support*: If you're isolated at work and in your personal life, you might be more stressed.
- *Work–Life Imbalance*: If your work takes up so much time and effort that you just don't have the time and energy to spend with family and friends, you may burn out quickly.

They have also identified four factors that may contribute to job burnout:

- You have a heavy workload and you work long hours.
- You struggle with work–life balance.
- You work in a helping profession, such as healthcare or Information Security.
- You feel you have little or no control over your work.

The consequences they have identified are as follows:

- Excessive stress
- Fatigue
- Insomnia
- Sadness, anger, or irritability
- Alcohol or substance misuse
- Heart disease
- High blood pressure
- Type 2 diabetes
- Vulnerability to illnesses

In addition to this, other factors can include:

- Exacerbation of other addictive behaviors (Ampelis Recovery 2021)
- Becoming more of a security risk, including increased susceptibility to phishing attacks (Tessian, Inc. 2022)

Why Do We Care?

We care because the people we rely on to keep our information secure are at a breaking point. Material for webinars and conferences is targeted at either bringing people into the workforce or bolstering the egos of people in management positions. Information Security professionals deal with a significant amount of toxicity from their coworkers, managers, and customers. Many of them feel marginalized, unheard, and ignored. Discovered issues are minimized because many people just don't want to deal with them. A negative use of motivation, better known as toxic positivity, ignores anything negative. This is often used to give the impression that all is well and build resistance to outside messaging that indicates that it isn't, even from within the same group or company. Social media builds on this and feeds it with its lack of actual social contact. Social media also gives a constant stream of messages exhorting and glorifying continual hustling and work with no time to rest. It's the 21st century equivalent of the use of "I'll sleep when I'm dead," the 1976 Warren Zevon song title co-opted by motivational speakers, except backed by algorithms that amplify that message and drown out everything else with the aims of engagement and profit.

COVID-19 and working from home inadvertently caused funnels that reduced social interaction outside of immediate work groups. This meant that these new funnels complemented the existing ones from social media. This also meant that the bad and toxic behaviors leaders propagated magnified significantly. Management, especially Information Technology and Information Security, have toxicity and trust issues.

It's been normalized to lie and underestimate how bad the situation is. It's also been normalized to give platitudes and empty promises of help. People no longer trust their leaders. They believe that they are going to be lied to and used to bolster the careers and egos of management so they can get a bigger bonus or buy a bigger Jet Ski. Reddit's /r/antiwork subreddit gives multiple examples of this daily, submitted by the recipients of bad management. Its popularity alone, with over 1.9 million users and 620 posts per day as of May 12, 2022, indicates just how big of a problem we face (Reddit.com 2022).

Our teams have had enough and are leaving. They're burning out faster than ever. At the best, many are just doing a good job hiding it while they crumble away inside. That takes energy, and they can't do it for long before large cracks show. We need to start fixing these issues before more crashes occur.

The corresponding crashing down of work–life barriers complement the internal crashes we're observing. We've operated under the illusion for years that work and personal lives were different. They are not, and we can observe many of the same behaviors between the two. Films and TV shows like Severance or Paycheck that dramatize extreme steps people take to segment their lives also show that totality is never possible. They also show that the race to totality has unintended consequences. The security community has, by and large, rejected those barriers, and instead embraced having multiple hacker subcultures that intermix work, personal lives, and culture along with people who just work in Information Security (Subculturelist.com 2022).

When I started writing this, I wanted to provide something other than the standard literature available on burnout and how to deal with it. One of my concerns was that no one really discussed the underlying cultural and business reasons why burnout is so prevalent or addressed business root causes. What I saw was an acknowledgment that it happens, not of its magnitude. Also, they were not covering information to the needed depth to take meaningful action.

We were not also linking it to addictive behaviors and larger issues. In this case, Information Security practitioners have more in common with business executives and leadership at a lower level as compared to other workers. The reasons why we need to care all map to burnout factors identified by the Mayo Clinic.

This is an area of security we do not address at our own peril. We focus on the employment shortage, tools, and trying to get senior leadership to listen. We don't focus on the issues our own teams have, and what they are experiencing. Our people are suffering and we're too focused on board-level metrics. We're not focused on how the people who are supposed to respond in the case of a disaster or data breach are doing when they are already overloaded. Much like the great people who work in Emergency Management, they are underappreciated and overworked. They also deal with a significant amount of toxic people in management positions who exist to support their own egos at the expense of others. Never underestimate how low someone, especially one who thinks their displays will impress a person in a superior role enough to recognize or promote them, will sink and plumb new depths.

Also never underestimate how some people enjoy being angry and outraged, and pushing others into being the same way. Never underestimate how some people think their way is the only way and will do anything to shut out differing perspectives. Social media and the internet have exacerbated our tendencies for anger and outrage, and this has also significantly contributed.

This is larger than just Information Security. It's more visible with this profession because its scope encompasses the entire business. We have a larger issue of our push toward earnings at all costs putting the information systems and support processes we have at risk, not just security teams. Ransomware proves that when we put these at risk, we endanger the business itself. We are at a critical point, due to the ignorance of what Information Security professionals are really facing by people in leadership positions. We cannot face this without understanding the scope of it, and its significance. We have ignored this for too long. People are leaving this field, and we're not talking about why or its origins.

This is not something we can look at in a silo by itself. We must start at the beginning and work through the major areas:

■ The cultural history behind our business culture
■ How we've built a society based on hero worship and conformance to ultra-tough norms
■ Storytelling, the monomyth, and enculturation into corporate culture

- How social media pushes the messages of Hustle Porn to normalize working impossible hours, hustling, and putting work above everyone else to meet goals
- Despite what we think, operating systems don't multitask well, and neither do we
- We communicate without anticipating noise
- Isolation and neurodiversity
- Addiction and how this can creep in and manifest
- The economic history of how we got here
- How businesses and corporations fail at business and excel at financial engineering
  - The Cycle of Disillusionment with Software and Technology, and other reasons why we're so burned out
- How we can structure ourselves and our initiative to address these concerns, reduce unnecessary issues, and build resilience with our team members

Understanding Information Security Pressures

How Is Information Security Different?

Information security professionals are significantly different than the rest of the IT workforce for multiple key reasons besides the presence of hacker subcultures:

- The scope of Information Risk Management requires them to interface with the entire business, as opposed to just their silo of IT and group of clients/customers.
- The scope of network security requires knowledge of all the core systems and how they interact with each other.
- The skill sets required for Information Security require a high variety of knowledge from multiple business areas:
  - Constantly expanding scope, with Cyber Insurance; Financial Systems Risk, and Healthcare Accreditation being added to an already overloaded workforce
  - Medical Device Security and Clinical Engineering added on top of this!
- The depth and breadth of skillsets required attracts highly intelligent people, especially from other fields.
- Neurodivergent people, who make up a significant portion of the IT and Infosec workforces, expend a lot of energy masking and trying to fit in.
- Cultural misunderstandings, especially with the multiple hacker subcultures.
- Information Security is often derided as being obstructionists to the business, starting with executives.
- Information Security teams are often asked to design defensive mechanisms and processes to address security risks created by legacy systems or other team members.
- They own responsibility for items they cannot control, specifically legacy systems and the behavior of systems that may be affected by security threats.
- They are often the first ones blamed when a security event happens.
- They are often not given the resources they need to address known risks, specifically in healthcare.
- High workload given low resource allocations.
- Organizational budgets, specifically in healthcare, don't even have the budgets to address basic IT, let alone security:

- Much like how people are forced to make the choice between healthcare and paying the rent
■ Security decisions being made based on wild promises, not actual results:
  - The false promises made by many blockchain vendors, especially about security
  - The false promises made by vendors that sell to non-Security executives using scare tactics and fear
  - Intelligent systems (AI/ML/Deep Learning)
■ Toxic leaders will make excuses to not address security issues:
  - Self-preservation efforts by leaders will trump efforts to modernize, make efficient, or streamline operations
  - Gaslighting on how the situation really is (thanks Christina!)
■ Outsourcing often lumps Information Security in with other IT functions, leading to ineffective security on top.
■ Unclear regulations, such as HIPAA, leave significant ambiguity for team members to even interpret how to do their jobs. The "Addressable" component of the HIPAA Security Rule caused many executives to interpret that it meant they did not have to address those issues if they gave a reasonable-sounding explanation. Meanwhile, Finance and Accounting can rely on Generally Accepted Accounting Principles (GAAP), FFIEC, and AICPA guidance, and the clinical care areas can rely on Joint Commission and their appropriate accrediting bodies to provide guidance.
■ Risk management is at odds with the risk-taking that management and executives want to do, specifically with emerging and unproven technologies that can present risks.
■ Risk management is at odds with executives that want to outsource to firms that are less secure to save money.
■ The entry-level positions can be monotonous, and lead to people feeling bored and unappreciated:
  - Buyer's remorse for some of the educational programs that promised "pew-pew maps" and put the people in Security Operations Center (SOC) analyst positions that did not resemble those at all
■ Higher stress positions, such as Digital Forensics/Incident Response (DFIR), security executives, or threat hunting, can sap energy and cause burnout.
■ Direct interaction with customers when they are having very bad days due to information compromise can also drain energy.
■ Whereas other positions are exposed to one or two areas of toxic organizational culture, Information Security sees more of it:
  - This includes the toxic positivity masquerading as management training and risk management in many departments, where so many will refuse to directly address risks to try and keep up an illusion of engagement.
  - This especially includes ignoring risks or making excuses why risks have not been addressed, that others just don't understand, and that all is well within the group. Any dissent from the message is treated as dissension from the norm and people get excluded and isolated because of it, even passive-aggressively.
■ The average tenure of CISO is 26 months, according to Catalin Cimpanu, in his article "Average Tenure of a CISO is 26 Months Due to High Stress and Burnout" (41Cimpanu 2020). This article further said that 88% of interviewed executives reported physical health issues, and half reported mental health issues
  - Constantly changing leadership causes uncertainty with the rest of the team

- − Burnout can lead to addiction or exacerbate existing conditions
- ■ Ethical concerns with:
  - − Unclear regulations
  - − Being asked to overlook known security issues
  - − Being asked to overlook attempts to circumvent laws and practices
  - − Being asked to secure systems that are insecure by design with no way to properly secure them
  - − Being asked to hide information through voluntary reporting such as risk assessments
  - − Being asked to ignore security issues because they don't contribute to a positive attitude or image of the organization or group (toxic positivity)
  - − Leadership pushing through projects and work designed to bolster their egos and make themselves look better without any thought for actual work or consequences

Why Are We Discussing Security as an Example?

Security is the proverbial canary in the coal mine. Canaries reacted differently to dangerous gasses and were a good sign to miners to get out of there if they died. The field of security has a fundamental component of utilizing and exploiting behaviors, conditions, and errors to induce systems and processes to behave in unintended ways.

Due to this, the field gets undue pressure. It also is visible to the entire business. Ransomware attacks just don't take down computer systems. They affect the entire business and the people that work there, especially healthcare providers. If a company doesn't realize how important security is before a ransomware attack, they do afterward. It's a fundamental building block that stretches across boundaries. In healthcare, the groups that bear the greatest brunt when cybersecurity issues occur are the caregivers and security teams.

This is the reason that security issues and burnout cannot be mitigated without looking at how companies manage, and what the factors are that got them there. We don't manage well. Executive suites are known for high turnover. IT executives are well-known for what is perceived as blind following of influencers and mantras such as "Digital Transformation" without knowing what they really mean. These influencers and mantras lead to an environment of toxic positivity where if something is perceived to not be fully part of what's being championed, it gets ignored or worse, actively campaigned against and insulted. Mimicry has been a way of life, especially in healthcare. Executives aren't trusted to understand or articulate what they are doing. People leave because they do not trust their organizations. Kelly Idell, Daved Gefen, and Arik Ragowsky, in the 2021 issue of *Communications of the ACM*, describe the most significant factor in turnover intentions to be distrust in the organization (Idell, Gefen, and Ragowsky 2021). Management passing down their internal fears and paranoia as part of corporate culture does not help the situation at all.

Management and executives are the ones responsible for communicating and relaying this message to others. Trust starts at the top. There are both formal and informal paths of communication. Discord between the two means that the organization is sending mixed messages. These breed mistrust, as Valerie McClelland discussed in her article "Mixed Signals Breed Mistrust" from *Personnel Journal* in 1987 (McClelland 1987). Hannah Drown, from Cleveland.com. discussed how mixed messaging has caused distrust of authority during the COVID-19 pandemic (Drown 2021). Distrust of the COVID vaccines has led to numerous deaths and injuries during the pandemic. Distrust in organizations has been a factor in the Great Resignation, according to

Christine Ro of the BBC in her September 16, 2021 article, "'Turnover Contagion': The Domino Effect of One Resignation," discusses this as an underlying one (Ro 2021).

The bifurcation between formal and informal communications, especially between those that are part of the "in crowd" that receives communication about heroic stories and corporate myths, and those who don't build distrust, isolation, and toxicity, in addition to the other factors, is discussed in this chapter.

Those who are most like, and favored by, the power structure in charge are ironically at higher risk of burning out because they are the ones most relied upon and given the most work to do. While they receive enculturation via the informal communication channels to exhort them onward, it can lead them to crash harder. Every part of a modern business relies on security. Whether or not they are like the power structure in charge, they have a significant amount of work to do, whether or not they receive informal support from the top.

Information Security team members often do not have the luxuries of being part of the "in crowd" in companies or being like the junior executives or young doctors being groomed for more senior roles. Yet they are given more work, less recognition, and are kept away from business decisions that matter. Dave Burg, Mike Maddison, and Richard Watson, in their summary of the EY Global Information Security Survey 2021, specifically point out three challenges (Burg, Maddison, and Watson 2021):

■ Cybersecurity is underfunded at a critical point when needed more than before.
■ Regulatory fragmentation is a growing issue causing more work and resource concerns.
■ Relationships with other business functions are deteriorating.

They further point out that security is not often consulted by organizations outside of IT, legal, or compliance, and that this has been a downward trend. Their 2020 survey indicated that 36% of security respondents were brought in at the Planning stage, and that dropped to 19% in 2021. Yet 55% of respondents indicated that cybersecurity is coming under more scrutiny today than at any other time in their careers.

According to these sources, we are setting up our security teams with less resources than before and without the inside track that the rising stars get. We are setting them up to fail by giving them more work, less resources, and less formal and informal support when they need it to develop. We consider them to be expendable. The lack of resources has caused many security personnel to become incredibly reliant on tools and software to give the illusion of effectiveness. This is to the point where many are considered "tools people" because they use them to automate a significant portion of their workflow to make up for the lack of resources. This is double jeopardy because executives will see this and think they are ripe for outsourcing and/or replacement.

One of the other challenges with outsourcing that we observe is that there is a significant cultural difference between IT managers and salespeople. The salespeople are more like executives in non-IT line areas, and often follow many of the same management philosophies. They are also trained how to communicate and interface with executives. Many career technologists do not receive this training. Homophily, which is the preference of associating with similar people that we will discuss later, may be a large factor in companies choosing to outsource. This is because often executives will want to surround themselves with people like them. This provides the benefits of having those people around while not being held directly accountable for operational matters or security concerns.

One of the other challenges that IT executives often encounter is falling into bad behaviors, which compounds these issues and makes them even worse. Lack of ownership, accountability, and playing political games to get ahead all amplify existing issues. The tendency to bully subordinates while at the same time trying to hold others, specifically the C-suite, accountable for making decisions, will damage credibility, trust, and the organization itself.

Psychological and Economic Factors – How Does Our Business Culture Encourage This?

Psychological Root Causes

One of the biggest concerns that I had with reviewing literature to better understand the root causes behind burnout is that no one discussed the reasons why we got here in the first place. How did we end up in this position? We can't take steps forward unless we understand how. This doesn't start with technology or security. This starts with understanding underlying human behaviors. We then can take them and align them with the personal values systems of managers. We then align that with homophily, which is the tendency for group members to hold a preference for relationships with those like them. The characteristics of hegemonic masculinity, especially that of its exemplars being celebrated as heroes to be emulated using storytelling as a technique, will also be discussed.

The reason for taking this approach is because we need to understand the root causes that drive human behavior that lead to burnout. Security issues don't start with security. They start with people and their personality traits. We need to understand people, their motivations, and the decisions they make before we extrapolate to security. Security is a manifestation of these.

The one area where we can start is by understanding the norms that drive behavior. We need to begin with these so we understand the basic behavioral characteristics that drive people. Thankfully, there is an accepted instrument that provides these. The Conformity to Masculine Norms Inventory (CMNI), developed by James R. Mahalik and numerous other colleagues, identifies 11 distinct characteristics of masculine behavior (Mahalik, Locke, Ludlow, Diemer, Scott, Gottfried, and Frietas 2003):

- Winning
- Emotional control/restraint
- Risk-taking
- Violence
- Power over other genders
- Dominance
- Playboy
- Self-reliance
- Primacy of work
- Disdain for homosexuals
- Pursuit of status

By themselves, these norms appear to be the archetype of almost every action hero. However, when combined with the findings in the article "Sectoral Ethos: An Investigation of the Personal Values Systems of Female and Male Managers in the Public and Private Sectors," by Richard W. Stackman, Patrick E. Connor, and Boris W. Becker (Stackman, Connor, and Becker 2006), they intersect with the characteristics of private sector managers. This study discusses the personal values systems of 884 public and private sector managers. This study also demonstrates that the personal values systems of male and female managers within an employment sector are not significantly different. The value orientations identified in this study were:

- Profit maximization
- Ability
- Aggressiveness
- Influence
- Power
- Success
- Change
- Risk

In addition, this article references three other studies and characteristics they uncovered, including Howard, Shudo, and Umeshima's article, which highlights:

- Individuality
- Competence
- Inner-directedness
- Family security

Frederick and Weber also found the following values:

- Personal achievement
- Ego satisfaction

Adler, Ichikawa, Apasu, and Graham found the following characteristics of North American businesspeople:

- Family security
- Delayed gratification
- Personal orientation over social orientation

What we've learned from this is that many of the same characteristics that apply to masculine norms also apply to the personal values of private sector managers. There is a significant mapping between them (see Figure 8.1 and Table 8.1).

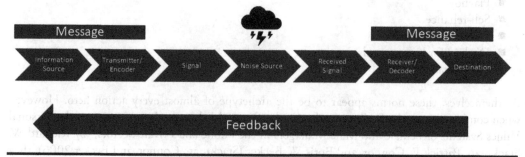

Figure 8.1 Shannon–Weaver communication model. Source: Adapted from Communication Theory (2014).

Table 8.1 CMNI Mapping to Personal Values Systems of Managers

CMNI	Stackman	Howard	Frederick	Adler
Winning	Profits, success, power			
Emotional control				
Risk-taking	Risk, change			
Violence	Aggressiveness			
Power over women	Power			
Dominance	Influence, ability, power	Competence		
Playboy				
Self-reliance	Power	Individuality, inner-directedness	Personal achievement	Personal orientation
Primacy of work				Delayed gratification
Disdain for homosexuals				
Pursuit of status	Influence, power	Family security	Ego satisfaction	Family security

Source: Mahalik, Locke, Ludlow, Diemer, Scott, Gottfried, and Frietas (2003).

Hegemonic Masculinity and Homophily

Tim Carrigan, Bob Connell, and John Lee describe hegemonic masculinity in their article "Toward a New Sociology of Masculinity" (Carrigan, Connell, and Lee 1985). They describe it as a movement where the dominant group can adapt to new circumstances without breaking down the socio-structural arrangements that actually give them power (Carrigan, Connell, and Lee 1985). Mike Donaldson, in his article "What Is Hegemonic Masculinity?", identifies several defining characteristics of this, which include (Donaldson 1993):

- Winning and holding of power by the dominant group
- Formation and destruction of social groups to support those
- The ability to impose a definition of the situation
- Setting the terms in which events are understood
- Setting the terms in which issues are discussed
- Formulating ideals
- Defining morality
- Persuading the greater part of the population, especially through the media

- Organization of social institutions in ways that appear "natural," "ordinary," and "normal"
- Usage of the state and its organs as instruments of punishment and enforcement for non-conformity
- Preservation of dominance

What these papers indicate is that the ones in charge of the system utilize it to keep themselves and their group members in positions of power through manipulation of social, media, state, and ostensibly corporate means to do so. Jodi Detjen, in her dissertation, "Masculinity and Leadership Inequities: An Examination of the Ways in which Masculine Cultural Norms Underlie the Barriers to Women's Leadership Acquisition," discusses the concept of homophily (Detjen 2021). This is when people are attracted to others based upon similar characteristics. This becomes replicated in groups (Detjen 2021). According to Detjen, there are two types:

- *Status-based*: based on ascribed status
- *Value-based*: based on similar values and beliefs

This status is often signaled via demographic indicators such as gender or race. Detjen cites Holgersson who calls this homosociality (Detjen 2021). She indicates that cultural norms and stories are defined and reinforced by the privileged group, and then used as gatekeepers. Detjen's dissertation reinforces the use of hegemonic masculinity as a management tool to keep the dominant groups in charge. Detjen further indicates that these biases become systemic and magnify the inherent bias inherent in the process, unless designed to avoid it.

How Does This Relate to Hero Worship and Hustle Porn through Storytelling?

Storytelling is a very important part of corporate life. Olja Arsenijevic, Dragan Trivan, and Milan Milosevic, in their article "Storytelling as a Modern Tool of Construction of Information Security Corporate Culture," discuss the importance of building and sharing narratives as a corporate communication tool (Arsenijevic, Trivan, and Milosevic 2016). Their article indicates that they:

- Build trust
- Promote norms, values, rules, and principles of organizational culture
- Promote a positive image of the company
- Enable targeted employee impacts
- Transfer knowledge
- Enable learning
- Create emotional connections
- Induce desired behaviors and actions
- Raise the level of effectiveness of communication
- Raise the level of effectiveness of organizational change

These stories and narratives have an important role in reinforcing organizational culture. A company that can effectively communicate through this method will likely have positive outcomes due to these positive impacts (Arsenijevic, Trivan, and Milosevic 2016). These stories are used to facilitate enculturation and reinforce primacy.

History has often ascribed heroic qualities to those who have demonstrated multiple characteristics of the CMNI as part of their success stories and removed mention of the negative ones. The perfect fictional example is that of Midge Kelly, the boxer from the celebrated Ring Lardner short story, Champion. Midge was an incredibly successful boxer who abandoned his family and partners to starve, did not pay debts, and used people to further his career. Yet at the end there were stories glorifying his life and achievements. When someone asked about it at the end, the response was "The people don't want to see him knocked. He's champion" (Lardner 1916). More modern examples include one public figure, who despite numerous personal bankruptcies, bad financial decisions, and personal issues is presented as the ideal man by numerous organizations and is considered a hero by many. Other examples include the great 19th-century industrialists and titans of business such as Andrew Carnegie, JP Morgan, and Thomas Edison, who despite the methods they used to get to the top, were lionized. As someone advances through the ranks, the story gets invented to demonstrate the myth of a meritocracy.

More modern stories of these revolve around social media, where numerous stories of people who put their heads down, grinded, and pushed hard made them multi-millionaires through marketing, sales, cryptocurrencies, or technology. The examples of Bill Gates and Mark Zuckerberg loom large over this. Elon Musk's work habits and quotations from his numerous tweets also repeatedly show up as memes. Before her downfall, Elizabeth Holmes' quotation "The minute you have a back-up plan, you've admitted you're not going to succeed" from a Stanford Business interview was plastered all over social media as a meme. Instagram's recent infestation of Peaky Blinders and The Rock memes, featuring motivational quotes superimposed over pictures of Cillian Murphy's character or Dwayne "The Rock" Johnson, often discussing self-reliance, hard work, and pushing hard, are also popular. So are memes challenging people to work out, read, swear off bad habits, and raise themselves from the position they're in because they keep failing. Even though it looks like positive motivation, it's negative because it reinforces that people are in a bad place and are less than those who work harder.

Social media has evolved to keep people algorithmically engaged, and constant exposure to these memes can change someone's thought patterns through constant exposure. This has led to the platforms evolving into constant dopamine hits. Julie Jargon, in a recent *Wall Street Journal* article, "Why Some Kids Seem Hooked on Social Video Feeds," discusses how one company's algorithm figures out what users like based on how much they watch each video, and serves up more of the same. This creates a dopamine hit or reward (Jargon 2022). Instead of a feed full of wide-ranging topics, it turns into dopamine hits for hustle porn or worse. This is not a unique tactic.

Social media also serves the purpose of signaling assertiveness and other CMNI characteristics, especially those of risk-taking, dominance, and pursuit of status. The constant stream of posts knocking down others that are not considered meeting traditional values, insulting others, and attempting to shut down others' points of view despite the facts shows this. It's not about the facts. It's about who signals dominating characteristics and values. The people who do this often want attention because they signal what they believe to be socially acceptable hatred. Many enjoy being in a position where they can actively or passively insult and demean others, with the evidence being the toxic environments of Facebook, Twitter, and LinkedIn.

Numerous exhortations of people who started from nothing, no matter how true or false they are, are shared on other platforms like LinkedIn. Alexis Ohanian, the co-founder of Reddit, called it one of the most toxic and dangerous things in the tech industry when he spoke at the 2018 Web Summit in Lisbon, Portugal (Petroff 2018). He elaborated by saying:

> It is this idea that unless you are suffering, unless you are grinding, unless you are working every hour of every day and posting about it on Instagram, you are not

working hard enough," he told the audience. "Do not let hustle porn win here. And do not let it infect your brain … It is such B.S.. Such utter B.S. And the worst part about it is it has deleterious effects, not just on your business, but on your personal wellbeing.

(Petroff 2018)

Many of these modern stories can be mapped along the Hero's Journey, which is a myth-based framework that numerous traditional adventure stories follow (Grand Valley State University 2021). Stories such as Lord of the Rings, Harry Potter, The Odyssey, and Beowulf all follow this. Due to its ubiquity, we consider it viable to explain many of the hagiographies of businesspeople that are used to exhort others to work more. It also explains how many in business are viewed as heroes by virtue of their completing quests.

These stories have three main parts:

- *The Separation* – Where the hero sets out on their journey, seeking adventure
- *The Initiation* – Where the hero commits to the task, goes through the trials, fights enemies, and goes through an ultimate battle before achieving their goal(s)
- *The Return* – The hero has finished what they set out to do, obtained the goal object, and has to return home, ultimately changed.

The framework has 12 steps, which can be explained using the following means using contemporary business examples:

1. *The Ordinary World* – This is the original world, which suffers from deficiencies that need to be corrected. The hero is lacking something, or something is taken from them (Grand Valley State University 2021). A modern example of this is an electrical vehicle company, whose purpose appears to reduce the reliance on fossil fuels and move toward a zero-emission future. The deficiency of the world is the waste of fossil fuels. One company's mission, which was to provide true personal computing to the masses in a world deficient of automation, can also be considered this. The mission of many modern security companies, which is that we exist in a world of uncertainty where anyone can be attacked, and that we are lacking the means to protect ourselves, is another one. This has been twisted by many politicians, however, to indicate an ideal world that needs to be corrected by the subjugation or removal of people who oppose the hero's message.

2. *The Call to Adventure* – The hero is given a challenge, problem, or adventure. It may appear as a blunder or chance encounter. This establishes their goal. The most famous call is that given to Steve Jobs and Steve Wozniak to build a true personal computer (MIT 2022). A security-related example is John McAfee. He was a programmer at Lockheed in California when he read about the Brain computer virus in the Mercury News. (He then wrote a program to find and eradicate it (Kelion 2013). He left Lockheed to found McAfee Associates to market and sell this software.

3. *The Refusal of the Call* – The hero must be set along the correct path. They must weigh the consequences and be excited by a strong motivation to move forward. Steve Jobs' frustrations with the Lisa project, specifically the expensive price (Bianchini 2019). Jobs' failure to run the Lisa project led him to be set along the path to the Macintosh, which was much lower-priced and simpler.

4. *Meeting with the Mentor* – The hero encounters a wise figure who prepares them for the journey. The figure gives advice, guidance, or an item, but cannot go on the mission. Mike

Markkula played the role of first investor and mentor. The example of Steve Jobs' visit to Xerox's Palo Alto Research Center to see the Xerox Alto and its Graphical User Interface (GUI) is an example because Jobs was not able to take this information back to the company he had founded.

5. *Crossing the Threshold* – The hero has committed to the task and enters the new world. A similar example from the business world is when people start new jobs or new projects. They are also often met by a threshold guardian (Grand Valley State University 2021). That guardian, in business terms, would often be their new manager or coworkers.

6. *Tests, Allies, and Enemies* – In this special world, the hero learns the new rules. This is through meeting people and getting information, along with a place where everyone gathers to discuss and meet (Grand Valley State University 2021). This is true through many fantasy books, role-playing games, including the Ultima ones from the 1980s and 1990s, and corporate life. The business example would be through new employee orientation, meetings, learning corporate norms, and learning what their tasks are going to be. The gathering place in fantasy novels is replaced by the coffee shop, team meetings, or instant chat channels.

7. *Approach to the Innermost Cave* – This is where the hero and their allies approach the place where their ultimate objective lies. This is often the land of the dead (Grand Valley State University 2021). In the case of the Lord of the Rings, it's Mordor. In the case of Jobs with the development of the personal music device, it was the land of failed portable music players such as the Diamond Rio that came beforehand. In the business world, this is often the beginning of a large project or initiative.

8. *The Supreme Ordeal* – This is where the hero faces danger, and it can be life or death (Grand Valley State University 2021). The dangers can be physical or psychological (Grand Valley State University 2021). In the Lord of the Rings, Frodo faced many of these on his quest. This can also be explained in the rush to get Intel chips into older computers to address the significant performance deficit from PowerPC chips (Dormehl 2021).

9. *Reward, or Seizing the Sword* – After surviving, the hero gets the object (Grand Valley State University 2021). This can be a treasure, weapon, knowledge, token, reconciliation, or accomplishment (Grand Valley State University 2021). In the business world, this can be the completion of a product or project. In terms of software, this can be similar to the completion of a new Linux Kernel or new Microsoft operating system revision.

10. *The Road Back* – The hero now has to deal with the consequences of their actions. They may still be dealing with challenges from the Supreme Ordeal (Grand Valley State University 2021). They now have to make the decision whether to return to the world or not (Grand Valley State University 2021). In the technology world, this can be described as those post-go-live bugs that organizations have to fix before the product ships, while dealing with the plans to release it and the challenges of doing so. Windows 2000, which shipped with over 63,000 potential issues, is an example of this (Foley 2000).

11. *Resurrection* – One final test is required for the purification and rebirth of the hero. It can also be a transformation (Grand Valley State University 2021). The best example from modern business would be the introduction of the Macintosh in 1984 with the Super Bowl commercial. This transformed computing to the graphical user interface paradigm. The second is the introduction of the popular smartphone in 2007, which changed smartphones from keyboards and styluses to finger-operated screens. It also transformed Steve Jobs from being just another tech CEO to a visionary leader capable of understanding more than just computers. A more common example is that big presentation that team members or leaders give

to the board, senior leadership, or at a conference that is used as the justification for their promotion or new opportunity.

12. *Return with the Elixir* – The hero returns to the ordinary world with the reward or elixir (Grand Valley State University 2021). In the case of technology, this is when it's released to the world and people can buy it, for example the crowded lines when new smartphones have been released prior to COVID-19. Nowadays, it's when people virtually queue up to buy a new Xbox or Playstation. A defeated hero who did not accomplish these tasks, much like the Gunslinger in the Dark Tower series by Stephen King, is doomed to repeat the lesson (Grand Valley State University 2021). Another example of the failed hero is the person who fails to ship a product on time, messes up the presentation, or does not meet the customer's requirements. They are set back to learn from the lesson and repeat it until they succeed. The excellent essay "On Cooling the Mark Out – Some Aspects of Adaptation to Failure," by Erving Goffman, from 1952, discusses how people on their way down, such as failed heroes, are indistinguishable from those on their way up (Goffman 1952).

Given the multiple examples from business stories here, specifically Steve Jobs and Apple, we can align many business narratives to this framework. The case studies often presented in MBA classes also align with this. They often show how companies succeed and benefit shareholders, or how they failed and had to go back and revisit the issue, if they were still able to.

We can also see how the narrative of the hero can be used as a participatory fiction tool to align people with stories and values. Building narratives around saving the company money, saving projects, or succeeding at all costs helps reinforce the norms of what it takes to succeed. Giving people the chance to participate in these stories and build their own as part of corporate tribal knowledge is an unspoken part of corporate messaging. Participation also helps build loyalty and helps people think they are being part of a heroic story.

A destructive and toxic version of this can be movements such as an Anti-Semitic conspiracy theory that has turned into interactive fiction (ADL, 2022). This has morphed into a narrative that draws people in with heroic goals of saving children and America from predators and evil people. It causes people to live in a live-action role-playing game that is not reality-based (Berkowitz 2021). People are so drawn into this that they ignore truth, reality, and even their own families because they are so drawn to participating in the heroic narrative. They will ignore their own basic needs. Their lives before this did not have much meaning to them, and this gives them that, even at their own expense and loss.

The use of the Monomyth and heroic narratives is commonplace with multiple scenario types, especially those meant to draw people into to support political agendas and authoritarian politics. People want to be the hero, and these stories give them the opportunity to participate in being one with minimum effort. People want to use being right as leverage against others to demonstrate superiority.

When I was discussing this chapter with one of my team members in a 1:1 meeting, one of the thoughts that came into my head is that the transformation of team members in a toxic environment can be thought of as "Cult for Business." This is because enculturating high-achieving people into the drama and toxicity of the workplace through interactive fiction and storytelling makes them part of something with similar drama and heroic goals twisted into something else. We discussed how people we had known transformed from high-achievement and idealistic to toxic and self-centered as a by-product of going through the funnel as they climbed higher in toxic organizations. The use of blame-shifting and negative connotations to entire groups or departments also occurs more than we're willing to admit in our workforces.

A very recent example of this was with a former Midwestern restaurant franchise executive. They openly discussed taking advantage of rising gas prices to lower wages and force employees to work more hours because they lived paycheck to paycheck (Blest 2022). The outputs from this email were the mass resignations of restaurant employees across their service region, and the executive's job loss due to those. This email was toxic and demonstrated how little they cared for team members or their well-being. What wasn't covered in the press was the email chain itself that circulated, with the text from the first email in it saying "Words of wisdom from XXXXX!" This indicates that leadership and those who felt that being obsequious to them would get them ahead saw this email as a lesson or story, and that its recipients agreed with the message. They thought they were doing a good heroic deed for them by agreeing with its message and starting to implement their vision. The negative connotation given to the entire workforce at the restaurant chain and subsequent agreement demonstrates toxicity at the highest levels of that organization. The numerous people who agreed demonstrates people who would throw their entire teams under the bus for a compliment from leadership. They also demonstrated characteristics of someone seeking attention by signaling assertiveness and dominance.

In Information Security, we've seen television shows such as Mr. Robot and marketing campaigns used to draw security professionals into heroic myths. An example of this has been the numerous advertisements used to convince executives that the big bad enemies are ready to pounce on your insecure protection and breach your network. They draw people in to think that they are fighting a war against Anonymous, criminals in hoodies, or various criminal syndicates. They give examples and scenarios that are far outside the norm of what really happens. Many of them also are meant to scare people. Only with the elixir of the tools can the security teams save the organization from whatever exaggerated threats combined with vendor data breach reports are used to sell the products.

The numerous reports from CISA, Verizon, IBM, or HIMSS paint a much less mythic picture, which is one of phishing messages, stolen credentials, and privileged access being the main gateways, not 0-day attacks. These narratives are dangerous, particularly in security, because they draw attention away from insider threats, privileged access management, combating phishing attacks, and addressing password/credential reuse. They also draw budgets away from initiatives and resources that would help improve security. Younger security professionals who have been sold the myths of Mr. Robot or Anonymous who see the reality of what security is may also become disillusioned at having to work in a Security Operations Center to start, instead of what's been popularized and sold, especially to get them into boot camps that may cost thousands of dollars to get their certifications.

Instead of resources that can help with reducing overload and burnout, you get heroic stories of pew-pew maps. When CISOs get together, they discuss tools. We don't discuss what our team members are really experiencing. We can do better here. As a former CIO I worked for once told me, people are only as good as they have to be to get the job done. For many of these successful attacks, they don't have to be very good. The sheer amount of credentials on Tor sites/The Dark Web that can be used to gain access to networks sans two-factor authentication is evidence of that.

What Have We Built?

While modern corporate messaging for all team members has messages of diversity, equity, inclusion (DE&I), and fairness, the messaging used in interpersonal and informal communications does not. The unwritten rules of hegemony are encoded in these. While people may end up in higher positions, they aren't necessarily part of the "in crowd." We need to address interpersonal communication and expectations based on long-standing expectations of management and leadership. We have to directly address the social media funnels or algorithmic traps that people get caught in that amplify bad behaviors and spread negativity.

Jennifer Berdahl, Peter Glick, and Marianne Cooper, in their *HBR* article "How Masculinity Contests Undermine Organizations, and What to Do about It" (Berdahl, Glick, and Cooper 2018), discusses how people who question the culture are marked as losers. People are afraid of losing status and being perceived as less than. Roger Jones, in his HBR article, "What CEOs Are Afraid Of," discusses that imposter syndrome, the fear of being found incompetent, is prevalent even in the C-suite (Jones 2015). Berdahl, Glick, and Cooper indicate that the resolution for this must come from top leadership setting values-based goals and enforcing them through modeling and behavior change.

What we've built, according to Arianna Balkeran, in her master's thesis from CUNY, "Hustle Culture and the Implications for Our Workforce" (Balkeran 2020), is a culture where people who hustle, enculturate themselves, and push themselves harder at the expense of everything else succeed to avoid being marked negatively. They receive more social support from their supervisors for hustling (Balkeran 2020). They are more likely to be engaged to receive the communications and storytelling needed to further enculturate them. The 1:1 meetings with top executives that top performers get contain much of this communication. We have more in common with our ancestors that swapped tales around the campfire than we realize.

We've also built a culture of self-perpetuating toxic environments. Social media makes it very easy to advertise companies that have "hustle porn" cultures and find candidates willing to work there. There is a subculture of people who believe that toxicity builds character, and they do find employment at these places. Much like the people that find self-actualization in outrage and anger, especially on social media, some find it in openly seeking out toxic and hostile places to work where they can express socially acceptable hatred. There are sadly numerous companies that still openly accept and facilitate this behavior. Not all of them have dramatizations starring Hollywood stars like Joseph Gordon-Levitt, Jared Leto, or Anne Hathaway. Often they're like the recent story out of Kansas , where an Executive Director at one company sent an email discussing lowering wages to get people who were already struggling financially to work more hours, to the agreement of many people in the management chain who considered it wisdom. When you go on LinkedIn or Instagram, there are numerous job postings from companies advertising hustle cultures. There are also numerous posts from both management and aspirational team members of these organizations actively and passively insulting and belittling others that don't buy into their mindset as lazy, not successful, and not deserving of what they have. The current sociopolitical environment encourages people who revel in cruelty and toxicity to signal each other based on who they support politically, and do so in a way to offend others and bring attention to themselves, often involving their businesses in their rants.

Information Security conferences such as DEF CON, BSides, and Black Hat often fill the void of informal communication with the lack of top management. They often facilitate top-level performers who have risen through the ranks to become senior figures in the industry educating the newer professionals on what they need to do. They serve a very important role in providing the leadership that many do not receive from their management or employers. Many members of the security community also hustle on their own of their own volition, not as part of employment, to advance, and use these conferences as part of that. They also provide ties from the outside into the multiple hacker subcultures, with DEF CON being the largest showcase and celebration of these worldwide.

Matt Plummer, in his *Harvard Business Review* article, "How Are You Protecting Your High Performers from Burnout?" (Plummer 2018), indicates three methods by which they advance. This is by being put in the hardest projects, being used to compensate for others that don't perform as well, and being asked to help on other projects unrelated to their work. This work bleeds over from standard hours, and results in people having to rapidly context-switch, removing their accuracy and making them less effective. Information Security is especially vulnerable to this because it's often used to patch issues from other teams (compensate), asked to help on other projects such as Cyber Insurance, and is often on the hardest projects with the least recognition.

Operating Systems Don't Multitask Well, and Neither Do We – How We're Like What We Create

When computing first started, we did not need to understand the complexities of multitasking or concurrency. We controlled everything on the computers ourselves and were able to shape them to our needs. Generations learned on the personal computers how to program and get results with running one program at a time. Later generations introduced the concepts of multitasking and multi-threading. The internet brought with it the concept of multi-connectivity and concurrency. Virtualization, containerization, federated technologies, and intelligent systems have exponentially increased complexity. While operating systems, compilers, and tools have evolved significantly to provide resilience to address many of the issues caused by misbehaving program code, we have not.

Computers have a limited set of registers to perform operations on, and the need to move addresses and values in and out of them. Calling functions requires us to push and pop operations on and off the stack. Multitasking and multi-threading in mainstream programs required the use of semaphores and locks to protect memory from being accidentally overwritten. Desktop computing and mainstream web apps consists of virtualization, containerization, and federated technologies require the use of trust and identity, along with security, to ensure a reasonable degree of accuracy between networked systems. Each of these operations comes with a serious performance penalty.

People are a lot like the systems we create and develop. Each context switch on a computer takes semaphores, locks, waits, and pushing and popping on and off the stack. The same happens when we are working on multiple tasks. We don't do this well. People are not efficient multitaskers. We often mess up, much in the way programming errors can cause issues on the stack. The difference is, instead of a computer locking up or a program not functioning, we don't function. Again, operating systems, compilers, and programs have evolved to address stack issues. We have not.

Buffer and Stack Overflows, which occur when someone overloads memory buffers and enables them to bypass protections to run their own code in privileged space, have tools to help address them. We need to utilize the human equivalent to not allow extremes that overload and break us to become more resilient. One bad task or distraction can cause a reaction like a badly written program on MacOS 7.5, Windows 98, or MS-DOS, and bring us down to a crash. A programming example is with the C programming language's string copying and concatenation functions, strncpy() and strncat(). These functions assume that programmers will be able to properly format string termination characters manually. As numerous exploits and crashes prove, this is not the case. Todd C. Miller of the OpenBSD Project created the functions strlcpy() and strlcat() to safely format strings and avoid potential mistakes that can lead to crashes or worse (Miller 2019).

Unlike the OpenBSD Project, we don't do enough to build these competencies. Similarly, memory leaks, where a program doesn't free up memory it has utilized, take up all resources eventually and starve others out. We can have tasks or distractions that take up so much of our memory that they crowd out others. Sometimes it is information we really need to know, and that causes cascading issues.

We, like computers, have one central scheduler and orchestrator, no matter how many processors we have. While we may be more efficient at some tasks, we still can focus on one task at a time. The significant increase in burnout we are observing comes with a price. Work–life boundaries have been destroyed. Phone distractions happened 46 times a day, according to Coleman and Coleman in their 2016 *HBR* article, "Don't Take Work Stress Home with You" (Coleman and Coleman 2016). The increase in connectivity since 2016 has doubtlessly increased this. We are doing more tasks with less focus and more errors. Olivia Goldhill, in her article "Neuroscientists Say Multitasking Literally Drains the Energy from Your Brain," cites Gloria Mark, a professor from UC Irvine, in indicating that when someone is interrupted, it typically takes 23 minutes and 15 seconds to go back to their work (Goldhill 2016). She further indicates that people will do two intervening tasks before going back to their original work. This can lead to a buildup of stress. Mark's further research, as defined in the article "Neurotics Can't Focus: An In Situ Study of Online Multitasking in the Workplace" (Mark, Iqbal, Czerwinski, Johns, and Sano 2016), also indicates that switching between different tasks results in a 50% increased completion time, the median duration of online screen focus was 40 seconds, neuroticism was associated with shorter online focus duration, and there is a factor of lack of control that significantly predicts multitasking.

Physicians, nurses, and caregivers have to deal with what is known as Alarm Fatigue. According to Sue Sendelbach and Marjorie Funk, in their article "Alarm Fatigue: A Patient Safety Concern," from the journal *AACN Advanced Critical Care*, 72–99% of clinical alarms are false (Sendelbach and Funk 2013). These alarms are distractions. Exposure to too many alarms results in desensitization. Patient deaths have occurred because of these. Distractions cause errors. When they occur to caregivers, they can directly lead to the worst adverse outcomes.

We take our top performers, give them significant amounts of work, and then expect them to deal with torrents of distractions and interruptions. We ask them to do more than the average worker. Top performers are socially linked more into the corporate narratives and storytelling and educated in norms and practices that still have associations with toxic behavior. We're taking the work, giving it to the best people, and continually trying to do more with less while piling on multitasking.

We Assume a Perfect Communication Environment and High Signal-to-Noise Ratio

Another way that we isolate people is through misunderstanding communication effectiveness (Figure 8.1). Communication, while being the medium by which entities directly message, is not perfect. Claude Shannon, in his 1948 article, "A Mathematical Theory of Communication," describes the path that it takes from source to destination (Shannon 2022). Communicationtheory .org adds in the concept of feedback from destination to source (Communication Theory 2014).

Noise is an integral part of communication according to Shannon, and feedback helps us understand whether the message was received correctly or not. The International Standards Organization (ISO) developed the Open Systems Interconnection (OSI) model to explain this (Figure 8.2).

The OSI model (Figure 8.3) discusses how to encode and decode messages through the Application, Presentation, and Session layers to form well-formed messages intended for users and groups. The Transport, Network, Data-Link, and Physical layers handle transmission and reception of well-formed messages across the physical layer. Network protocols such as TCP/IP make explicit whether there are many recipients of a message. The Transport and below layers, which map to Signal, handle error checking and correction across the communications link. The Application layer determines if there are adequate resources to process the messages, while Presentation formats them and Session determines the recipients.

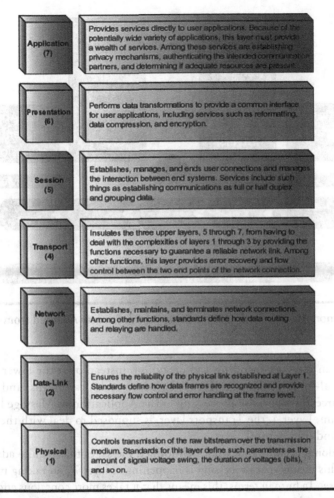

Figure 8.2 OSI model. Source: www.fcc.gov/help/public-safety-tech-topic-2-internet-protocol -ip-based-interoperability.

Corresponding to this, the upper bound to a link is C, according to Shannon's Theorem, which is a mathematical formula that gives the upper bound to the capacity of a communication link in bits per second (Free University of Berlin 2022):

$$C = B * \log_2(1 + S/N)$$

where:

   C = Upper-bound capacity in bits per second
   B = Bandwidth of the line in bits per second
   S = Average Signal Power
   N = Average Noise Power
   S/N is expressed as db * 10

We assume that noise is just communications noise or error correction. Well-formed messages that interfere with the intent of our messaging or distract from it are also noise and

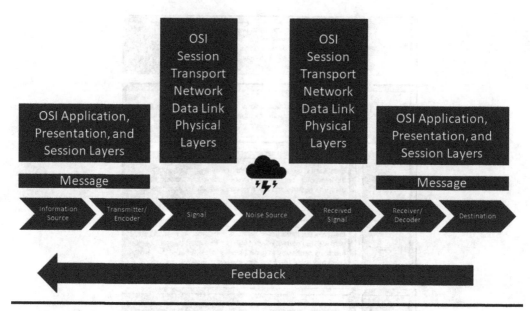

Figure 8.3 Shannon–Weaver with OSI model overlaid. Source: Adapted from Communication Theory (2014).

take resources from message encoding and decoding, instead of at the lower levels. Our calculations for signal-to-noise ratio (SNR) are based on technical means, and do not directly address the resource costs of noise processing at the Application or Message layer. According to the OSI diagram, Layer 4, the Transport layer, is supposed to deal with the complexities of error correction and flow control.

What we are doing is communicating well-formed messages that require additional processing for validity. This means that processing is happening at layer 7 and taking more resources to decode and interpret. In human terms, this means that it takes more conscious effort and energy to understand and comprehend messaging that otherwise would be handled by unconscious reflexes.

An example of this is a minimum Wi-Fi 6 (802.11ax) connection, which according to Intel has a minimum max link rate of 600 Mbps (Intel, Inc. 2022). According to Meraki Networks, now part of Cisco, a minimum 25 db SNR is needed for voice communication, which is a ratio of approximately 316 (Cisco Meraki 2020). That means for every 316 good messages we receive, we get one bad one. Taking the above equation and plugging in the numbers for a 600 Mbps Wi-Fi connection with a 25 db signal-to-noise ratio, we get:

$$C_{phys} = 600000000 * \log_2 (1 + 316)$$

$$C_{phys} = 600000000 * \log_2 (317)$$

$$C_{phys} = 600000000 * 8.31$$

$$C_{phys} = 4.986 \text{ Gbps}$$

That's pretty good for a modern Wi-Fi connection. That's a lot of raw data. However, it's actual unprocessed data at layer 4, not layer 7, which requires more processing power.

It also gives the wrong picture because it assumes the physical and messaging SNRs are the same. They're not, and the noise increases due to too many well-formed messages that are just noise that people must process. As noise increases, the overall upper-bound capacity of the link can decrease. The signal-to-noise ratio can approach 0 and take the overall effective capacity down to that. We assume all useful data, and not distractions or non-useful data that layer 7 processes. We posit that the realistic SNR, given the data we receive, is much lower due to distractions. We assume physical and messaging SNR are the same. What we really see is very different.

For example, if a user gets 1 useful message for every 100 distractions, which is not an unrealistic ratio given inboxes, Slack, social media, and messaging, that Wi-Fi connection's realistic upper-bound capacity is:

$$C_{act} = 600000000 * \log_2 (1 + 0.01) \left(0.01 \text{ being the real SNR} \right)$$

$$C_{act} = 600000000 * 0.014$$

$$C_{act} = 8.4 \text{Mbps}$$

We can calculate the percentage effectiveness by dividing C_{act} into C_{phys} and multiplying by 100.

This means that the overall effectiveness of our connection is 0.16% of what the physical connection is. Distraction takes something that's super high-performing and can remove 99.84% of its effective capacity. That means you have given someone the effective bandwidth of a computer running 10 Mbps ethernet through distractions. You've turned an Intel Core i9 or AMD Threadripper into a 386 that can barely run DOOM.

A computer is considered a computer by many if it can run DOOM, which was state of the art in 1993. Distractions interfere with communications, decrease the real SNR, and remove that capability. Now imagine how this affects the people you work with, or the Information Security teams that must deal with torrents of information given little guidance. You're lucky if you can load that first level of DOOM given the real SNR, and you're isolating them through distractions and torrents of data.

What these equations show mathematically is that we're reducing our team's mental bandwidth capacities significantly. When you factor in the distractions from multitasking and their equivalents to badly written code, what we're doing to our people is the logical equivalent of running MacOS from 1992 on a Mac Studio from 2022. It's written for the wrong processor, uses a small fraction of available resources, and code or misinformation that assumes it has all the resources that can take over and stomp over existing protections, in addition to reducing effective bandwidth to next to nothing. It's a combination of overflow and Denial of Service attacks on them. No wonder they feel isolated, burned out, and exhausted, and more susceptible to worse, including misinformation. The bigger question is what arbitrary code is running in our minds due to these overflows because we expend too much energy processing them at higher levels? What sorts of messaging are we allowing to play in our minds and subliminally influence us due to overload?

This real signal-to-noise ratio has the effect of turning us into zombies, much like the ones in The Walking Dead. The actual usage of our brainpower is roughly the same as them after the virus restarted enough of their brains to function. We've tried to use technology to augment our capabilities. It accomplished the opposite. We've given our top performers, such as physicians, security team members, caregivers, and executives, a torrent of noise and are asking them to figure out how to deal with it themselves. The noise weighs them down so much that they are less effective. Shannon's Theorem only serves to show how badly. Alarm Fatigue also does, with much more tragic consequences.

Isolation – How Virtually Isolating Team Members Causes Paranoia

What we also do with our top performers is isolate them from the rest of the organization in their own bubble. Every group within an organization has its own culture, stories, and social organization. Top performers are scheduled and given time with the senior leadership and high-performing executives. There is a definite difference in how high performers, including those who have been chosen by senior leadership for additional opportunities, socialize. The communications they have are much different than those that other teams have with their members. They are exposed more to the corporate narratives, storytelling about corporate heroes, and are part of the "in crowd." They also get overloaded with messaging and bandwidth. The online equivalent was described in O'Callaghan et al.'s article, "Down the (White) Rabbit Hole" (O'Callaghan, Greene, Conway, Carthy, Cunningham 2014). This is because these people, much like YouTube users that have been exposed to extreme propaganda, end up in an ideological bubble. In this case, it's one that espouses the winning at all costs philosophy that can be endemic to top management, and its norms. High performers are also exposed to the fears and behaviors of executives more than a workforce that may just see a smiling face at large department meetings.

Roger Jones, in his *HBR* article, "What CEOs Are Afraid Of," discusses the five key fears that executives face (Jones 2015):

- Impostor syndrome – Fear of being found incompetent
- Underachieving – Which can lead to risk-taking to overcompensate
- Appearing too vulnerable
- Political attacks by colleagues
- Appearing foolish – This limits their ability to speak up and have honest conversations

These five top fears resulted in the following dysfunctional behaviors:

- A lack of honest conversations
- Too much political game-playing
- Silo thinking
- Lack of ownership and follow-through
- Tolerating bad behavior

The fallout from these behaviors can include:

- Poor decision-making
- Focusing on survival rather than growth
- Inducing bad behavior at the next level
- Failing to act unless there's a crisis

What we have observed is that the top fears, combined with the isolation caused by reliance on top performers, causes silos to appear in companies, even if unintended. Leaders depend on who they can rely upon to deliver so that they don't appear to be underachieving, vulnerable to attacks, or politically vulnerable because they can't deliver. They're willing to put up with jerks and bad behavior if it means they deliver something on time that makes them look better than anyone else. This is considered a calculated risk and is a common occurrence in business. Leaders isolate themselves, and through their actions, they pass on narratives and stories to their top performers that propagate their behavior. In many cases, it's paranoid, neurotic, and inward-focused. The top people are exposed to more of the worst behavior, and the dysfunctional behaviors make it to the next generation of leaders. We're left with the leaders and their acolytes, much like the relationship between bullies and their "friends" that egg them on so they don't face their wrath. In the business world, its bullies, acolytes, and the management consultants, investment bankers, and accountants that enable them. When you combine this with effective bandwidth isolation through a very low signal-to-noise ratio and information overflows, it's a recipe for disaster.

The best way to explain this to outsiders is that executives are like parents, and their direct reports are like children. While many good activities and practices get passed from parent to child, so do the ones you don't want. These can include when you use profanity or adult phrases or tell them negative things about people. You have no control over what your children will do afterward.

The product of these fears is often corporate conspiracy theories, much like a "Cult" For Business" we discussed earlier. Passing down the paranoia is the best phrase that explains this. We fall into these patterns because we see so many of them outside of work. We repeat the same patterns we see in school, with our friends, and on social media. Much like the negative cues from social media, we see mistrust, false information, and the paranoia of senior leadership funneled down to their teams.

Significant resources are expended upon political game-playing, stories, and myth-building. Isolation and silo thinking can cause many unfounded thoughts. Political games lead to stories, lies, generalizations, and infighting. A very low SNR reduces effective bandwidth and crowds out the ability to effectively process messaging to a fraction of what it needs to be while taking up resources. Business degrades into conspiracy-mongering instead. We cause significant negative stress where we don't have to, and cause others to expend energy otherwise used to combat their issues. We can cause downward spirals for team members feeding corporate conspiracy theories and myths to sate our own egos and make ourselves look better. The lack of energy to keep up the façade draws it from other areas, combined with significant resources to process distractions and messaging, and makes people more susceptible to negative cues and effects like addiction.

Neurodivergence and Masking Drains Energy

Neurodivergent people, specifically those with autism spectrum disorder (ASD), often expend significant energy masking, which is compensating for behaviors not considered neurotypical (Stanborough 2021). This has three stages, according to Healthline. These are motivation, which is the realization that they need to hide their differences, act differently, or change characteristics about themselves. They spend significant time and energy in the second stage attempting to act or pass as neurotypical by teaching themselves how to do so. The third stage, consequences, is the feeling drained and exhausted by trying to conform to those different standards of behavior.

Imagine always being an actor on stage and feeling that you must always perform. Imagine being given a script and being told to prepare for a performance without preparation. Imagine always feeling left out and left behind because you don't pick up on social cues, or don't know how people interact socially. I want you to think about having to script and think about everything you do with interacting with your own family because those interaction types are new to you and different. This also makes dealing with workplace bullies and toxic management significantly harder, as you can't pick up on the social cues or messaging that you're dealing with very bad people.

Bullies and toxic management will also take advantage of neurodiverse people to advance their agendas and provide self-satisfaction.

Think of being in a meeting and on camera and having to turn it off and decompress for an hour because you're spent. Think of always feeling like an outsider, like you've always made someone upset, and walking on eggshells because you think your behavior disgusts people. Imagine having compassion for people and being afraid of negative consequences at the same time.

Impostor Syndrome, which according to Verywellmind.com is an internal experience of believing that you are not as competent as others think you are, is significantly magnified (Cuncic 2021). When you know that you're throwing up a mask for others, the fear of being found out or considered a phony is magnified a hundredfold. The stress of that is devastating. Dealing with that drains much more energy. Trying to be the perfectionist, superhero, expert, natural genius, or soloist, according to that article, along with masking, expends additional energy.

Being successful or recognized at something can also have two possible outcomes. You can be happy because you've never been recognized or accomplished, you come off the wrong way that people think you're egotistical, and you fail to follow recognized logical paths and fly off the rails. You can also have the impostor syndrome kick in and cause you to retreat into yourself with embarrassment. Having a response considered normal is difficult to achieve.

The energy spent acting and performing and being worried about negative consequences, while dealing with the existing resources and fitting into the business hierarchy, drains people and leaves them significantly more susceptible than those who keep up just one façade. Imagine keeping up multiple, inception-type levels of them at the same time. Unlike the movie, the time and perception don't increase with the levels. It decreases and requires significantly more agile and time-sensitive reflexes to keep up. The energy expended here causes significant stressors and can break neurodivergent people more easily. Burnout happens because the energy they spend engaging is significantly more than the average team member, and resources drain that much more quickly.

Our Actions Reduce Energy and Make People Susceptible to Addiction

Addiction, according to Barrows and Van Gordon, is generally understood to be a dysfunction of brain circuitry involved in reward processing and motivation (Barrows and Van Gordon 2021). Garland describes it as a negative effect to stressor stimuli that combined with a drug-related cue leads to addictive behavior (Garland 2016). This biases attention toward drug-use action schema, which biases attention toward relevant stimuli and increases cravings (Garland 2016). People who have some degree of control try to suppress this and control it. When self-control resources are exhausted, they self-medicate, which strengthens the habit through negative reinforcement conditioning (Garland 2016).

While drug and alcohol addiction are prevalent in technology and management, other types of addiction, such as gambling, social media, and gaming are also. We are putting many people who already do not have good responses to stress that manifest in one or more addictions in a position where they are more susceptible. We are also putting them in positions where they are already suppressing emotions and trying to meet norms while dealing with fears such as impostor syndrome, the constant stimuli of multitasking, and the fear of failure. We're taking people already using much of their energy to maintain themselves and draining whatever is left, increasing their risk. In security terms, this is a very poor compensating control.

The best reward for good work, some say, is more work. As our current culture indicates, this is often the case. The best reward for stress is not more stress. However, this is also the case. We need to address resiliency with our top performers to restructure reward processing and help them build the coping mechanisms they need to deal with these increased levels that leave them more susceptible. Barrows and Van Gordon cite Garland and Howard in discussing five key mechanisms at play for this:

- Enhancing executive functioning
- Increasing dispositional mindfulness
- Attenuating stress reactivity
- Decreasing cue reactivity
- Reduction of thought suppression

According to them, mindfulness-based interventions (MBIs) such as acceptance and commitment therapy (ACT) can be used to address these behaviors.

Addiction is the opposite of flow, which is a state where people are fully immersed in an activity and highly focused on what they are doing with positive feedback, according to Kendra Cherry in her article, "What Is a Flow State?" (Cherry 2022). It can occur when people are entirely focused on the task at hand. Flow has four components, which are to

- Set clear goals
- Eliminate distractions
- Add an element of challenge
- Choose something they enjoy

Addiction is the opposite of flow because it is a negative effect, a distraction as it is a response to stress, and is a negative coping mechanism that facilitates ignoring all other wants and needs. It looks like flow; however, it's negative feedback that leads to self-destruction instead of something positive.

We take our best people and isolate them through work, constant stimulation, and multitasking. We put them in environments where they are constantly communicated to through storytelling and alternative means above others. We put them under significant pressure. We expose them to leaders that can pass on their fears and bad behaviors. We leave them more susceptible to addiction through increasing their stressors and decreasing resources they can use for self-control. We gauge them against norms that reward keeping emotions hidden, being aggressive, taking control, and competency at the job that have roots in being tougher than everyone else. We drive them toward coping mechanisms, negative flow, and addiction. We do a poor job of Information Risk Management. The algorithms and dark patterns behind keeping people engaged on social media do an excellent job of keeping people addicted to that feed or other negative stimuli and in a negative flow.

If we use addictive methods and processes as part of engagement or management, it's going to the dark side. Forever it will dominate your management destiny and burn people out. A toxic manager, do not be like that. Because you can inadvertently become the cause of addiction in others.

We are setting our best up to fail hard at work and at life for the purpose of a higher corporate net income and the glory of others. The top performers often work until they break. When they crash, they crash hard. The crashes not only affect their work lives, they affect them personally. We are leaving broken people in our wake trying to fix broken software and solutions that often suck.

How Does Software Development Relate?

They can suck because they do not have to be good. We have put solutions in place that are good enough to be sold. We have managed to build incredibly complex systems. However, according to Sam Williams in his article, "A Unified Theory of Software Evolution," quoting Meir Lehman, we build systems that eventually become unmaintainable and overly complex (Williams 2002). The lines of code in Windows 11, Mac OS, or the Linux Kernel bear this out. We have a cycle where some pieces of software, such as Enterprise Resource Planning or other critical business software has a very long lifecycle. However, other technologies last about six to eight years (Mitopia Technologies 2022). In Williams' article, he discusses how IBM, in a race with Bell Labs, was attempting to demonstrate that they could be as productive at writing lines of code (Williams 2002). What Meir Lehman found upon further analysis was that they were spending more time writing code for new features, and less time fixing issues in it.

What happened with OS/360 repeated itself several times with other operating systems. The most memorable time was when Microsoft stopped development of Windows Longhorn to address root causes of numerous security issues in 2003–2004, resulting in Windows XP Service Pack 2 and Windows Server 2003. Windows Vista and Server 2003 were delayed fixing many issues caused by feature creep. Apple utilized the Snow Leopard release of MacOS, 10.6, to stabilize and make the OS more performant. Linux has aggressively removed older architectures and unsupportable platforms from the Linux Kernel, with the most important being i386 support, as it prevented effective implementation of symmetric multiprocessing (Larabel 2012).

We have a cycle in this industry where we build software to meet needs, try to get it to work with everything else, have it become incredibly complex, and then throw it away for something else because it's too complex, clunky, or unmaintainable enough to meet needs. We prioritize features over good code and efficiency. While there are numerous exceptions to this rule, sadly, this is the norm for a lot of software, especially security software. The standard three-year contract is often how long a software product lasts due to many of these issues (Figure 8.4).

We take our best and brightest and have them attempt to configure and use these platforms to protect businesses. We build our businesses on bad software that our customers don't like. The negative stimuli alone burns people out. It's about looking good and selling. Yet we continue to do this. What is the economic reason for us to continue down this path? We are effectively doing more with less.

The Cycle of Software Disillusionment captures the cycle that much software goes through. Clients determine that they have a need, and then go through a selecting and purchasing process that is long, opaque, and often oversells clients on what the products can do. Integrating into the network then takes significant investment. Interoperability with security software, much like healthcare software, is nearly nonexistent and requires significant custom and professional services work that isn't disclosed beforehand. If someone tells you that software integrates easily, it's to get the sale by saying anything that sounds good. Major security companies, like healthcare companies, are more than willing to charge significant amounts for professional services for interfaces that require constant maintenance and additional operating costs. These interfaces are often not able to be secured well and present numerous security issues. Customers become disillusioned when they realize they've been oversold and overpromised. They have further concerns when they realize what the real operational costs are. They finally realize that they have not met their needs, and the cycle starts again.

This is good for software companies when they need to have a consistent sales cycle. According to Mito Systems, the average lifecycle of software is six to eight years (Mitopia Technologies 2022). With security moving faster because of significant changes, and enterprise contracts moving to an annual and multi-year model (Tunguz 2016), this cycle can play out in one to five years. The rapid change in

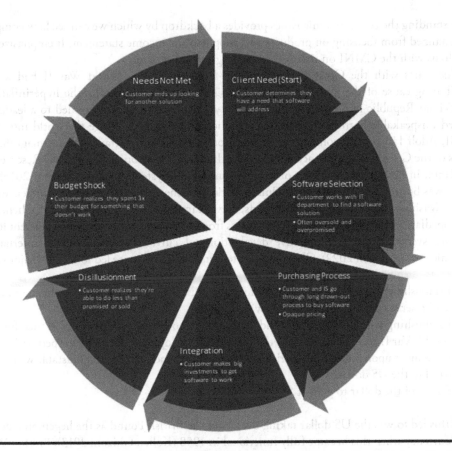

Figure 8.4 The cycle of software disillusionment. Source: Parker, Indiana University Health (2022).

Endpoint Protection solutions from vendors such as McAfee, Symantec, and Trend Micro to newer entrants like Crowdstrike, Blackberry, and VMW over the past three years demonstrates this.

The rapid churn of security companies and Mergers/Acquisitions in this space bear this out. According to Momentum Cyber, M&A in the cyber space has incrementally increased from US$5.9B in 2011 and $4.7B in 2012 to $28.1B in 2019 and $20.0B in 2020, the latter affected by COVID-19 (Momentum Cyber 2022). There has been an increase in financing activity from $0.9B in 2011 to $10.7B in 2020, according to the same source. There is significant pressure to get out and sell to capture that money and growth, and the 745% growth in market cap and 192% increase in public cybersecurity companies can also increase the pressure to put something out, even if it's a bad product. Cybersecurity sells on the public markets, which have become the major economic force driving our economy and have become a major focus of all business. With the pressure to sell, why bother putting out products that work well together?

What Are the Economics behind This?

We need to look at the economics behind this that emphasize doing more with less that led to this focus. How did the US and world economies change from the Great Depression until now? How did we evolve to value profits at all costs, especially with people? What steps did we take?

Understanding the core economic issues provides a backdrop by which we can see how companies have changed from focusing on products and people to the income statement. It emphasizes how and aligns with the CMNI on risk-taking and aggressive behavior.

This starts with the Great Depression and Bretton Woods. World War II had a major contributing cause of the massive economic crisis in Europe, specifically the hyperinflation in the Weimar Republic in Germany. The instability and economic depression led to a leader who enacted unspeakable horrors and murdered millions, and helped plunge the world into World War II, Adolf Hitler. The competitive devaluations and restrictive trade policies magnified the effects of the Great Depression, according to Sandra Kollen Ghizoni of the Federal Reserve Bank of Atlanta, in her article "Creation of the Bretton Woods System" (Kollen Ghizoni 2013). This system was born out of the July 1944 United Nations Monetary and Financial Conference at the Mount Washington Hotel in Bretton Woods, New Hampshire. According to James Chen, in his Investopedia article, approximately 730 delegates from 44 nations met to create an efficient foreign exchange system, prevent competitive devaluations of currencies, and promote international economic growth (Chen 2022). Ghizoni indicated that this conference led to several major results:

- Establishment of the International Monetary Fund (IMF) to monitor exchange rates and lend reserve currencies to countries with deficits.
- Establishing what is now the World Bank Group to provide financial assistance for post–World War II reconstruction and the economic development of less-developed nations.
- Agreement upon member countries to keep their currencies fixed but adjustable within a 1% band to the US dollar.
- Fixing of the dollar to gold at $35 an ounce.

What this led to was the US dollar taking over from the British Pound as the hegemonic de facto world currency when this became fully functional in 1958 (Kollen Ghizoni 2013). Bretton Woods was a precursor to the Yalta conference in 1945, where Churchill, Roosevelt, and Stalin made major steps toward deciding the post–World War II order in Europe. It was one of the signature events that led to the post–World War II economic boom that put the United States at the top of the world order.

Staying there was another issue. As the world grew stronger, specifically West Germany and Japan, the US dollar became devalued against the Mark and Yen. Douglas A. Erwin, in his article, "The Nixon Shock after Forty Years: The Import Surcharge Revisited for the National Bureau of Economic Research (NBER)" (Irwin 2012), discusses the events that led to this. The devaluation of the US dollar against these strong currencies led the Nixon administration to close the gold window and establish an immediate 10% tariff on imported dutiable goods to force countries to renegotiate their exchange rates. This led to the abandonment of much of Bretton Woods and floating exchange rates, meaning arbitrage across different currencies was now possible. The abandonment of the gold standard and reestablishment of new exchange rates due to West Germany and Japan's growing economic dominance helped start a series of events that persist to this day. Arbitrage was now a major driver.

Prior to the 20th century, most countries utilized some sort of gold standard or backing by a commodity. As international trade and finance grew in scale and scope, however, the limited amount of gold coming out of mines and in central bank vaults could not keep up with the new value that was being created, causing serious disruptions to global markets and commerce. Fiat money gives governments greater flexibility to manage their own currency, set monetary policy, and stabilize global markets. It also allows for fractional reserve banking, which lets commercial banks multiply the amount of money on hand to meet demand from borrowers. Though like many institutional mechanisms, fiat money poses challenges to economies and governments including susceptibility to inflationary pressure (Douville, 2022).

The 1970s were not known for good economic times. New York City famously almost went bankrupt. Stagflation, the gas crisis, and numerous other issues led to serious economic issues. The U.S. left the gold standard in 1971, which threatened the concentration of wealth and power that the gold standard enabled (Douville, 2022). The Revenue Act of 1978, which was signed by President Carter and took effect on January 1, 1979, was a complex tax cut bill (The New York Times 1978). However, as Tim Stobierski from Northwestern Mutual noted, section 401(k) of the Revenue Act allowed employers to avoid being taxed on deferred compensation (Stobierski 2018). In 1981, the IRS introduced rules that allow employees to fund their 401(k) plans through payroll deductions. This had a major effect as the previous method for deferred compensation, pensions, are considered long-term liabilities according to the CFA Institute (CFA Institute 2022).

This had the effect of shifting pensions from corporate liabilities to third parties, greatly increasing the potential valuation of companies. These 401(k) plans also invested in stocks to get the desired returns for their customers. The Investment Company Institute indicates that $7.3 trillion in assets are held in 401(k) plans as of 2Q 2021 (ICI 2021); 63% were held in equity securities at year-end 2018. Medina, De La Cruz, and Tang, in their OECD report, Owners of the World's Listed Companies, note that institutional investors owned 72% of the stock in US-listed companies at the end of 2017 (Medina, De La Cruz, and Tang 2019). Institutional investors held the stock on behalf of themselves and their customers. They also held 41% of public equity worldwide, equivalent to US $31 trillion (Medina, De La Cruz, and Tang 2019).

The continued investment in 401(k) funds drives the market and growth, and major investment firms demanding higher returns on investment due to their ownership leads to companies needing to maintain continual growth. The Economic Recovery Tax Act (ERTA) of 1981 and Tax Reform Act of 1986, led to this. The ERTA accelerated depreciation tax reduction, cut the top tax rate to 50%, and reduced the capital gains tax. It also reduced taxes on windfall profit gains (CFI 2021). The Tax Reform Act of 1986 lowered the top individual tax rate to 28% and the top corporate tax rate to 34% from 46% (Nellen and Porter 2016). The net capital gains tax went from a 60% deduction to ordinary tax rates. These changes meant that corporations kept more money.

The Gramm-Leach-Bliley Act of 1999 (GLBA) repealed part of the Glass-Steagall Act, enacted in the wake of the Great Depression, which disallowed banks from offering investment and insurance services (Kagan 2022). It also removed the ban on simultaneous service by any securities firm officer, director, or employee at a member bank. The most damaging part of GLBA was the failure to give any agency the authority to regulate investment bank holding companies (SEC 2008). Banks voluntarily reported into the SEC. Self-regulation failed, as the bankruptcies of Lehman Brothers and Bear Stearns proved. This deregulation led to unchecked bank consolidation and greater risk-taking by them, two of which were the financial giants that went bankrupt. The mortgage-backed securities based on bad paper also were a symptom of this.

At the same time, the Double Irish tax tool, where corporations would incorporate in Ireland, use their Irish location to license intellectual property, and then indicate the command and control decisions were happening in the Bahamas, and turned Ireland into a tax haven (Holland 2021). In 2010, it shielded $100 billion annually in corporate profits (Holland 2021). One company was exposed as having $252 billion in untaxed cash (Paradise Papers Reporting Team 2017). The use of tax havens abroad, including Bermuda, St. Kitts and Nevis, and Dubai, also has helped hide true origins of money and the flow of it worldwide.

The Tax Cuts and Jobs Act of 2017 reduced the federal corporate tax rate from 35% to 21% (York 2018). This further increased corporate profits. The tax provisions in them indicate a maximum tax rate of 13.125% on repatriated overseas profits (Pomerleau 2018). This means that publicly traded corporations that rely on increasing net income have incentive to locate their profits in the United States, and little, if anything else.

With this, there's been an increase in telecommunications and data bandwidth. While corporations have had advantageous structures for moving around finances, they have also now had the bandwidth to move the actual technology work to different places for outsourcing. According to the International Telecommunications Union, there has been a 931x increase in internet bandwidth in India from 2007 to 2020 (ITU 2020). Over that same time, Viet Nam had a 1,106x increase, and China had a 119x increase. This means that more work was able to be moved there due to the significant increase in bandwidth. Outsourcing becomes a lot easier with the large amount of bandwidth available.

Additionally, large container ships that can hold over 20,000 20-foot equivalent units (TEUs) have been used since 2015 (Zeymarine 2020). These allow cargo to be shipped from Asia more cheaply than before and makes it more advantageous to outsource manufacturing. This is because the shipping costs are now low enough to provide a cost advantage by making products overseas and shipping them over given these massive ships. The Ever Given accident that blocked the Suez Canal for several weeks was an example of disrupting the global supply chain because of this. Also, it takes significant logistics to move all the TEUs from one of these ships to their ultimate destination. If there is one spot on the supply chain that has an issue, the overall delivery gets impacted. This includes the transportation of goods to and from the ports, the people and equipment that put them on the ships, the roads and rails to and from the warehouses, and the transportation from warehouses to stores and customers. All these need people, and they don't come as cheaply as Hiro Protagonist, the main character of Snow Crash. Neal Stephenson prophesized a dystopian America in that novel that did four things well: music, movies, microcode (software), and high-speed pizza delivery (Seiden 1993). Getting it from the ports to the pizza store is much different than getting it to the customers directly. Our current environment does not deviate that much from this, or idiocracy for that matter.

The gig economy, as exemplified by a number of modern day "unicron" companies, uses contractors who are paid per job to drive people in their own cars and deliver items they need. These contractors are not employees. They have to pay for their own insurance, cars, and supplies. This is a return to piecework, and their parent companies have not been considered liable for them as full employees. This has caused issues in countries such as the UK, where the courts upheld a ruling classifying UK rideshare drivers as employees (Lawrence 2021). The Biden administration's Labor Task Force in the United States issued a report in February 2022 that considers gig economy workers misclassified as independent contractors instead of employees, and seeks to remedy that (Desrosiers, Kupetz, and Roseman 2022).

We can also look at the economic model of Bootleggers and Baptists, which Diane Lim discusses in her article "'Bootleggers and Baptists' – How Crony Capitalism Has Captured Regulatory Policy for Centuries." This article discusses how groups with self-interests, such as bootleggers of liquor, align themselves with groups that share the same interests, and provide a cloak of respectability, such as religious groups that wish to see no liquor sold, to meet their shared ends (Lim 2015). What this does is that it actually reduces the ability to produce outcomes in the public interest by creating ones to protect the "bootleggers." In this chapter, this aligns with the existing hegemony. A more recent example of this is when a company uses smear tactics of planting news stories and false op-ed columns in newspapers across the United States through a consulting/lobbying firm to drive down the number of users. This is to maintain dominance in their platform space and was exposed as utilizing political means to do so (Associated Press 2022).

The leveraging of social groups as proxies to change policy to benefit private industry through the selective enforcement or loosening of controls is ubiquitous in America. The book *Bootleggers and Baptists: How Economic Forces and Moral Persuasion Interact to Shape Regulatory Politics*, by Adam Smith and Bruce Yandle, goes even further and demonstrates multiple scenarios (Smith and Yandle 2014). This has been successfully leveraged multiple times to give the cloak of legitimacy and morality to initiatives that diminish the public benefit and enrich private companies using religious and moral organizations as a proxy, including the removal of public benefits.

What these examples show is that over the past 51 years, there has been a movement to reduce corporate taxes and liabilities to increase profits. Retirements have been switched to third parties who invest in public companies. This can cause pressure on these companies to constantly find ways to increase earnings per share and reduce expenses. The constant reduction in liabilities over the past 50 years combined with outsourcing labor and manufacturing to cheaper places and using increased bandwidth and shipping capacity to reduce costs demonstrates this. The use of independent gig economy contractors to further reduce labor costs to the bare minimum also does. Software and the rapid sales cycles of it that don't meet customer needs are a microcosm of our economic push to make money at all costs. This encapsulates the management philosophy discussed in this chapter, and its underlying norms and values. We have switched from collaboration and quality to making money, being aggressive, and taking risks at all costs. People are just an additional cost to be shed so that our numbers can look better and our standing and income will improve. Security products make money, get investments, and are hot. Dollars matter, not the product itself, if earnings increase and cost of goods sold decreases according to analyst expectations. Security itself doesn't improve, and some can make the argument it's significantly regressed. The same can be said for healthcare IT, which has roughly the same issue with a large portion of products.

How Do We Misuse Metrics?

Organizations will often use metrics to hide issues. Metrics often use the law of averages for ones such as average project completion time, average service desk call length, or average time to repair a PC. The book *Noise*, by Daniel Kahneman, Olivier Sibony, and Cass R. Sunstein, discusses how averages can make an organization such as a court have an average sentencing time for defendants well within acceptable metrics and mask excessive sentences given by one judge.

When you look at this in terms of financial disasters, the 2008 one fell victim to averages. That was when mortgage-backed securities that were based upon very shaky finances, the equivalent of "junk bonds," were repackaged with higher-grade ones and resold on the market. This was one of the causes of the financial market crash. The average ones were high-grade, and the junk ones brought the value of the rest down and caused a cascading crash. To go back even further, the Tulip crisis of the 1600s was predicated upon the use of margin derivatives and the anticipation of paying 5–15% of value now for something whose price will always increase. Paying the rest of the money to sell the asset and make money off the top with tulips is like what people did with real estate.

Averages help leaders who want to hide their mistakes. They are good at telling the story of heroics and heroic myths by presenting a narrative that is biased toward the good outliers and hiding the bad in their noise. They are not good at finding issues so they can be resolved. Averages keep people grinding ahead looking to increase numbers rather than addressing root causes.

Good metrics use standard deviations to find and resolve issues. The standard deviation, according to Robert Niles, is a statistic that tells you how tightly all the various examples are clustered together around the mean (Niles 2022). It's the unit that measures the distance from it. If something is one standard deviation or more below the mean, it means that it likely significantly lags. Conversely, if it is one or more above the mean, it is significantly higher performing. This can be calculated for a series of numbers and metrics using Excel or a similar tool.

How this can help manage is by taking appropriate metrics and looking for the ones that are one standard deviation or more below the mean. This will identify problem areas to work on and resolve. One of the recommended metrics is resolution time of these issues. Another is to conduct risk analysis, such as failure mode and effects analysis (FMEA) or other ones, directly on processes to determine areas of high risk in process implementations. Using the metrics of issues discovered during process risk analysis and time to resolve presents a much more accurate picture of the environment and complications.

How Do We Start Down the Right Path?

How Do We Do This Right? How Do We Resolve These Issues?

This chapter has gone to some very bleak and negative places. It identified traditional narratives being used as dark patterns to keep people engaged, and how they lead to negative behaviors that can cause addiction and burnout. The goal of this section is to provide the tools needed to do it right and break the dark patterns that burn us and our teams out. These extend way past where we have traditionally gone in the past.

To do this, we need to focus on creating that environment of psychological safety. The March 2022 *Harvard Business Review*'s IdeaWatch section had a part on teams called "To Create Psychological Safety, Share Negative Feedback about Yourself" (Harvard Business Review 2022). This was about a study where team leaders were divided into four groups to gauge psychological safety. The first was instructed to ask team members about their performance. The second was instructed to discuss development areas from their own performance reviews. The third was asked to do both, and the fourth neither of these. After a year, the groups were surveyed on psychological safety. The second group had significant gains. The reasons, as explained in a series of subsequent interviews, were because it took time to build that level of safety needed and build the virtuous cycle. We need to build that credibility. This is not going to be something we can build in a quarter. It takes years to do this right, and if you rush it, it will backfire on you. Hustling and appearing busy is the exact opposite of this.

Start with identifying the organization's core values. According to Northern Ireland Business Info, they are the standards that guide the way you do business. They encompass what your business stands for, influence the culture, and determine the how and why you do things (Invest Northern Ireland 2022). These are also the main methods by which you articulate and elaborate how you do business in the frame of the corporate mission. The mission statement is also key, however, it needs the core values to support it.

The concern is that many companies don't live or demonstrate them. We've all seen the headlines. Chery Bachelder, in her blog entry, "The Honor Code of Business" (Bachelder 2016), discusses how one bank not following their values caused significant corporate discord, damaged relationships, and damaged team member morale. Team members see when this happens. Leaders often do not think they do, and people will leave companies over this.

We also need to get over ourselves and the need to continually dominate and win to look better. Of the 11 norms identified in the CMNI, values aren't there. Focusing on winning, dominance, and pursuit of status, while they are associated with profits, success, power, influence, and ability, can backfire if we don't live our values and burn people out. People see themselves as being used to make others look better and say they are living the values at the expense of making life better for the executives on top.

We need to check ourselves on negativity and paranoia. We're passing down our paranoia and insecurities to our teams, and it's causing them stress they do not need. It can push them over the edge. We can do better than this. Good leaders model resilience and values and demonstrate those, not the negatives. Even one or two leaders doing this can poison an entire organization, as examples in this chapter demonstrate. We need to check each other and our teams. There are aspiring leaders who do this, and it's because they follow what they've been taught formally and informally. We must make sure they don't propagate it or other behaviors they may have learned elsewhere.

Defer to the security and technical expertise on your teams. You don't know security as well as they do. Just because it's in a publication that you read, most likely written by product marketing teams, does not mean that you know as much as someone that has been a practitioner for years. Let them participate and lead to provide that expertise instead. One of the ways that senior executives can lose credibility is by bringing out the same old tired numbers and statistics about security and data breaches. Your team knows more. Let them demonstrate it and give them the opportunities and guidance to do so.

Set initial expectations with excellent service built in like the Big 4 consultants with your engagements. Service delivery is critical. It also helps with planning out resources and needs. This is important to set borders and service expectations first. Doing it cheaply and with a plan to focus just on lower price signals that you will race to the bottom and don't value your team. According to Brian Jackson's article, "How a Race to the Bottom Hurts Your Business's Bottom Line," there are three negatives that this will destroy your business and the people within it, and potentially burn them out (Jackson 2021):

- This kills innovation and the environment team members need to feel engaged and safe enough to do so. In the security world, this means not creating good solutions to meet customer needs and putting solutions in place that won't adequately protect or work well. If the security is bad, either clients will get hacked or they will remove solutions themselves because they kill business.
- It means the brand is cheap. Being cheap means that you sacrifice quality. Brian's article cites a previous *Harvard Business Review* flowchart that indicates the overwhelming answer is No if you want to compete on low prices. You have to find the right price for your value or you won't be valued. And you won't value your employees.
- It puts your future at risk. You won't be able to invest in it. If you raise prices (or wages), you'll lose customers and business.

Going back to the Cycle of Software Disillusionment, a lot of companies compete on price only and don't discuss value. When they see that a competitor is offering more features at the same price, they try to promise more features, or they cut out security and integration as line items to make the price more attractive. When they do, they pull resources from all over the company to try and make it work. When this happens, often so does failure. The client is left with a negative opinion of the software and likely bad security. The next thing that happens is they don't renew. However, the salesperson that made that deal thinking of their commission check is already on to making promises they can never keep at their next employer, and the security team that was brought in last-minute is left holding the bag. After enough times of having their employer treat them like a zero-cost add-on that doesn't provide value, they burn out and leave.

The same story has repeated numerous times with companies that ship products with numerous known bugs and issues. The previously referenced OS/360 story is true to the mantra that if you try and do everything, you will do nothing well. Focus on core features that align with the mission and values, and make sure that they are done correctly. If you are cheap, try to squeeze features in, and forget security, you will develop bad products. Your security team will burn out, and they will leave. The rest will follow. You will damage them in the process. We discussed what people and multitasking operating systems have in common earlier. People take a long time to switch back and forth between tasks, like an operating system having to swap items in and out of memory. We add time and increase the probability of errors. Like OS/360, these will hit critical mass. Sometimes a single task or distraction overloads us. Computers can be rebooted and services restarted. We can't.

Stand up for your security team. There are a lot of people who place blame on others to make themselves feel better and assert their dominance. If you give in to their blame and actions, it means you don't care about your team and will sell them right out to make yourself look better in front of a bully. Don't be an acolyte or a toady that blindly cheers on toxic behavior. If you acquiesce to them it means you don't value yourself or your team, and only care about yourself and getting ahead. It also means that you don't care about security, only looking good for people that could care less about you because they only care about themselves.

Build real stories. Don't buy into trying to build corporate myths and legends to support your ego and team. Customers don't care about that. They would rather have the time spent meeting their needs and building something they want, not a foundation for your ego. If you align with mission and values, your team will want to build something together. Let what you build and support be the story, and your authenticity and vulnerability support it. Forget the stories and attempts to bolster egos and focus instead on how to improve. Security is about building, collaborating, and improving processes. It's not about building monuments.

Treat your team members like people, not resources. This is a simple lesson. You cannot parallelize sequential processes and make a baby in a month. We don't multitask well, and neither do many business processes. We have other responsibilities outside of work. Trying to draw on addictive patterns and behaviors to keep people engaged will spectacularly backfire. We need to think of our team members' well-being first. Figure out what they can do within the time they have, and only reserve anything over full resourcing for when it's really needed. Communicate with them and try to demonstrate appreciation for the work they do. Help them do better and improve as part of that. Reduce the distractions to an absolute minimum. Remember, a high SNR reduces bandwidth and mental capacity significantly! You don't buy a 12th Generation Core i9 to run DOOM at the same speed as your parents' 386/40 in the garage. Yet when you employ these patterns, this is the level of effectiveness you have.

If you have neurodivergent team members, work with them to understand them. Give them the space to not have to mask themselves continually. Communicate about their needs frequently. Make sure they don't feel like outsiders. Make that extra effort to include them and explain norms. Watch out for workplace bullies and toxic leadership attempting to use, take advantage, or push around your team members, especially your neurodiverse ones.

We can use two important psychological techniques, which are coping techniques and resiliency, to help our team members become more mindful and address the cues that can cause negative feedback and burnout. One of the techniques that works well is a coping technique for anxiety. According to the Mayo Clinic Health System, this exercise can work.

This is also one we have put in materials we've developed for our team and is like one a team member uses with their children. The exercise is to sit quietly, look around, and notice (Mayo Clinic Health System 2020):

- Five things that you can see
- Four things that you can feel
- Three things that you can hear
- Two things that you can smell
- One thing that you can taste

The other technique that we use is based on ACT, as discussed by Stephen C. Hayes in his book, *A Liberated Mind* (Hayes 2020). We like ACT because it teaches how to mindfully address events that can cause negative cues and feedback. ACT also aligns with values, which are important to ground positive actions on (Hayes 2020). While we could likely spend an entire book writing about its benefits, we recommend those who are interested to read this book and speak with an expert in the field first.

The techniques from ACT that we use are as follows (Hayes 2020):

- Cognitive diffusion – A way of encapsulating and viewing events and memories indirectly to better understand them mindfully
- Self-pivot – Pivoting from self-interest to a greater connection with others
- Acceptance – Learning from our pain
- Presence – Living here in the now
- Values – Giving ourselves greater meaning by purpose and choice
- Action – Committing to change and a greater sense of life competence

They help combat the challenges from addictive behavior and give people the space to mindfully make changes that align with a good set of values. Being able to address and redirect negative cues and distractions helps keep the people we care about engaged. We want to help make people more resilient to a hostile environment and distractions, and to be able to mindfully address them. Distractions and multitasking, especially caused by anxiety, notifications, and negative thoughts, can derail someone, make them less productive, and give them negative cues that cause negative cycles and burnout. We don't want that for our team members. Security is challenging enough. We need to build up our people to address these challenges as part of leadership, instead of building a fire that burns them out.

Giving our security people tools to cope and be more resilient gives them time back. They need this time to think and absorb their environment and stimuli to understand what is going on and process it. When you don't give people time to process items, especially with multitasking, they will make unforced errors. We need to enable positive flow, not negative flow. The latter can lead down a very dark path for many, and their failure will be on us.

Another part of that is that we need to not force paperwork or processes on our teams when we need them to fix something. One of the major challenges of working in security are the byzantine processes, procedures, and forms that people create for the purpose of minimizing the work they must do. What security teams see is work avoidance through bureaucracy. While that might have suited Dr. Sheldon Cooper on the Big Bang Theory when he wanted to get his way by drowning people in paperwork, it doesn't help fix issues. As Joan Osborne asked in the song Dracula Moon, "What if the cure is worse than the disease?"

The numerous labyrinths that security teams need to traverse to address changes openly discourage people from finding, reporting, or fixing issues. Bad management that openly threatens penalization or arrest for discovering and reporting issues does not help. An example is when one public official threatened a reporter in October, 2021 that disclosed a data breach on a website (Krebs 2022). Overly cautious change management or complex processes only help the teams that want to minimize their work. They do not align with values. Neither do threats for disclosing security concerns.

Leadership's job to reduce burnout is to facilitate addressing these issues and preventing other teams from hiding behind change management, help desk tickets, and toxic management. A security leader handles the process management and prevents their team from having to play these games while appropriately prioritizing work. These issues cause distractions and negative cues, which have multiple negative effects on their work and lives. Our team members are smart, and Dr. Cooper is a fictional character who admits that the purpose of his complex contracts and agreements is to benefit him. Don't facilitate people like Sheldon. Cut them off and don't give into their games. They do it for their own benefit and to make themselves look good.

We all work together on a team, although some would like to believe that they work in silos and that their work prioritizes all others. We need to better understand that we all work together and plan for that. A team that doesn't plan and always imposes on others will cause multitasking and distractions, and will lower the quality of work. They can easily turn an entire organization toxic. We need to avoid the tragedy of the commons and watch for abusers. Abusers ruin companies if you let them. It only takes a few to destroy the work of many. The mistrust built by this also prevents security teams from working effectively.

Accountability and transparency with issues, especially with the ones that matter, also help. Addressing abusers of the commons takes communicating transparently about responsibilities, workload, and needs. It also takes being open to change, not blaming others, and being vulnerable. Addressing issues with these behaviors, instead of through bullying and intimidation, sets a model for leaders to follow, and helps build better relationships instead of more of the same toxicity. Security needs to espouse these values and collaborate, as opposed to being demanding and condescending.

Clear and transparent communication is a must. Unclear communication and being secretive leads to paranoia and suspicion. We have to be direct, answer questions, and align with the values in what we do and say. We must do this with our own teams and with others. There is often suspicion of outsiders, and we need to break this down. Propagating paranoia will not advance us. Meeting needs and being clear and open with our customers will.

Collaboration is also important. While our teams are important, they are not the whole company. No one group is the entire company. Cliques and exclusivity can ruin an organization. We actively must search for and mitigate these. Including high achievers and people like the leaders in special groups is this. We need to focus on working together and actively catching ourselves when we practice these behaviors. Align with the values, clearly communicate, and work toward shared goals that demonstrate a better way.

We're expected to model the values. They are a moral and corporate benchmark. We constantly must appeal to them and who we are here for, not for someone's advantage. While we don't agree with it, there are people in management or leadership positions who will twist rules to their advantage. Extreme cases and toxic environments can cause corporate conspiracy theories.

We need to further build resilient defenses against burnout through reasonableness. We need to avoid chaos and overloading. The security frameworks that we speak of and espouse are good for addressing technical and management controls. They're not designed for people, their needs, or resource management. We don't get good technical or management controls without those.

The goal of this is to devise a plan against burnout that encompasses a framework like the ones we gauge the effectiveness of our management and technical controls. We burn ourselves out to be compliant with NIST, HIPAA, PCI-DSS, management demands, other frameworks, or buzzwords. We don't have one to protect ourselves against burning out doing so. One of the major causes of burnout is management. We want to provide a way to incorporate the lessons learned into better management techniques.

Protection and Environmental Considerations Framework

The purpose of this framework is to provide more than just another book chapter leaving an open-ended solution for a problem. Information Security, and Information Technology team members in general, need repeatable processes and procedures for addressing initiatives and projects. Instead of just having yet another framework that covers security or process management, this covers overall risk management of people, processes, and technologies. It also covers financial management those other frameworks do not.

Leadership needs to take a more active role in reducing burnout. The best way to make this happen is to map out what's needed to integrate anti-toxicity measures, values-based alignment, acceptable norms, team member resilience, and multiple levels of planning to integrate existing security frameworks into a plan that they can utilize.

With that, let's get into the details below:

1. *Statement of Values.* Each initiative has a statement of values and how they align with the greater organizational values. We need to understand why we are completing initiatives, and why they are important.
2. *Establishing Acceptable and Unacceptable Norms.* Writing down and discussing the acceptable and unacceptable norms and expectations of team members. We recommend using the norms in the CMNI as unacceptable.
3. *Set Team Member Roles and Expectations.* Clearly map out who will have what primary roles, what their expectations are, and how performance will be initially evaluated.
   a. Align the work with team members' development plans.
   b. If your team members don't have one, make one for them. We cover this later on.
4. *Establish Internal Communications and Norms.*
   a. Set expectations that the use of heroic myths and extensive informal communication as a form of enculturation is forbidden.
   b. Set expectations that overly distracting people is not allowed.
   c. What methods will be used for this communication?
   d. What communications will go out to the project team?
   e. What communications will go to leadership?
   f. Who are the involved stakeholders?
   g. Who are our collaborators?
   h. What communications will go to them?
   i. How do we minimize the absolutely necessary?
5. *Establishing a Clear Stance against Toxicity.* Project executive leadership and sponsorship discusses toxic behavior and provides a means for team members to report it in.
6. *Developing and Reinforcing Coping and Resilience Mechanisms.* Discussing the use of these mechanisms with team members that need this behavior reinforced and providing them the options for further assistance.

7. *Understanding the Customer.* The OCTAVE framework emphasizes interviewing people at three layers of the organization: Executive Leadership, Line Management, and Team Members (Alberts, Behrens, Pethia, and Wilson 1999). Structured interviews that help answer the following objective types are a must to align with them to greater increase the probability of success:
   a. Understand what they hope to get out of the initiative, and their objectives.
      i. What are their goals?
      ii. What are their needs?
   b. Understand their business.
      i. What do they do?
      ii. How do they do it?
      iii. How are they structured and how do they operate?
   c. Understand their market segment and constraints.
      i. What drives them?
      ii. What are the norms and practices in their area?
   d. Understand their values.
      i. What are their values?
      ii. How do they align with the rest of the organization and with you?
      iii. How can you build shared values?
   e. Understand their metrics.
      i. What defines success for them?
      ii. How can you help contribute to improving their success?
   f. Understand their financial requirements.
      i. How do they report?
      ii. How do they budget?
      iii. How do you align with these to provide the information they need with a minimum of distractions and potential errors?
   g. Understand regulators and regulations.
      i. Who evaluates and regulates them?
      ii. What are they looking for? Examples are the Joint Commission (JC) or College of American Pathologists (CAP) in hospitals.
      iii. What standards do they look for?
      iv. What are they being audited on?
   h. Understand their security and security requirements.
      i. What are their existing policies, standards, and processes?
      ii. How do they provision systems and users?
      iii. What happens when a user transfers departments?
      iv. How do they log/audit?
      v. How long do they retain logs for?
      vi. How long do they need the data from the systems you are working on for? How to delete it?
      vii. How do they destroy data and systems when no longer needed?
      viii. How do they respond to events?
      ix. What is their staffing?
      x. How do they manage third-party risk?
      xi. Who is managing security for them?
      xii. What are their needs for success?

8. *Defining Technical and Architectural Standards.* Leveraging what we learned in Understanding the customer, define appropriate technical and architectural standards to follow for the initiative.
9. *Defining Security Frameworks.*
   a. What security frameworks and guidelines, such as PCI-DSS, IEEE, ISO, HIPAA, or NIST, apply to this?
   b. What ones do we have to use?
10. *Scope the Project.* Using knowledge gathered in the previous steps, build a project scope to understand how to build the project plan.
11. *Build the Project Plan.* There is significantly more material out there about building project and management plans, and we recommend using those in conjunction with the previously gathered materials to do so. Incorporate security framework requirements into here.
12. *Develop Financial Plans.* Map out the initiative costs and anticipated post-go live costs according to the plans.
13. *Develop External Communication Plan.* Answer the following questions:
    a. Who are the stakeholders that need to be communicated to?
    b. What information do they need to know?
    c. What are the best methods for conveyance?
    d. How often do they need to be communicated with?
    e. Who is assigned to communicate?
    f. Who needs to approve communications?
    g. How do we reduce it to what's absolutely necessary?
14. *Develop Operational Plan.* Build the plan for ongoing success by using knowledge acquired in previous steps to map out roles, responsibilities, and allocations for managing and maintaining what's being developed. Leverage automation technologies as much as possible to reduce the probability of errors. Incorporate security framework requirements here.
15. *Find Potential Failures.* Use FMEA to run through the operational plan and define where there can be areas of improvement. The Institute for Healthcare Improvement has significant resources available at http://www.ihi.org/resources/Pages/Tools/FailureModesandEffe ctsAnalysisTool.aspx to implement this on your own (IHI 2017). Use this step to identify exception conditions so that you can monitor for and develop plans to address them.
16. *Refine Plans.* Refine Project and Operational Plans based upon findings from FMEA or similar analysis. Incorporate lessons learned to build a much better plan to iterate on.
17. *Develop a Security Plan.* Build an ongoing security management plan that addresses their needs discussed before which plugs into what they have, augments it, and helps them respond to exception conditions better. Security frameworks need to be incorporated into this step.
18. *Develop Exception Management Plans.* Work with the customer to develop plans on how to manage and address exceptional conditions or failures, both security and non-security related. With the use of intelligent systems such as artificial intelligence or machine learning, knowing how to deal with these cases is critically important as well. Manage alarms and alerting to minimize false alarms and distractions.
19. *Develop Ongoing Support Plan.* Plan out resources and teams needed to successfully maintain the service, and what they will be doing.
20. *Develop Controls Tests, Management Metrics, and Testing Plans.* Identify controls and thresholds required for meeting successful operational management and audit criteria. Develop a plan to periodically test and measure them, and steps to refine them if below thresholds. Identify how the customer wants them and present them in the format they need.

    a. As part of those control tests, identify metrics that fall one standard deviation below the others to better focus on where to improve first.

21. *Develop Annual Review Plan.* Have a plan to review user entitlements and access at least annually, which is one of the most common requirements across regulations such as HIPAA.

22. *Dedicate Time to Engineering.* Look at the time schedules of technical resources and engineering on the team and reduce their distractions in coordination with other leadership. Don't over-allocate resources as they will get burned out and accomplish much less.

23. *Ensure Team Members Get Time Off to Not Work and Not Be Distracted with It.* People need time to recharge. They need time to exercise. They need time to eat somewhere other than their desk.

24. *Make Sure to Reduce Overall Distractions to a Minimum to Increase Your Team's Bandwidth.*

The Roots of Burnout

Burnout in Information Security has multiple root causes. Most of them don't have to do with security. They start with a firm grounding in psychological norms and how they extend to management. The toxic narratives that many members of management create, specifically around heroic myths, and trying to build themselves up to look good, create a parallel culture inside organizations. This culture is focused on propagating their hegemony and ensuring others like them reap the rewards and stay on top. The people in charge want to make sure that only those like them are the ones who get ahead, and they use multiple formal and informal means to enforce this.

This culture takes groups, starting with the best and brightest, isolates them, gives them significant distractions, and uses informal communication, heroic myths, and stories to keep them grinding ever harder toward ever-increasing results and earnings at the expense of overload and burnout. This can result in multiple negative cues and the magnification of addictive behavior. Both the incredibly hard work and addictive behavior can lead to crashes and burnouts. People who burn out and leave because of this are often derided as not having character, not being tough enough, or weak.

Information Security teams often have very talented people who must understand multiple facets and layers of the business to be effective. Unlike the "best and brightest" they want to transition into positions of power, they make the work environment significantly more challenging in multiple ways, specifically by not including them as part of organizational management. Leadership challenges with management skills, not understanding security, hiding information, value engineering, and not providing an environment of psychological safety all contribute. This can also lead to significant negative cues, burnout, and exacerbate addictive behaviors. We also do not prepare people to deal with these numerous distractions and give them the tools they need to build this environment of safety, especially our team members that are neurodivergent.

This takes active work, starting with senior leadership, to address. The current security frameworks only deal with compliance and methods to meet their requirements. What we propose in this chapter is a management framework that intertwines required security frameworks and draws some pieces from the original CMU OCTAVE framework to do so. The goal here is to provide leaders a better way to implement the non-technical tools they need as part of an overall toolset, not just a set of disjoint processes that contravene each other. We need to have better methods to address this because leadership in general has not.

Burnout is real. Why do we push people over the edge? What can we do to mitigate this? It's not enough to pick up a book or go to therapy and think the problem is solved. It's not enough to go to a 12-step program and work through the steps continually without understanding the deeper meaning behind them other than trusting a higher power. Blind trust only goes so far. However, it doesn't help you move ahead meaningfully and with better comprehension.

There's always one layer down that can be examined to get a better understanding of why these events occur. In writing this chapter, one of the items that came out was cynicism on how, despite multiple initiatives, we have a regressive management culture that is based upon the same heroic myths told around the fire countless generations ago. It is designed to only keep the ones who can survive the overload. We've evolved in numerous ways since then. We can do better than taking our best and putting them in business cultures that burn them out and exacerbate significant negative behaviors. We need to be leaders and improve this. Part of that is taking negative experiences and using them to provide the people reading this with tools to create positive ones for their teams from them.

You are not perfect and will not be perfect. You will make errors, some of them egregious. You need to own them, address them, and move forward. You're not an Instagram meme of Cillian Murphy or The Rock with someone else's hustle quotations pasted in. You're not some millennial bragging about their 24/7 hustle and grind lifestyle on LinkedIn. You're you, and no number of memes is going to address these issues the way that concerted action will.

What Can You Do for Others?

You are a leader and can help others out. The situation with Information Security and burnout has been poorly understood by organization leadership, either by choice, unwilling ignorance, or willful ignorance. Whether it occurred or not is not the question. It has. We have a choice to do the right thing by the people who report to us. If you're a C-suite executive, CIO, CISO, Director, or Manager, this is aimed directly at you. We have too many people in leadership positions who aren't doing what's needed to help their teams. This is accelerating burnout with them and decreasing the quality of their lives. We're going to break this out by group, so you can tailor your responses. The most important thing to keep in mind is that you need to evolve as a leader to avoid further damage.

It's never too late to fix ourselves so we can do better by others. We let ourselves fall into the trap of trying to satisfy others and fit ideals that we read about on social media, saw at a conference, or watched a video about. We need to realign ourselves to what our people need, and what we need to mitigate risks. We don't need to demonstrate competence by our conformance to measures that ultimately do not communicate value or matter.

The days of advertising job postings requiring five years of experience and a CISSP for minimum pay and maximum hours need to come to an end. We're asking too much for too little, and we're putting our team members in a position where they will fail.

Align with values. Stop with the heroics and trying to build or reinforce heroic myths. Build on shared values and align with them. Also, no product you buy is going to turn you into a hero, stop people with obscured faces in hoodies, or save you from well-funded FSB members. Stop trying to benefit yourself at the expense of others and benefit your team instead.

Team members need a realistic path forward. This means that you need to have career plans that cover training courses, conferences, certifications, and requirements for advancing in their careers. Instead of hiring warm bodies to watch monitors and pew-pew maps all day or night, we need to treat them like valued people. Approach their careers with a continually updated plan that incorporates:

■ *Career Path* – Where they are, where they want to be, and what they need to get there. Have the difficult discussions about what exactly is required to advance. Define what is needed and make it quantitative.

■ *Individual Development Plans* – Develop customized individual plans for team members that include technical training, leadership training, presentation/speaking experience, writing experience, and higher education.

■ *Goals and Actions* – Always set definitive SMARTEST goals with actions and requirements that have intent and definitive actions for team members to take afterward. Unclear direction can cause minds to race and inadvertently burn out.

  – **Specific**: Are your goals targeted, with numbers or other measurable metrics behind them?

  – **Meaningful**: Do they motivate you enough to make you want to push yourself to achieve them?

  – **Achievable**: Are these goals realistic, concrete, and able to be completed within a defined time frame?

  – **Relevant**: Are they relevant to your life, work, and what you want to improve about yourself? Do they align with any of the 12 qualities you want to improve?

  – **Time-bound**: Are you able to set an exact date for goal achievement? You need to make them measurable.

  – **Empathic**: Is the goal considerate and understanding of the needs of others, or is it focused just on one person? Does it focus on putting you above others because of something you know or perceive puts you above them, or understanding them better as equals?

  – **Synchronized**: Are they aligned with the specific qualities you want to improve, and/or with further improving your positive qualities?

  – **Tracked**: Are they written down, reviewed, and measured on a daily, weekly, and monthly basis toward their goal?

■ *Mentoring* – Ensure that your less experienced team members have mentorship opportunities available to get the formal and informal communication channels they need to learn to succeed from more experienced team members in multiple areas.

■ *Reverse Mentoring* – We need to have our senior team members hear what is really going on from other team members. Having less experienced team members work with them to help them understand their perspectives and get feedback can help them improve. This can also help leverage informal communication models to better inform others of the organization's norms.

■ *Advancement Preparation* – Provide the leadership training and planning needed for team members to advance. Don't just give them TED videos to watch or books to read. They can do that on their own at home. Give them definitive training on management and leadership before they need it so they don't fail at their next job.

■ *Required Certifications* – Define the ones needed for the job and why they are specifically needed.

■ *Conferences* – Which ones are worth the time to attend, and what courses/training they have there that is of benefit.

■ *Training Courses* – What courses are worth pursuing at a reasonable pace that will get them where they need to be.

■ *Coping Techniques and Resilience* – Give them the tools they need to be able to mindfully address challenges they have.

■ *Distractions* – Reduce distractions to allow them to concentrate on what they need to, not constantly barraging them.

■ *Communication* – Check in to make sure that they can address negative cues using Coping Techniques and Resilience. You also need to make sure that your communications are value-based, meaning heroic myths and preferential communication to high achievers to enculturate them need to go away. You can use the BRAVE method we've developed to help monitor for better communications:

- **Bias**: Is what we are saying or thinking something that can be considered prejudiced against someone or something else?
- **References**: We need to always remember to reference others, not ourselves. No one knows our frame of reference. Always communicate with an outward, empathic frame.
- **Appropriateness**: Is it appropriate and aligned with social norms and values? Is it something that may make someone feel uncomfortable?
- **Values**: Is what we are communicating aligned with our values?
- **Emotion/Empathy**: Are we showing an excess of emotion? Are we showing empathy to who we are speaking with?

■ *Communication* – Make sure we are communicating effectively, without malice, and with the intent to improve in line with our values. We can use the THINKER method to gauge this:

- **T: Is it True?** Is this a fact or is it an opinion?
- **H: Is it Helpful?** Does it help others to say this? Does it help the situation? Or does it just make you look good?
- **I: Is it Inspiring?** Does this improve the situation?
- **N: Is it Necessary?** Do we really need to say this?
- **K: Is it Kind?** Is this something that could unnecessarily hurt someone or give someone a bad opinion of you for communicating it?
- **E: Is it Empathic?** Does it show empathy toward who you are communicating with?
- **R: Is it Relevant or Repeating?** Is this an appropriate response to the situation at hand? Are we repeating something else someone said? Does it align with our values? Did we think of what someone else said or did we blindly repeat or repost it like a social media meme?

■ *Implicit Biases and Microaggressions* – Get over the urge to try and find the people most like you. Don't use myths, communication, and microaggressions to try and filter out people unlike you.

■ *Employee Assistance Programs* – Make these available to team members and erase the stigma behind them by talking about their benefits. There is too much of a stigma associated with these, and we must erase that.

■ *Gratitude* – Always show gratitude and be thankful for your team members. Don't take one day a year because a group is being honored to show it. Those days mean nothing if you don't do it every day for everyone. Always find something good to do that benefits others. Words and days mean nothing to people without committed actions to demonstrate appreciation and gratitude daily.

■ *Alumni* – Your former team members are your best advertisements. You need to maintain great relationships with ones that leave for better opportunities.

■ *Errors* – You will not be perfect. You will mess up. Own it, apologize, and do better.

Conclusion – We Can Do Better, One Step at a Time!

We have a very realistic problem in security that affects us at all levels. We need to do a better job to address it. The goal here is to make sure that we build in the mechanisms at multiple levels to combat this before we contribute to the destruction of our teams, their customers, and the organizations that we work for. Security is at the forefront because of visibility. Addressing these issues here first can help address them for the rest of the organization, and hopefully healthcare. We're here to help people live better lives, not destroy them by being toxic jerks. We need to look into the mirror to see what we can improve before we can improve it.

Take that first step. Read through this. There is no stigma in seeking help. Everyone has low points in their lives and seeking assistance for the causes is not a sign of weakness. Trying to pretend everything is OK and continually hustling toward impossible goals will hollow you out inside and leave a shell that easily cracks and crumbles. Just keep moving. It's a hostile, crazy world. If you keep moving one step at a time, you'll make progress. It's not easy. It takes a lot of introspection and work; however, you and others are worth it. Don't let narcissistic tendencies take over.

References

"401(k) Plan Research: FAQs." Investment Company Institute. Investment Company Institute (ICI), October 11, 2021. https://www.ici.org/faqs/faq/401k/faqs_401k.

"About Tesla." Tesla. Tesla, Inc. Accessed May 12, 2022. https://www.tesla.com/about.

"Burn-out an 'Occupational Phenomenon': International Classification of Diseases." World Health Organization (WHO), May 28, 2019. https://www.who.int/news/item/28-05-2019-burn-out-an -occupational-phenomenon-international-classification-of-diseases.

"Burnout and Addiction: What's the Link?" Ampelis Recovery, April 7, 2021. https://ampelisrecovery.com /burnout-and-addiction/.

"Business Values." What are Company Values? Invest Northern Ireland. Accessed May 13, 2022. https:// www.nibusinessinfo.co.uk/content/what-are-company-values.

"Chairman Cox Announces End of Consolidated Supervised Entities Program." Press Release: Chairman Cox Announces End of Consolidated Supervised Entities Program; 2008–230; Sept. 26, 2008. Securities and Exchange Commission (SEC), September 26, 2008. https://www.sec.gov/news/press /2008/2008-230.htm.

"Countdown to Make Anxiety Blast Off." Mayo Clinic Health System, June 6, 2020. https://www.mayocli nichealthsystem.org/hometown-health/speaking-of-health/5-4-3-2-1-countdown-to-make-anxiety -blast-off.

"Cybersecurity Almanac: 2021." Momentum Cyber, Inc. Accessed May 13, 2022. https://momentumcyber .com/cybersecurity-almanac-2021/.

"Economic Recovery Tax Act of 1981 (ERTA)." Corporate Finance Institute (CFI), February 3, 2021. https://corporatefinanceinstitute.com/resources/knowledge/economics/economic-recovery-tax-act-of -1981-erta/.

"Failure Modes and Effects Analysis (FMEA) Tool: IHI." Institute for Healthcare Improvement, 2017. http://www.ihi.org/resources/Pages/Tools/FailureModesandEffectsAnalysisTool.aspx.

"Hacker Culture." Subculturelist.com. Accessed May 12, 2022. http://subcultureslist.com/hacker-culture/.

"Highlights of the Tax Law Signed by President Carter." The New York Times, November 9, 1978. https://www.nytimes.com/1978/11/09/archives/highlights-of-the-tax-law-signed-by-president-carter -individual.html.

"IdeaWatch." *Harvard Business Review* 100, no. 2, 2022.

"Know the Signs of Job Burnout." Mayo Clinic. Mayo Foundation for Medical Education and Research, June 5, 2021. https://www.mayoclinic.org/healthy-lifestyle/adult-health/in-depth/burnout/art -20046642.

"Non-Current (Long-Term) Liabilities." Chartered Financial Analyst (CFA) Institute. Accessed May 13, 2022. https://www.cfainstitute.org/en/membership/professional-development/refresher-readings/ noncurrent-long-term-liabilities.

"Qanon." Anti-Defamation League (ADL). Accessed May 12, 2022. https://www.adl.org/qanon.

"Shannon and Weaver Model of Communication." Communication Theory, July 10, 2014. https://www
.communicationtheory.org/shannon-and-weaver-model-of-communication/.

"Shannon's Theorem." Free University of Berlin. Accessed May 12, 2022. http://www.inf.fu-berlin.de/lehre
/WS01/19548-U/shannon.html.

"Signal-to-Noise Ratio (SNR) and Wireless Signal Strength." Cisco Meraki, October 5, 2020.
https://documentation.meraki.com/MR/WiFi_Basics_and_Best_Practices/Signal-to-Noise
Ratio(SNR)_and_Wireless_Signal_Strength.

"Software Evolution – Mitosystems – Mitopia Technologies." Mitopia Technologies. Accessed May 13,
2022. https://mitosystems.com/software-evolution/.

"Statistics." International Telecommunications Union (ITU). Accessed May 13, 2022. https://www.itu.int
/en/ITU-D/Statistics/Pages/stat/default.aspx.

"Steve Jobs and Steve Wozniak." Lemelson. Massachusetts Institute of Technology (MIT). Accessed May
12, 2022. https://lemelson.mit.edu/resources/steve-jobs-and-steve-wozniak.

"Subject Guides: The Monomyth (the Hero's Journey): The Hero's Journey." The Hero's Journey – The
Monomyth (The Hero's Journey) – Subject Guides at. Grand Valley State University, June 10, 2021.
https://libguides.gvsu.edu/c.php?g=948085&p=6857311.

"Subreddit Stats." Subreddit Stats – statistics for every subreddit. Reddit. Accessed May 12, 2022. https://
subredditstats.com/r/antiwork.

"The Psychology of Human Error 2020." Tessian, Inc., March 29, 2022. https://www.tessian.com/research
/the-psychology-of-human-error/.

"What is Wi-Fi 6?" Intel, Inc. Accessed May 12, 2022. https://www.intel.com/content/www/us/en/gaming
/resources/wifi-6.html.

Alberts, Christopher J., Sandra Behrens, Richard D. Pethia, and William R. Wilson. "Operationally
Critical Threat, Asset, and Vulnerability Evaluation (Octave) Framework, Version 1.0." Carnegie
Mellon University – Software Engineering Institute, September 1999. https://resources.sei.cmu.edu/
library/asset-view.cfm?assetid=13473.

Arsenijevic, Olja, Dragan Trivan, and Milan Milosevic. "Storytelling as a Modern Tool of Construction
of Information Security Corporate Culture." *Ekonomika* 62, no. 4 (2016): 105–14. https://doi.org/10
.5937/ekonomika1604105a.

Bachelder, Cheryl. "The Honor Code of Business." *Serving Performs*, November 1, 2016. https://www
.cherylbachelder.com/the-honor-code-of-business/.

Balkeran, Arianna. "Hustle Culture and the Implications for Our Workforce." *CUNY Academic Works*,
June 8, 2020. https://academicworks.cuny.edu/bb_etds/101/.

Barrows, Paul, and William Van Gordon. "Ontological Addiction Theory and Mindfulness-Based
Approaches in the Context of Addiction Theory and Treatment." Religions 12, no. 8 (2021): 586.
https://doi.org/10.3390/rel12080586.

Berdahl, Jennifer, Peter Glick, and Marianne Cooper. "How Masculinity Contests Undermine
Organizations, and What to Do about It." Harvard Business Review, November 2, 2018. https://hbr
.org/2018/11/how-masculinity-contests-undermine-organizations-and-what-to-do-about-it.

Berkowitz, Reed. "Perspective | Qanon Resembles the Games I Design. but for Believers, There Is No
Winning." The Washington Post. WP Company, May 11, 2021. https://www.washingtonpost.com/
outlook/qanon-game-plays-believers/2021/05/10/31d8ea46-928b-11eb-a74e-1f4cf89fd948_story.html.

Bianchini, Riccardo. "The Different Fate of Apple's Lisa and Macintosh (and Why Design Matters)."
Inexhibit, November 1, 2019. https://www.inexhibit.com/case-studies/different-fate-apples-lisa
-macintosh-design-matters/.

Blest, Paul. "An 'Atrocious' Email Caused a Mass Resignation at a Kansas Applebee's." VICE, March 29,
2022. https://www.vice.com/en/article/4aw8p9/kansas-applebees-leaked-email.

Burg, Dave, Mike Maddison, and Richard J. Watson. "Cybersecurity: How Do You Rise above the Waves
of a Perfect Storm?" EY, July 22, 2021. https://www.ey.com/en_gl/cybersecurity/cybersecurity-how
-do-you-rise-above-the-waves-of-a-perfect-storm.

Burns, Tiffany, Jess Huang, Alexis Krivkovich, Ishanaa Rambachan, Tijana Trkulja, and Lareina Yee.
"Women in the Workplace 2021." McKinsey & Company, September 27, 2021. https://www.mckin-
sey.com/featured-insights/diversity-and-inclusion/women-in-the-workplace.

Carrigan, Tim, Bob Connell, and John Lee. "Toward a New Sociology of Masculinity." *Theory and Society* 14, no. 5 (1985): 551–604. http://www.jstor.org/stable/657315.

Chen, James. "Bretton Woods Agreement and System: An Overview." Investopedia, March 21, 2022. https://www.investopedia.com/terms/b/brettonwoodsagreement.asp.

Cherry, Kendra. "What is a Flow State?" Verywell Mind, February 17, 2022. https://www.verywellmind.com/what-is-flow-2794768.

Cimpanu, Catalin. "Average Tenure of a CISO is Just 26 Months Due to High Stress and Burnout." ZDNet, February 12, 2020. https://www.zdnet.com/article/average-tenure-of-a-ciso-is-just-26-months-due-to-high-stress-and-burnout/.

Coleman, Jackie, and John Coleman. "Don't Take Work Stress Home with You." Harvard Business Review, July 28, 2016. https://hbr.org/2016/07/dont-take-work-stress-home-with-you.

Cuncic, Arlin. "What is Imposter Syndrome?" Verywell Mind, November 23, 2021. https://www.verywellmind.com/imposter-syndrome-and-social-anxiety-disorder-4156469.

Desrosiers, Alex, Ariella Kupetz, and Lauren Roseman. "White House's Labor Task Force Takes Aim at Gig Economy in New Report." Fisher Phillips, February 8, 2022. https://www.fisherphillips.com/news-insights/white-houses-labor-task-force-aim-gig-economy.html.

Detjen, Jodi. "Masculinity and Leadership Inequities an Examination of the Ways in Which Masculine Cultural Norms Underlie the Barriers to Women's Leadership Acquisition." Temple University Libraries, May 19, 2021. https://dx.doi.org/10.34944/dspace/6503.

Donaldson, Mike. "What is Hegemonic Masculinity?" *Theory and Society* 22, no. 5 (1993): 643–57. https://doi.org/10.1007/bf00993540.

Dormehl, Luke. "Today in Apple History: Apple Chooses Intel over Powerpc." Cult of Mac, June 6, 2021. https://www.cultofmac.com/484394/apple-intel-over-powerpc/.

Douville, Sherri "Why Techno-Libertarianism Clashes With Medicine." Medium, (July 9, 2022), https://sherridouville.medium.com/why-techno-libertarianism-clashes-with-medicine-fe92a9928a9b

Drown, Hannah. "Booster Shot or Not? Mixed Messaging Creates Distrust during COVID-19 Pandemic." Cleveland.com. Advance Local, September 27, 2021. https://www.cleveland.com/coronavirus/2021/09/booster-shot-or-not-mixed-messaging-creates-distrust-during-covid-19-pandemic.html.

Foley, Mary Jo. "Bugfest! Win2000 Has 63,000 'Defects'." ZDNet, February 14, 2000. https://www.zdnet.com/article/bugfest-win2000-has-63000-defects/.

Garland, Eric L. "Restructuring Reward Processing with Mindfulness-Oriented Recovery Enhancement: Novel Therapeutic Mechanisms to Remediate Hedonic Dysregulation in Addiction, Stress, and Pain." *Annals of the New York Academy of Sciences* 1373, no. 1 (2016): 25–37. https://doi.org/10.1111/nyas.13034.

Goffman, Erving. "On Cooling the Mark Out." *Psychiatry* 15, no. 4 (1952): 451–63. https://doi.org/10.1080/00332747.1952.11022896.

Goldhill, Olivia. "Neuroscientists Say Multitasking Literally Drains the Energy Reserves of Your Brain." *Quartz*, July 3, 2016. https://qz.com/722661/neuroscientists-say-multitasking-literally-drains-the-energy-reserves-of-your-brain/.

Hayes, Steven C. *A Liberated Mind: How to Pivot toward What Matters*. New York: Avery, An Imprint of Penguin Random House LLC, 2020.

Holland, Stephen. "How Ireland became One of the World's Biggest Tax Havens." Independent.ie, December 4, 2021. https://www.independent.ie/regionals/sligochampion/business/how-ireland-became-one-of-the-worlds-biggest-tax-havens-41117761.html.

Idell, Kelly, David Gefen, and Arik Ragowsky. "Managing it Professional Turnover." Communications of the ACM. Association for Computing Machinery (ACM), September 1, 2021. https://cacm.acm.org/magazines/2021/9/255039-managing-it-professional-turnover/.

Irwin, Douglas. "The Nixon Shock after Forty Years: The Import Surcharge Revisited." 2012. https://doi.org/10.3386/w17749.

Jackson, Brian. "How a Race to the Bottom Hurts Your Business's Bottom Line." Kinsta®, October 20, 2021. https://kinsta.com/blog/race-to-the-bottom/.

Jargon, Julie. "TikTok Brain Explained: Why Some Kids Seem Hooked on Social Video Feeds." The Wall Street Journal. Dow Jones & Company, April 2, 2022. https://www.wsj.com/articles/tiktok-brain-explained-why-some-kids-seem-hooked-on-social-video-feeds-11648866192?fbclid=IwAR32E_fiG-FWc6Bd3o_xyDtYxS9292BTnqZU-ok0OlZWqtL3cAg8w9MF4Xy4.

Jones, Roger. "What CEOs Are Afraid Of." Harvard Business Review, February 24, 2015. https://hbr.org /2015/02/what-ceos-are-afraid-of.

Kagan, Julia. "The Gramm-Leach-Bliley Act of 1999 (GLBA)." Investopedia, February 18, 2022. https:// www.investopedia.com/terms/g/glba.asp.

Kelion, Leo. "John McAfee: Addict, Coder, Runaway." BBC News. October 11, 2013. https://www.bbc .com/news/technology-24441931.

Kollen Ghizoni, Sandra. "Creation of the Bretton Woods System." Federal Reserve History, November 22, 2013. https://www.federalreservehistory.org/essays/bretton-woods-created.

Krebs, Brian. "Report: Missouri Governor's Office Responsible for Teacher Data Leak." Krebs on Security, February 22, 2022. https://krebsonsecurity.com/2022/02/report-missouri-governors-office-responsible-for-teacher-data-leak/.

Larabel, Michael. "Linux Kernel Drops Support for Old Intel 386 Cpus." Phoronix, December 12, 2012. https://www.phoronix.com/scan.php?page=news_item&px=MTI0OTg.

Lardner, Ring. "Champion by Ring Lardner." Metropolitan Magazine, October 1916. http://martinhillortiz .blogspot.com/2015/07/champion-by-ring-lardner.html.

Lawrence, Cate. "Uber Drivers Officially Recognized as 'Employees' Rules UK Supreme Court." TNW | Shift. The Next Web (TNW), December 8, 2021. https://thenextweb.com/news/uber-is-just-another -example-of-the-problems-of-new-mobility-models.

Lim, Diane. "'Bootleggers and Baptists' – How Crony Capitalism has Captured Regulatory Policy for Centuries." Committee for Economic Development of the Conference Board. Committee for Economic Development of the Conference Board, August 25, 2015. https://www.ced.org/blog/entry/ bootleggers-and-baptistshow-crony-capitalism-has-captured-regulatory-policy.

Mahalik, James R., Benjamin D. Locke, Larry H. Ludlow, Matthew A. Diemer, Ryan P. Scott, Michael Gottfried, and Gary Freitas. "Development of the Conformity to Masculine Norms Inventory." Psychology of Men & Masculinity 4, no. 1 (2003): 3–25. https://doi.org/10.1037/1524-9220.4.1.3.

Mark, Gloria, Shamsi T. Iqbal, Mary Czerwinski, Paul Johns, and Akane Sano. "Neurotics Can't Focus." Proceedings of the 2016 CHI Conference on Human Factors in Computing Systems, 2016. https://doi.org /10.1145/2858036.2858202.

McClelland, Valorie. "Mixed Signals Breed Mistrust." American Psychological Association, 1987. https:// psycnet.apa.org/record/1987-23621-001.

Medina, Alejandra, Adriana De La Cruz, and Yung Tang. "Owners of the World's Listed Companies – OECD." OECD Capital Market Series. OECD, 2019. https://www.oecd.org/corporate/Owners-of -the-Worlds-Listed-Companies.pdf.

Miller, Todd C. "Strlcpy, Strlcat – Size-Bounded String Copying and Concatenation." strlcpy(3) – OpenBSD manual pages, January 25, 2019. https://man.openbsd.org/strlcpy.3.

Nellen, Annette, and Jeffrey A. Porter. "30 Years after the Tax Reform Act: Still Aiming for a Better Tax System." Journal of Accountancy, October 1, 2016. https://www.journalofaccountancy.com/issues /2016/oct/tax-reform-act.html.

Niles, Robert. "Standard Deviation." Robert Niles. Accessed May 13, 2022. https://www.robertniles.com /stats/stdev.shtml.

O'Callaghan, Derek, Derek Greene, Maura Conway, Joe Carthy, and Pádraig Cunningham. "Down the (White) Rabbit Hole: The Extreme Right and Online Recommender Systems." Social Science Computer Review 33, no. 4 (2014): 459–78. https://doi.org/10.1177/0894439314555329.

Petroff, Alanna. "Reddit's Alexis Ohanian Warns 'Hustle Porn' Is 'Most Toxic, Dangerous Thing' in Tech Industry." Yahoo! Finance, November 6, 2018. https://finance.yahoo.com/news/reddits-alexis-ohanian-warns-hustle-porn-toxic-dangerous-thing-tech-industry-140033929.html.

Plummer, Matt. "How Are You Protecting Your High Performers from Burnout?" Harvard Business Review, June 21, 2018. https://hbr.org/2018/06/how-are-you-protecting-your-high-performers -from-burnout.

Pomerleau, Kyle. "The Treatment of Foreign Profits under the Tax Cuts and Jobs Act." Tax Foundation, May 3, 2018. https://taxfoundation.org/treatment-foreign-profits-tax-cuts-jobs-act/.

Press, Associated. "Facebook Reportedly Resorting to Smear Tactics against TikTok." MarketWatch, March 30, 2022. https://www.marketwatch.com/story/facebook-reportedly-resorting-to-smear -tactics-against-tiktok-01648674227?mod=newsviewer_click.

Ro, Christine. "'Turnover Contagion': The Domino Effect of One Resignation." BBC Worklife, September 16, 2021. https://www.bbc.com/worklife/article/20210915-turnover-contagion-the-domino-effect-of -one-resignation.

Seiden, Mark. "Where Pizza Deliverers Rule." Wired. Conde Nast, March 1, 1993. https://www.wired.com /1993/03/where-pizza-deliverers-rule/.

Sendelbach, Sue, and Marjorie Funk. "Alarm Fatigue." *AACN Advanced Critical Care* 24, no. 4 (2013): 378–86. https://doi.org/10.4037/nci.0b013e3182a903f9.

Shannon, Claude E. "A Mathematical Theory of Communication." Harvard University. Accessed May 13, 2022. https://people.math.harvard.edu/~ctm/home/text/others/shannon/entropy/entropy.pdf.

Smith, Adam C., and Bruce Yandle. *Bootleggers & Baptists: How Economic Forces and Moral Persuasion Interact to Shape Regulatory Politics*. Washington, DC: Cato Institute, 2014.

Stackman, Richard W, Patrick E Connor, and Boris W Becker. "Sectoral Ethos: An Investigation of the Personal Values Systems of Female and Male Managers in the Public and Private Sectors." *Journal of Public Administration Research and Theory* 16, no. 4 (2006): 577–97. https://doi.org/10.1093/jopart/ mui059.

Stanborough, Rebecca Joy. "Autism Masking: To Blend or Not to Blend." Healthline Media, November 19, 2021. https://www.healthline.com/health/autism/autism-masking#definition.

Stobierski, Tim. "401(k) Basics: When it was Invented and How it Works." Northwestern Mutual, March 30, 2018. https://www.northwesternmutual.com/life-and-money/your-401k-when-it-was-invented -and-why/#:~:text=Despite%20their%20popularity%20today%2C%20401,being%20taxed%20on %20deferred%20compensation.

Team, Paradise Papers Reporting. "Paradise Papers: Apple's Secret Tax Bolthole Revealed." BBC News, November 6, 2017. https://www.bbc.com/news/world-us-canada-41889787.

Tunguz, Tomasz. "What is the Optimal Contract Length for your SaaS Startup?" tomtunguz.com, October 20, 2016. https://tomtunguz.com/optimal-contract-length/.

Williams, Sam. "A Unified Theory of Software Evolution." Salon.com, April 8, 2002. https://www.salon .com/2002/04/08/lehman_2/.

"World's Largest Container Ships." Zeymarine, January 19, 2020. https://zeymarine.com/worlds-largest -container-ships/.

York, Erica. "The Benefits of Cutting the Corporate Income Tax Rate." Tax Foundation, August 14, 2018. https://taxfoundation.org/benefits-of-a-corporate-tax-cut/.

HOW CYBERSECURITY ENABLES DEPLOYMENT OF ADVANCED TECHNOLOGIES

III

HOW CYBERSECURITY ENABLES DEPLOYMENT OF ADVANCED TECHNOLOGIES

Chapter 9

Security Frameworks as a Foil for Larger Management Issues

Mitch Parker, Brittany Partridge, and Allison J. Taylor

Contents

This was initially planned as a security chapter. However, as our team worked through how healthcare organizations truly affect security changes, a higher-level notion came to life. It became clear that no leader in healthcare can solve security issues unless the underlying management issues are addressed first. There are numerous books out there (and a separate chapter in this book) that explain security frameworks and make you look competent in front of your board or C-suite.

DOI: 10.4324/9781003348603-12

173

This is not one of them and would consider that role an insult. This chapter is for most healthcare information technology (IT) organizations, many of which are centralized due to electronic medical records (EMRs) and other digital transformation initiatives. There are many forward-looking organizations out there. However, there are also a lot aspiring to reach an effective state. This chapter is for those who want to move forward.

In literature, there is the concept of a foil. According to the MasterClass staff, a foil is a supporting character who contrasts with the main protagonist by having a contrasting personality and set of values (MasterClass Staff 2022). Having them together is meant to highlight the attributes of the protagonist. They are there to shine the spotlight, according to that article, without needing to be in conflict. Examples of these include Watson, who is the foil to Sherlock Holmes, Captain America, and Iron Man in the Marvel Cinematic Universe movies, and Scottie Pippen and Dennis Rodman, who were foils to Michael Jordan on the late 1990s Chicago Bulls championship teams in NBA Basketball (Medellin 2020).

More recently, and applicable later in this chapter, is the example of neuroscientist Dr. Amy Farrah Fowler, played by real-life neuroscientist Mayim Bialik, PhD, from the Big Bang theory. Her character can be interpreted as a foil for the curmudgeonly Dr. Sheldon Cooper, highlighting both how he attempts to be a better and more sensitive person, and highlighting his disregard for others, including his wife. However, it is also her partnership with her husband, Dr. Cooper, on Super Asymmetry, that led to a positive feedback loop based on complementary learned knowledge applied differently which resulted in a Nobel Prize. These characters, while being fictional, provide a perfect example of how a character foil can highlight positive and negative characteristics, and how they can work together to make incredible discoveries that can improve the knowledge and understanding of many.

In healthcare, we can consider mobile technologies to be the character foil for information technology in general. We can also consider it a driver for continued improvement and accelerated development in new areas by providing complementary techniques to improve technologies and their usage.

IT in general has had a disconnect from the rest of the organization. Information management techniques and processes began evolving well before the development of digital computers and associated technology management disciplines. Mobile technologies are a foil to demonstrate the lack of coherence between existing frameworks and processes within health systems. They also demonstrate the lack of maturity of the IT field as compared to established ones within other disciplines. This presents significant issues because IT needs to evolve to become a strategic resource within the organization, yet it has no measurable guidance or demonstration of effectiveness. Instead, frameworks such as the NIST Cybersecurity Framework or CHIME's Most Wired levels have historically been utilized to demonstrate efficacy. These alone do not provide the complete model needed for overall effectiveness, and mastering the models themselves may give a false sense of security. Meanwhile, the root causes of many of our issues persist.

A message to executives reading this:

> You may think you understand what is going on within your organization. However, many factors come into play. One of the biggest is thinking that you can apply a cybersecurity or management framework and make your organization significantly more secure. You may think that people in the clinical departments will think this is a good idea and better than what they may already have. In researching this chapter, we realized that it doesn't work that way. What we end up doing is not fixing the environmental factors that cause many of the security issues we face, and we just keep heaping

more of the same security issues on already overburdened security and clinical staff (especially!). What's the point of discussing security frameworks if you don't address the environment that causes the burnout issues in the first place?

Security will fail if you do not plan well and manage better than everyone else. It is an outcome of good processes. Using it to continually provide cover for other processes is the technical equivalent of old cartoons or comedy sketches where the overburdened repairman would continually try and fix all the broken pipes and eventually fail. Rube Goldberg contraptions belong as memes, not as actual solutions.

Use this chapter to build your own program to improve your organizations and leverage the meta-framework and structure you require to succeed. We want to reduce the chaos of the IT organization and replace it with the certainty all stakeholders need.

Dismantling the Definitions of Frameworks

There is much discussion about frameworks; however, what we see as security frameworks, such as the NIST Cybersecurity Framework, ISO/IEC 27001/2, or OWASP, are more control sets or guidance. According to Merriam-Webster, a framework provides skeletal, structural, or open work frame (Merriam-Webster 2022). While these frameworks provide that for security controls, they don't provide enough detail to implement specific controls in measurable ways. There's a lot of subjectivity with them, and this leads to confusion and distraction. Also, you need to have a good framework and structure to effectively manage your organization.

Many of the disciplines within healthcare already have standard-based practices that they use to manage themselves and benchmark against each other. Pharmacies have the American Society of Health-System Pharmacists – AHSP (https://ashp.org). Radiologists have the Radiological Society of North America – RSNA (https://rsna.org), Health Information Management has the American Health Information Management Association – AHIMA (https://ahima.org). AHIMA also covers certifications for numerous revenue cycle positions. Laboratories have the College of American Pathologists – CAP (https://cap.org). Healthcare Technology Management and Clinical Engineering have the Association for the Advancement of Medical Instrumentation – AAMI (https://aami.org). AAMI also participates in standards development with the American National Standards Institute (ANSI), the International Standards Organization (ISO), and the International Electrotechnical Commission (https://iec.ch). Healthcare Finance has HFMA – the Healthcare Financial Management Association (https://hfma.org). For hospital accreditations, we can look to the the Joint Commission (https://jointcommission.org). Nursing and physician organizations have numerous associations and practices, and an entire book could be spent discussing them. There are separate subcultures with doctors, nurses, and other professional caregivers.

Many of these organizations have been around much longer than the concept of a dedicated data processing organization, well before IT, and long before technology was pervasive. AHIMA was founded in 1928, HFMA in 1946, and AAMI in 1967 (AHIMA 2022; HFMA 2022; AAMI 2022). RSNA was founded in 1915 (Purdue University 2022). The Joint Commission was founded in 1951 (TJC 2022). The Healthcare Information Management Systems Society (HIMSS), founded in 1961, is a relative newcomer (HIMSS 2021). HIMSS started out as the Hospital Management Systems Society (HMSS) before pivoting to its current form (HIMSS 2021).

What this means to us is that we have numerous organizations/practices/subgroups that have developed years of processes and procedures for management of their specialty areas. These groups have been around longer than IT. They have been organizing longer than IT. And they have been managing technologies as an evolution of existing processes, as opposed to managing technology and strategy as means to ends in themselves. They do a better job of it and are more credible because local programs align with their practices. What we have discovered in our experience and research is that these organizations and their associated departments and practices have 12 common characteristics, known as the 8 P's and 4 C's:

1. Example **Programs**, benchmarks, data, and documentation based on evidence-based **Practices and Standards**
2. Standard evidence-based **Protocols**
3. Standard evidence-based **Processes**
4. **Peer Accreditation and Review** of the first 4 P's
5. **Periodic Review** of Programs, Protocols, Practices, and Processes
6. **Professional Culture**, Associations, and Networking – Defense against fabulists, who are people that falsify and embellish their credentials and history to appear credible:
   i. Continuing education
   ii. Certifications/Licensure
   iii. Credibility – Are these people who they say they are? Have they been vetted?
   iv. Cultural norms
7. **Peer-Reviewed Publications** discussing research, improvements, applications, and explanations of Programs, Protocols, Practices, and Processes

What makes IT especially challenging in healthcare organizations is that there are, in fact, numerous hegemonic professional subcultures under one roof attempting to operate as one organization. Often it looks like three children standing on each other's shoulders wearing a trench coat trying to impersonate an adult and operates just as efficiently. If we are not cognizant of these professional cultures and their nuances, any enterprise initiatives that we attempt to undertake will fail. An example of this is from Michelle Finneran Dennedy's book, *The Privacy Engineer's Handbook*. She discusses the cultural impact of privacy, specifically between the United States, Europe, Japan, and China, and how each has an impact on data collection and practices (Dennedy, Fox, and Finneran 2014). The same is true of the Professional Cultures that we have within the healthcare umbrella. We're going to refer to these characteristics throughout the rest of this chapter as the P's and C's.

We need to mind our P's and C's when we are dealing with healthcare and cannot think that we can put in an enterprise management approach without being cognizant and intentional in addressing them. One of the areas where Epic Systems has succeeded has been in enculturating this into their development process. According to Bernie Monegain, in her article "A Look Inside Epic's Design and Usability Teams," developers are required to spend a minimum of 4 days, up to 18, testing their applications in the field (Monegain 2016). This helps Epic address these P's and C's in their software development processes. It's not perfect by any means; however, it's something.

We need to recognize the P's and C's in all of our processes. There's no way that we're going to address Diversity, Equity, and Inclusion (DE&I) without recognizing the various subcultures within healthcare. If we don't address the subcultures within frameworks and strategic plans, we're not going to address DE&I. You're not going to address cultural factors behind burnout. We're not going to be meeting their needs by understanding how their organization is structured. This means we won't be credible, and Shadow IT will be more prevalent than us. We're just going to be another framework that they ignore because it doesn't help them improve their business.

Shadow IT exists, particularly in healthcare, because central IT organizations don't have the structure or understanding of these organizations to properly support them. Many times, it comes down to credibility, reliability, and trust. These organizations do not trust that there are solutions in place from the central IT organization that meet their needs. They often do not believe that an understanding of their organization and culture exists, and that centralizing will inhibit that. Many of these departments have been organizing and cataloging information since before modern digital computers, such as in the case of AHIMA, 1928. It's not Shadow IT if the processes started before computers existed, and an entire set of credible information demonstrating success exists. We've all been organizing information before IT was prevalent. IT has the issue of not recognizing, in general, the Dewey Decimal Systems that evolved beforehand that incorporated human factors better. IT has also not evolved to understand these cultures and adapt to meet their needs.

When it comes to information risk management, there are frameworks already in place for quality that have also been used for Information Security. These include Just Culture, Failure Mode and Effects Analysis, Comprehensive Unit-based Safety Program (CUSP), and numerous quality improvement frameworks such as the National Database of Nursing Quality Indicators (NDNQI). Medicine has the Institute of Medicine (IOM) framework, and there are many others, especially when you get into nursing and other care areas. The cybersecurity frameworks are incomplete compared to the robustness of what is already used, and often don't fit into existing organizational frameworks. We need to align with these so that we can all do better.

Fraud, Fabulists, and the Blockchain – Why Tech Will Never Replace Professional Organizations

These P's and C's also address fraudulent products and the credibility of product sellers. Products that have been recommended or approved by these professional organizations or their practitioners have met their standards for rigor and credibility. An example is Crest toothpaste, which has the seal of the American Dental Association on many of its products available in stores. There is a history behind this that stretches back before any of us were born.

In the last 19th century, there were numerous people selling patent medicines, which were unregulated medications that claimed to cure diseases, and caused more harm than good due to the use of ingredients such as alcohol, cocaine, or opium (Hagley Museum 2017). The Pure Food and Drug Act, passed by Congress in 1906, allowed for public health action against unlabeled or unsafe ingredients, misleading advertising, quackery, and false representation (Hagley Museum 2017).

With COVID-19, there have been numerous people who have attempted to wind back the clock and sell miracle solutions. A disgraced television personality, best known for spending the charitable donations sent to their organization and spending time in federal prison for multiple fraud charges, was fined $156,000 for selling their supplement, Silver Solution, as a cure for COVID-19 (Bertram 2021) (Salter 2021). A family in the United States was charged with selling a toxic industrial bleach as "Miracle Mineral Solution," a cure for COVID-19, cancer, autism, and numerous other serious medical conditions (USDOJ 2021).

These products have had significant support on social media, where private Facebook groups emerged to urge parents to poison their autistic children with this type of bleach to cure them (Zadrozny 2019). One streamed support for their Silver Solution online. One media personality infamous for their talk show and numerous conspiracy theories was estimated to make $25 million a year from selling their dietary supplements of questionable value (Brown 2017). Many enjoy continued support on numerous fringe social media sites.

Social media has been used by anti-vaccination, extremists, and anti-government groups that oppose vaccinations and COVID-19 mandates. An example of a group signaling false credibility by attempting to impersonate a professional organization is a group posing as a legitimate medical association and offering telemedicine services, who also oppose vaccinations and credible medical treatments for COVID-19 (Fiore 2022). This was later revealed to be an extremist political group, not a medical one (Bergengruen 2022). The continued popularity, especially on social media, of a disgraced former physician, whose fraudulent and retracted study falsely linked childhood vaccinations to autism, continues to contribute to outbreaks of childhood diseases once thought suppressed, such as rubella.

Physicians and medical professionals act as a control against fraud. However, social media, especially due to organizations employing physicians while impersonating credible groups, has threatened the value of their word. People attempting to sell supplements often do so by slamming "big pharma" or greedy doctors. Millions of people are buying these products and spreading disinformation. This has caused real-world harm.

The technology companies, for the most part, have let this fester. Some argue they have grown into massive disinformation platforms. Snake oil peddlers have figured out how to buy ads and get preferred placement online, and willing voices to listen on extremist podcasts (Hannigan 2021). The big tech companies have let disinformation campaigns reign for years. Doctors are not going to trust Big Tech given how snake oil solutions have been allowed to exponentially propagate, which has effectively knocked progress on the front against their use back over 100 years.

A major part of what healthcare professional organizations do is provide risk management and defense against fabulists and fraudsters. Big tech companies are used to enabling them for their financial benefit. This has undermined the work that many healthcare professionals have done to help improve the lives of others. Attempting to replace or augment risk management processes in healthcare, such as credentialing, based upon these companies' technologies will be met with serious doubt. Big tech companies are like live television coverage 20 years ago, where someone could arbitrarily claim they were an important and relevant person during breaking news and then shout out "Howard Stern," taking advantage of the media's appetite for insider information to bypass checks and balances. Treating extremist political groups that happen to employ doctors on social media or questionable supplement peddlers with the same apparent credibility as the American Academy of Pediatrics gives tech companies a serious credibility issue within professional organizations and subcultures.

A Sputnik Moment

On October 4, 1957, the Soviet Union launched the first artificial satellite, Sputnik-1. This was the shock that jolted the US space race into high gear (US State Department 2022). The hegemonic power that had just tipped the scales in World War II, the United States, was just eclipsed. This kicked off a space race that eventually resulted in the moon landing in 1969. While President Eisenhower downplayed this event, it directly led to significant funding increases for the space program (US State Department 2022). Brent Maddin, who was a doctoral student at Harvard at the time, discussed how this directly led to the passing of the National Defense Education Act, which increased funding for scientific and technical education (Powell 2007). Resources poured in, to not let the United States fall behind the Soviets and cede space to communism. The United States ended up with an advantage in space that has persisted to this day. According to Statista, as of January 1, 2021, there are 3,372 artificial satellites in orbit around Earth (Salas 2021): 1,897 were of US origin, 176 were Russian, 412 were Chinese, and 887 were from other countries (Salas 2021). It can be argued these days that an event of such magnitude would be blunted by numerous distractions, namely social media. We are so distracted that it would take a significantly more impactful event to have a similar effect to motivate the gigantic effort that led to men on the moon.

So far, COVID-19, despite millions of deaths and the effects of long COVID, has not led to a similar moment nor had similar effects. Think about it. One satellite launch led to decades of effort and innovation that have led to a hegemonic American advantage in space, even today. It gave entrepreneurs the raw data and scientific foundations they needed to build their own programs and supplant ones that some argue to be mired in bureaucracy and appropriations. One satellite has had more effect on our society than millions dead and sickened from a pandemic. What will be our moment to motivate us to fix healthcare IT so that we can see the same types of effects with healthcare?

We need to build discipline and rigor around IT to be considered credible like Surgery, Radiology, Pathology, or Nursing and its many subspecialties such as Critical Care or Nephrology. Even engineering is considered more credible. Each of these has its own requirements for data and IT analysis based upon their existing P's and C's. We want to have something that recognizes all these involved subcultures and supports them, rather than attempt to make them all fit a mold. In terms of conferences and education, we need to be like DEF CON and its inclusion via villages and broad themes, rather than ones that try to make everyone fit into a narrow theme and focus points. If we do the latter, it just won't work.

We don't have the same focus or mission, and we are nebulous because of it. We see much of the same with how organizations treat IT. They don't understand it, think it can do anything, and that a tech guy can work miracles to do whatever the organization needs to get ahead. They look at it tactically, like someone hooking up a router or television. They don't look at it as supporting the mission or requiring the same level of management that a design and construction or clinical project has. The critical path has never flowed through technology because it's considered an appliance and secondary to the mission. Technology was managed, and in many cases is still managed, the same way organizations manage their boilers, food service, or physical plant. Local affiliates are given significant latitude to make decisions, and this reflects in the tech used. As Dr. David Feinberg, the CEO of Cerner, said in the April 22, 2022, episode of This Week in Health IT, it's provider-centric (Russell and Feinberg 2022). Each practice is given high latitude and autonomy to make decisions, based upon decades of practice in some places. This is much like how physician staffs in health systems have significant local governance and different policies and practices at the hospital level, even though they may all be employed by the same parent organization. Hospitals and health systems have built an analogue of this with technology. It has never been a consideration because it played a secondary role to managing the individual subcultures within the organization. With the advent of electronic medical records and the pervasive and mobile use of technology in patient care, it now is primary. While many organizations have grasped this, many more have not, and this affects all of us. The EMR reflects this fragmentation. The continued use of numerous bespoke applications and services to support clinical operations also reflects this. Decades of practice before the first general-use digital computer, ENIAC, in 1947, support this mindset.

A Modern-Day Guild System

Healthcare isn't retail or supply chain, where solutions can be duplicated and scaled much more easily. It is still based upon processes and methods that haven't changed much in the past century. Healthcare is based on centuries of methods and traditions that are a modern analogue of the medieval guild system. It's very localized, and it's also very subjective. The most successful organizations know how to build and align around this subjectiveness to optimize the entire process. They've involved physicians and organizational subcultures through the entire process and discussed what is needed to be successful. They've had clinicians do the co-development and selling of digital transformation, which adds credibility.

Health Systems are loose confederations of medical practices that share numerous common support, administrative, finance, and staffing services, and many of those services, including the practice of medicine, are performed by other organizations under contract. Your competitors for heart surgery and emergency services are your providers for OB/GYN services or patient transport. You may employ an organization like TeamHealth or Vituity for your Emergency Department services. If you're in a rural area, the local community doctors are likely also the ones providing services at your Critical Access Hospital. The same techniques that work in finance, retail, or manufacturing do not work here for addressing corporate structures. If anything, it resembles real estate, where according to the article "LLCs: Best Biz Structure to 'House' Multiple Properties," real estate investors often create Limited Liability Companies (LLCs) for each property they own to shield themselves from liability (Akalp 2022).

Shedding the Expense Item Mindset

Many organizations are also stuck in the mindset that IT is an expense and a support cost like flowers or environmental services. IT has not been included in strategic plans. They brought in operational management for the CIO role as a service to the practices. They managed it tactically based on organizational and practice needs. They asked their teams to integrate systems and patch them together to save money and make them work for minimal cost. What this has done is turned teams into ones that perform day-to-day maintenance work and spend little time on innovation, digital transformation, and addressing systemic issues. One can easily make the same argument for many health systems in general, and that the truly successful ones are the ones that have figured out how to strategically manage technology and concurrently manage their practices.

IT has been asked repeatedly to drop costs like the same groups that serve food, environmental services, or physical security. This has resulted in numerous outsourcing companies coming in and offering these services for the least amount of money possible. This has also resulted in numerous hacks, custom-designed, and insecure systems that are holding together the operational and process infrastructures of health systems while simultaneously, exponentially increasing their risks. IT in many organizations keeps the lights on and is considered ancillary to the patient care process. This is a way of thinking that both the pervasive use of technology and its ever-increasing risks demonstrate is no longer valid.

IT's relegation has resulted in years of systems designed for tactical purposes that are not easily able to strategically expand. Instead of strategically investing in system-wide alternatives, health systems minimally invested in tactical and practice needs, often paid for by the practices themselves. This has led to organizations having numerous stovepipe systems all over the place. Businesses depend on these for daily operations, and these cobbled-together systems are what is running a significant portion of healthcare. This worked when these were the only systems. Now everyone else has caught up and we need to too.

These solutions did not scale, required significant work to maintain, did not meet the security standards of the software or services the organizations would have purchased had they made a strategic investment. They were not held to the same criteria as clinical systems, laboratory systems, or the Enterprise Resource Planning (ERP) system that requires ever more stringent guidelines to provide accurate results according to Generally Accepted Accounting Principles (GAAP), Institute of Internal Auditors (IIA), and American Institute of Certified Public Accountants (AICPA) standards. The systems that support IT in health systems often resemble their Byzantine corporate structures.

Many organizations went out and purchased top-of-the-line electronic medical records systems from Epic or Cerner, thinking that the sheer presence of a well-developed and managed system would percolate all over the rest of the IT department, and demonstrate a commitment to world-class security while improving the corporate structure.

In the words of Maury Povich when presented with an envelope of polygraph results, "That Was a Lie." It didn't work for ERP systems, and it's not going to work because of your EMR, no matter how good the big vendors are at protecting their core systems. A number of health systems all proved that incredibly well with the ransomware attacks that crippled them.

Even with top-of-the-line EMRs and ERP systems in place, these systems sat on top of legacy systems, hardware, and processes. According to Sinno, Gandi, and Gamble, in their Becker's Health IT article, "8 Problems Surrounding Meaningful Use," government incentive programs only covered 20–25% of the implementation costs, which required significant IT investments beyond the systems themselves (Sinno, Gandhi, and Gamble 2011). You also cannot expect departments to switch processes and systems on the fly and lose years of organizational knowledge and muscle memory. What we often end up with, despite the best efforts of our organizations to digitally transform and provide services that providers and patients want with electronic medical records (EMR) systems and patient portals, are islands of well-maintained EMR and ERP systems being fed by numerous insecure legacy systems and stovepipe applications designed for use cases of medical staffs of local hospitals or practices. They do not scale, and patient medical record security is an afterthought.

The Lack of a Security True North and Supporting Programs

What this also gives rise to is several different methods and ways by which organizations try to claim compliance with the HIPAA Security Rule. The "gold standard" for this is not clear, and HIPAA does not have a certification and enforcement structure, or a quantitative instrument that organizations can follow with the same level of assurances that accounting systems have. Signaling credibility with security is difficult for many organizations to accomplish. Many just rely on the word of the vendor that they've done a risk assessment or are "HIPAA compliant." There is no HIPAA certification from CMS that would guarantee that.

Jessica Davis, in her SC Media article "'Voluntary practices' in healthcare insufficient for its dependence on legacy tech," discusses how many healthcare organizations do not have security programs and are not even aware of the resources such as Health-ISAC that exist (Davis 2022). These "cyber-poor" organizations do not have the support that the financial services sector enjoys, including Federal Financial Institutions Examination Council (FFIEC) cyber guidance that is shared with numerous others. The New York State Department of Financial Services, for example, recommended on February 6, 2014, that all NYS-Chartered Depository Institutions join the Financial Services ISAC (FS-ISAC) to promote cybersecurity and share information (Lawsky 2014). Christian Dameff, MD, in the same article, recommended to the US Congress that entities need to be incentivized to do so (Davis 2022). This is similar to the Meaningful Use incentive program that brought electronic medical records to hospitals and practices. The top-down approach applied in finance needs to happen with healthcare.

To move from a system built on stovepipes, wires, and local connections requires excellent strategic management skills. This requires a metrics-driven approach to digital transformation and security. If an IT organization wants to deem themselves worthy of the funding that other strategic components of the organization seek, they need to demonstrate effective quantitative management skills and mind the P's and C's. This starts with building the discipline needed to understand and build toward goals and leveraging a framework and structure to do so.

It's no longer enough to look at the incredibly good technology advisory firms or Big 4 consultancies and say their word is enough to convince CEOs to make massive strategic technology investments. Much like the practices will provide peer-reviewed evidence-based results and data to support their investments, practices, and processes, we need to do the same. We need to partner with and incorporate them as part of a larger strategic framework that includes their goals for success and plans to improve their practices. We need to fit together several key areas in one plan to be credible to the leaders who are staking their careers on us:

- Practice Programs, Standards, Practices, Data, and Metrics
- An applicable Information Technology Maturity Model and Framework
- An applicable Information Security Management Plan Framework
- An applicable Security Culture Framework
  - Leverage the Security Culture Framework for the ITMM and Framework as well
- Detailed operation management plans that incorporate Risk Management as part of the plan itself at multiple levels

Just presenting Electronic Medical Record Adoption Model (EMRAM) or Most Wired isn't going to be enough to get the practices to buy into your vision. Plaques don't make money, and most patients won't care about that, especially if the technology does not work for them the way they need. Without good solid planning, partnership, and discipline, claiming CHIME Most Wired or HIMSS EMRAM Level 7 will make you look less credible. If the practices don't buy into the plan and feel validated by it, they will be disengaged, and it will cause issues. In Rick Blizzard's article for Gallup, Nurse Engagement Key to Reducing Medical Errors, he indicated that a key Gallup study finding of outcomes at more than 200 hospitals was that nurse engagement was the number one predictor of mortality variation across hospitals (Blizzard 2021). Technology is no good if it's not a cultural fit and causes disengagement, even if it's a billion-dollar investment. The framework aimed for needs to be secondary to engagement and understanding customer needs. It is with that concern that this framework has been designed to address long-range strategic plans built to improve organizations, not just have a narrow focus on one component such as IT or security. You won't have either if you don't have buy-in from the practices.

The strategy we mention has several major components, which are summarized below and detailed throughout this chapter:

- Understand Organizational Goals and Strategies
- Mind Your P's and C's
- What Signals Are We Sending?
  - Signaling Theory Applied to Healthcare Information Security as a Foil for Highlighting Less Credible Frameworks
- Management Past the Book – What the Thought Leaders Didn't Tell You

- Determine the Right Enterprise Metrics
- Perform an Enterprise Risk Assessment, Score the Risks, and Perform Gap Analysis
- Are These in Alignment with Our Goals and Strategies?
- Picking the Risks to Address
- More Than Words – Digital Transformation That Matters
- Tracking This across the Organization
- Feedback to the Enterprise Strategic Plan – A Virtuous Cycle

Understand Organizational Goals and Strategies

Our goal is to avoid the numerous traps that have befuddled our predecessors and caused organizational disengagement and chaos before us. We want to have a framework and strategy that lead to lowered risks. However, just presenting frameworks without context, especially given the disengagement that we are seeing with providers now, is not going to help.

We need to admit we need to change to understand and support organizational goals and strategies. This means that we need to have full comprehension of what goals the leadership is trying to accomplish. The issue IT has here is that we attempt to provide solutions instead of listening to problems and understanding needs. People just want to make the big problems go away and use Big Tech to do so. They want to use the easy button and have someone else do it. The problem is that it's often IT that promulgates this mindset first.

An example of this has been with Blockchain, a technology that while it does have uses, has had numerous applications that have failed because of the attempts to find problems to solve using it. Jesse Frederik, in his article "Blockchain, the amazing solution for almost nothing," discusses how it is a solution in search of a problem (Frederik 2020). He gave the example of a Dutch town, Zuidhorn, that implemented Blockchain technology as a method of implementing a municipal poverty aid package for children. The software won awards and brought recognition to the town. However, the developer admitted the software could have been written without it. What it did was bring attention to old, staid processes and consider them targets of innovation and change. The solutions that other Dutch providers put in place did not have this technology; however, they were inspired to think of solutions because of it.

Our job as IT executives is not to try and build ultimate solutions based upon analyst buzzwords, media hype, or awards. It is to provide an understanding of what's possible, and to use these new technologies as Character Foils for the technology challenges they face. As excellent translators of what the new technologies are, without attempting to solution everything, and allowing the customers to tell you what they're thinking, has two advantages. First, this gives customers a different way to explain their goals and strategies to you. Second, it lets us understand what these goals are by using these as Character Foils to highlight what areas you need to focus on to develop the IT strategy.

In healthcare, the ultimate example of Solutioning has been with the Electronic Medical Record itself. In Dr. Atul Gawande's November 12, 2018, article, "Why Doctors Hate Their Computers," he discusses how the implementation of a billion-dollar EMR system at the hospital he worked at led to more work, less time spent on patients, more physician burnout, and less meaningful collaboration due to technology bloat (Gawande, Mukherjee, and Groopman 2018). He cites studies that indicate that there were epidemic levels of burnout among clinicians, including 40% screening positive for depression, and 7% reporting suicidal thinking. This is before COVID-19 made their situation exponentially worse. He also discusses how decisioning around the EMR was taken out of the hands of the people performing the care and given to administrators who did not.

We must do better than this. Dave Fornell, in his article "7 reasons clinicians are leaving jobs in the era of the great resignation," indicates that 60% of clinicians want to leave their jobs because they feel burned out (Fornell 2022). In many cases, this was linked back to inefficiencies with their hospital EMR. Thirty-five percent strongly disagree with the statement that their organization has done a great job implementing, training, and supporting the EMR, and are strongly dissatisfied with it.

Doctors and clinicians do not like the technology solutions we have implemented. The EMR, despite being a technological achievement on many levels, has made life worse for clinicians. They are burned out. They are leaving. They are becoming disengaged. This has and will continue to have effects on patients.

We need to admit the truth of EMRs not working as designed and realize that the same effect can happen to mobile technologies if we let it. We need to start by understanding the goals and strategies of the organization and use technologies as a foil to better understand intentions and characteristics. Don't put it all on the Blockchain and expect the awards to make up for losing your clinicians to burnout. Also, you're putting alerts on devices that clinicians use without likely consideration for Alarm Fatigue, which ECRI Institute considers a major health technology hazard, according to Lauren Dubinsky, in her *HealthCare Business News* article "From alarm management to AI, patient monitoring gets a facelift" (Dubinsky 2020). This article also predates COVID-19.

Is it any wonder that organizations sanction unofficial IT projects to fill the gaps and provide actual solutions for their clinicians? Shadow IT, which according to Cisco is the use of IT-related hardware or software by departments or individuals without the knowledge of IT or security, is more prevalent than we think (Cisco 2020). This can encompass the cloud, software, and hardware. When EMR changes affect how physicians can see patients, as Dr. Gawande hinted at multiple times in his article, you will have undesired effects. One of them is that you are taking people who are often much smarter than the IT teams that support them, have been using technology longer, are highly motivated, and giving them negative motivation to work around the IT solutions. They're not being listened to, being overridden by people who have no agency in how the technology is used in the patient care environment, and writing off IT to do their own thing.

There is a personal reason why I became interested in mobile healthcare technologies. In 2009, I was attending my cousin's Bat Mitzvah in Westchester County, NY. One of my other cousin's husband, a pediatric oncologist and researcher at a large East Coast academic healthcare institution, showed me his smartphone and all the custom applications his team had written for mobile device usage. I took this knowledge back to the teams I worked with at my old employer to warn them. I told them that the doctors were going to see this technology at conferences and with their friends in the field, and we would too. Initially I was blown off. Then a smart tablet came out in 2011, and Stores on the Main Line and at King of Prussia in suburban Philadelphia had a lot of visits from doctors who were suddenly interested in them. Healthcare became very mobile, very quickly, as the clinical chairs started discussing how to use iPads to take care of patients. MedCityNews covered part of a BYOD program my previous employer established, especially the part where we worked with each interested clinical department as much as possible (Baum 2014). We went to each discipline and spoke with them. Each one had different answers about how they intended to use iPads in their areas. Each had their own interpretations of how the technology was to be used. However, we all had the shared objective of improving the organization despite the technology in place. Part of the appeal of mobile technologies is because of the dissatisfaction with existing ones.

Mind Your P's and C's!

Where organizations have fallen flat is in understanding cultures. Healthcare is not retail. It's not an insurance company. It's not finance, banking, or any kind of traditional business. It's a loose confederation of individual practices. We discussed the 12 characteristics of these practices and subcultures before:

Example **Programs**, benchmarks, data, and documentation based on evidence-based **Practices and Standards**
Standard evidence-based **Protocols**
Standard evidence-based **Processes**
Peer Accreditation and Review of the first 4 Ps.
Periodic Review of Programs, Protocols, Practices, and Processes
Professional Culture, Associations, and Networking:
    i. Continuing education
    ii. Certifications
    iii. Credibility
    iv. Cultural Norms
Peer-Reviewed Publications discussing research, improvements, applications, and explanations of Programs, Protocols, Practices, and Processes.

This section is going to go into detail about what this means to us and why we need to mind our P's and C's. It is important for IT professionals, especially those coming from outside of healthcare, to understand the current state and how this differs from other industries. Too often we hear "It's different," with no explanation. The result is often very frustrated IT staff or executive leadership that get burned out and leave because no one is there to explain it to them.

Most areas of healthcare have associated professional organizations that have existed long before digital computers. They have significant resources available to all organizational members in their disciplines. There are eight P's – Programs, Practices and Standards, Protocols, Processes, Peer Accreditation and Review, Periodic Review, Professional Culture, and Peer-Reviewed Publications – and four C's under Professional Culture – Continuing Education, Certification, Credibility, and Cultural Norms – which we need to discuss to get an understanding of the practices we deal with, and how we can better align our plans to help them reach their goals. After we discuss the P's and C's, we will then discuss what we can do to improve.

These professional organizations provide full references on example Programs that clinicians can use to structure their own nascent ones. Through a mixture of peer-reviewed publications, conference programs, papers, and textbooks, a program administrator or clinician can have many of the resources they need to establish, structure, and maintain a practice focused on delivering care to patients using well-established and proven methods.

These methods include methods, benchmarks, and data on effective Practices that can quantitatively demonstrate their effectiveness. When a clinician approaches IT about using a new device or method that they heard about a clinician at a large academic health system using at a conference, it often comes with data demonstrating significant improvements as part of a paper or presentation. Healthcare practices are very much focused on using the scientific method to demonstrate clinical effectiveness. Drs. Espinosa-Brito and Barmúdez-López focus on this model in their Rapid response to Doctors are not scientists in the *British Medical Journal*, discussing how clinical care is the scientific method applied to patient care, rather than just the "application" of methods called "the truth" (Espinosa-Brito and Barmúdez-López 2004). They also discuss how scientific–technological advances of the past 60 years changed the technique, and not the clinics. The Practices predate the technology used, and not the established knowledge they are built upon. It's not Shadow IT if the processes existed before computers did.

Protocols, which according to the National Cancer Institute, are detailed plans of scientific or medical experiments, treatments, or procedures (National Cancer Institute 2022). These are the standard methods that clinicians follow. It is important to understand what these are, because the use of technology in both intended and unintended ways will affect how the clinics operate. These will need to be reviewed as part of any strategic plan because they will change.

According to Gartner, Business Processes are event-driven, end-to-end processing paths that start with customer requests and end with customer results. These often cross boundaries (Gartner 2022). Healthcare has numerous process flows, and one of the purposes of the EMR is to help coordinate crossing those boundaries across multiple areas. Understanding these processes, especially given that communication across the practices has been affected by EMRs given the examples in Dr. Gawande's article, is also critical.

Peer Accreditation and Review, according to the article "Clinical peer review in the United States: History, legal development, and subsequent abuse," by Dinesh Vyas and Ahmed E. Hozain, indicates that the Joint Commission on Accreditation requires hospitals to conduct these reviews (Vyas and Hozain 2014). The Health Care Quality Improvement Act (HCQIA) was passed by Congress in 1986 and provides a solid base for peer review as a means of quality improvement. The American College of Surgeons (ACS) began using this process in the early 20th century. Many of these peer reviews leverage chart reviews; however, site visits, either in-person or virtual, are not uncommon. Joint Commission, ACS, and other accrediting organizations use these. This means that in addition to the internal reviews done by practice and health system leadership, peer reviews using clinicians from other institutions also happen regularly. HIMSS uses peer reviewers as part of reviewing for Electronic Medical Record Adoption Model Levels 6 and 7 maturity (Rayner 2015).

Periodic Reviews also happen. Each practice area, depending upon their areas of specialty, is also reviewed by state and federal regulatory agencies, health departments, and even groups such as the Drug Enforcement Administration (DEA). Each practice must manage their own reviews, as each specialty has different reviewers and groups, even if the same agency covers multiple practices. Each practice will also have differing data requirements to submit to these agencies. Healthcare practices are very similar to finance and community banks in the number of audits and auditing agencies that audit and review their practices, protocols, and standards. Many healthcare administrative practices do not have to undergo the scrutiny that an Emergency Department or surgical program does by numerous agencies, peer reviewers, and accreditation organizations. The Federal Financial Institutions Examination Council (FFIEC) also provides standard guidance for cybersecurity assessments using their Cybersecurity Assessment Tool (Clement 2017). This is much like how the various disciplines have their own standard processes and reviews. The closest many IT organizations come to this is with recurring Payment Card Industry-Data Security Standards (PCI-DSS) attestations, Health Information Trust (HITRUST) Common Security Framework certification reviews, or ISO 27001/2 audits. While there are other tools and frameworks out there, they are not certified against instruments the way these three are. This will be discussed later.

Professional Culture, according to the *Mayo Clinic Proceedings* article "Healing the Professional Culture of Medicine," is the shared and fundamental beliefs, normative values, and related social practices of a group that are so widely accepted that they are implicit and often no longer recognized (Shanafelt, Schein, Minor, Schein, and Kirch 2019). Physicians, and many other clinical groups, have their own distinct professional cultures. This is deeply rooted in the history of the profession or organization, according to the article. We consider each practice and specialty to be representative of unique professional cultures by virtue of the manifestations of the three key levels identified:

■ Artifacts/Symbols – The visible manifestations such as actions, behaviors, heroes, and rituals
■ Espoused Values – What the values and priorities are claimed to be and manifested through mission and value statements, communications, and external communications
■ Tacit Assumptions – What we truly believe and value, the unwritten rules that drive our behavior

There are four artifacts of Professional Culture that demonstrate these norms, which we refer to as the 4 C's:

■ *Continuing Education* – Specialties provide continuing education and training that both provide professional training and further enculturation using Artifacts and Symbols.
■ *Certifications* – The International Certification & Reciprocity Consortium defines this as representing the achievement of a level of professional competency agreed by the international community as qualified to practice effectively (IC&RC 2022). This can also be a prerequisite to be licensed to practice in some localities. These certifications are key and important for identifying credible professionals in clinical practice and support, such as AHIMA's certifications.
■ *Cultural Norms* – The American Psychological Association explains these as societal rules, values, or standards that delineate accepted and appropriate behaviors within a culture (APA 2022). These set minimum expectations for participants.
■ *Credibility* – According to Oxford Reference, this is the quality of meriting belief or confidence, consisting of perceived trustworthiness, and perceived competence or expertness (Oxford Reference 2022). A major part of credibility is vetting out the fraudsters and fabulists.

These professional cultures also manifest in numerous professional organizations, such as the American Medical Association, American College of Surgeons, and AHIMA. These organizations provide the 4 C's and networking opportunities.

One of the artifacts that represents credibility, and represents a summary of these items, is Peer-Reviewed Publications. The United States Geographic Survey (USGS) defines this as a scholarly publication with a supporting peer-review process (USGS 2022). It is further defined as scrutiny of work by qualified peers possessing appropriate education or expertise that have no stake in the outcome of the review or publication of the work. USGS requires at least two scientific peer reviews for their products. Oxford Academic requires review by an editor and at least two reviewers (Oxford Academic 2022).

What does all of this equate to? This means that practices have significant cultures, subcultures, practices, and standards that have existed long before technology, and will likely outlast the current technology and business architectures in place. They are culturally separate organizations that operate in confederation with each other to provide patient care and services. The most appropriate comparison is that of the United States under the Articles of Confederation.

The question then becomes how we best support these organizations as part of a strategic framework. What steps do we have to take to be able to do so? Approaching them about joining the enterprise is the wrong way to do so. It makes it about yourself and your accomplishments and sends the wrong signals. It doesn't validate them, meet their needs, or give them a path forward.

The opinion of many doctors, especially from Dr. Gawande's article, is that EMRs made things better for administration and billing, and not for them. Technology has made their lives worse, and it's been implicated in burnout. Approaching them and offering more of the same is going to have worse effects, and it's going to make you look not credible.

There is a three-step set of methods we can use to combat this that we call Align, Improve, and self-Determination (AID):

■ *Align* – Align the goals of the strategic plans and initiatives with those of the practices. Align the strategic plan with their cultures. Align processes with their long-established protocols and processes. Use tools such as the A3 problem-solving format methodology to help clarify and align issues and goals (Healy 2020).

■ *Improve* – Leverage Protocols, Benchmarks, Best Practices, and Processes from their peer-reviewed sources as initiative goals for each area.
   – Leverage the Awareness, Desire, Knowledge, Ability, and Reinforcement (ADKAR) model to address change management (Prosci 2022).
   – Leverage Failure Mode Effects Analysis (FMEA) to understand their processes and how changes to them may cause failures (IHI 2017).
   – Examine P's and C's to determine if existing ones inadvertently lead to bias, and if so, how to leverage technology to avoid them. A discussion in Ars Technica characterized how existing processes and data can work with intelligent systems algorithms to potentially amplify bias in outcomes where they are applied (Hutchinson 2022). It also discussed how to modify processes to look at the overall processes before searching for the right algorithms.

■ *Self-Determination* – Give them access to the data and analytics they need to be able to leverage Practices, Protocols, Processes, and Benchmarks to perform their own analyses, conduct their own quality improvement programs, and report to their institutional, government, and peer reviewers. Work with them to be their own better self-monitors and incorporate these improvements into what exists. Help them make something better that they can take a larger part in sustaining.

We need to be credible in how we approach change and these new strategic plans with our customers. This is not something we can push down to them and expect everyone to buy into. Especially after COVID-19 and EMR challenges, we need to be very responsive to customers and understand what they are looking for. We also may not be aware of the individual cultures that a health system consists of. Minding our P's and C's and keeping AID in mind will help us build better strategic plans with all stakeholders, rather than pushing something down that is clearly not effective.

This is more than including clinical committees in technology design. This is about building plans for them that include them and their needs as part of the outcome. We've focused on outcomes that benefit the organization, not the confederation members. People want to feel validated and respected, especially within their professional cultures. Pushing frameworks and processes on top of them without respect for the culture and practices of their organizations will cause you both to fail. Chapter 11 of *The Security Culture Playbook* discusses how numerous organizational transformation programs over the past 50 years have failed because of the lack of understanding of (Carpenter and Roer 2022):

■ The organization's current culture
■ Future goals

- Purpose of the culture change program
- Inadequate resources
- Failure to handle change resistance
- General lack of cultural understanding and empathy

Our healthcare device and software manufacturers have hired Chief Medical Officers with the purpose of providing feedback and interpretation of the professional cultures and environment their software and devices will be used in back to the engineers who design the hardware and software. The Big 4 routinely hire doctors to help them provide better services to their hospital and health system customers. The Chief Medical Information Officer's role in healthcare organizations has been as the physician information technology leader, as defined by Leviss, Kremsdorf, and Moihadeen in their JAMIA article "The CMIO – a new leader for health systems" (Leviss, Kremsdorf, and Moihadeen 2006). Perhaps leveraging the CMIOs to help organizations use their P's and C's better as part of a larger strategic plan would benefit healthcare organizations and signal a positive change. They're not meant to be an island; however, sadly in many organizations they are without the support they need. With the CMIO also comes the Chief Nursing Informatics Officer (CNIO). Considering the single largest user group of clinical systems in hospitals is often nurses, they need to have their say as well to make sure that the systems meet their needs.

What Signals Are We Sending?

Going back to the question about signals, we are operating in a field of constant information asymmetry. There is a lot of information that we have as IT professionals that our customers do not. We do not effectively communicate this to our practice customers, who due to the professional standards of their respective organizations, have rigorous standards for evaluating credibility. Even Healthcare Finance has HFMA to communicate these standards. However, we've been disorganized in communicating about IT. We're going to use the concept of signaling theory from economics to describe characteristics of the information asymmetries we observe and add in criteria to better illustrate measures of credibility. We'll also use an example from television to do so.

The super asymmetry theory was discussed in the Big Bang theory. It fictionally posited to be an observer-relative opposite of supersymmetry, which proposes that every particle identified in the standard model has a supersymmetric partner (Lincoln 2019). Super Asymmetry, while being fictional, proposes that certain particles have observer-relative opposites.

While this is from a TV show, and not based on proven scientific theory, it is related to the concept of information asymmetry, which according to the Corporate Finance Institute is unequal, disproportionate, or lopsided information (CFI 2020). Like the concept of super asymmetry, information asymmetry is observer relative. We can observe how our customers know a lot less than we do about technology and security, and that they rely upon us to help them make complex decisions that their businesses rely upon. The article "Signaling Theory: A Review and Assessment," from the January 2011 *Journal of Management* by Brian L. Connelly, S. Trevis Certo, R. Duane Ireland, and Christopher R. Reutzel, discusses how signaling theory describes behavior when two parties have access to different information (Figure 9.1) (Connelly, Certo, Ireland, and Reutzel 2010). It discusses Communication of the Signal and its Interpretation. The example they give of information asymmetry is from Spence's 1973 article, which discusses how potential employers lack information about the quality of job candidates. They use the signal of higher education to signal their quality and reduce the information asymmetry present.

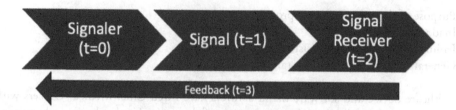

Figure 9.1 Signaling theory time graph. Source: Inspired by Connelly, Certo, Ireland, and Reutzel (2010).

This article identified four key concepts in signaling theory:

- Signaler – An insider who is in possession of information that others do not have.
- Signal – The deliberate communication of positive information to communicate positive organizational attributes.
- Receiver – Outsiders who lack information about the organization who would like to receive it, despite potentially conflicting interests.
- Feedback – Responsive signaling from the receiver designed to improve signal interpretation.

The way they fit together on a time graph is shown in Figure 9.1.

This article also discussed the concept of noise causing the signal to be misinterpreted (Figure 9.2), so we incorporated it between Signal and Receiver.

One of the other constructs we needed to look at was environment/results. The work we do in signaling is also dependent upon the environment the signal was communicated in and the results of the signaling. The paper discussed four stages and eight key constructs that can be utilized to model signals and messaging (Table 9.1).

What we've done is identify one additional stage and four constructs that help us better assess quantitative or qualitative assessments (Table 9.2).

The reasons we are discussing signaling is because of the current situation with Healthcare Information Security. People use compliance with established security standards such as the Health Insurance Portability and Accountability Act (HIPAA) Security Rule, HITECH Act, HITRUST, or the NIST Cybersecurity Framework to signal quality and assurance of safe data handling. The paper indicates that if a signaler does not have the underlying quality associated with the signal, however they believe that the benefits outweigh the costs, they will be motivated

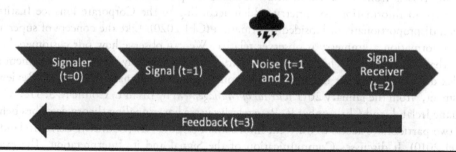

Figure 9.2 Signaling theory time graph with noise, which can cause the signal to be misinterpreted by the signal receiver. Source: Inspired by Connelly, Certo, Ireland, and Reutzel (2010).

Table 9.1 Key Signaling Theory Constructs from Paper Former Signal Theory Term Definitions and Constructs

Signal Stages/Constructs Stage/Actor	Construct	Definition	Representative Actors					
			Top-tier Academic Healthcare CIOs	Accredited Certification Firms (ISO, UL, NIST, HITRUST, AICPA)	Law Firms	Self-Assessments	Third-Party Non-certified Assessors	Representative Professional Organizations
Signaler	**Honesty (genuineness, veracity)**	Extent to which the signaler actually has the unobservable quality being signaled	High	High	High	Low	Low	High
	Reliability (credibility)	The combination of a signal's honesty and fit	High	High	High	Low	Low	High
	Institutional credibility							
Signal	**Signal cost**	Transaction costs associated with implementing a signal	Low	Low	High	Low	Low	High
	Observability (intensity, strength, clarity, visibility)	Signal strength, not accounting for distortions and deception	High	High	High	Low	High	High

(Continued)

Table 9.1 (Continued)   Key Signaling Theory Constructs from Paper Former Signal Theory Term Definitions and Constructs

Signal Stages/Constructs Stage/Actor	Construct	Definition	Representative Actors Top-tier Academic Healthcare CIOs	Accredited Certification Firms (ISO, UL, NIST, HITRUST, AICPA)	Law Firms	Self-Assessments	Third-Party Non-certified Assessors	Representative Professional Organizations
	Fit (value, quality)	Extent to which the signal is correlated with unobservable quality	High	High	High	Low	Low	High
	Frequency (timing)	Number of times the same signal is transmitted	Low	High	High	Low	High	High
	Consistency	Agreement between signals from one source	High	High	High	High	High	High
Receiver	**Receiver attention**	Extent to which receivers vigilantly scan the signaling environment	High	High	High	Low	High	High

(Continued)

Table 9.1 (Continued) Key Signaling Theory Constructs from Paper Former Signal Theory Term Definitions and Constructs

Signal Stages/Constructs — Stage/Actor	Construct	Definition	Representative Actors — Top-tier Academic Healthcare CIOs	Accredited Certification Firms (ISO, UL, NIST, HITRUST, AICPA)	Law Firms	Self-Assessments	Third-Party Non-certified Assessors	Representative Professional Organizations
	Receiver interpretation (calibration)	Amount of distortion introduced by the receiver, and/or weights applied to signals by the receiver	Low	Low	Low	High	High	High
Feedback/environment	Counter signals (feedback)	Responsive signaling from the receiver designed to improve signal interpretation	High	High	High	Low	High	High
	Distortion	Noise that can be introduced by the signaling environment, external referents, or other signalers	Low	Low	Low	High	High	Low
Environment/results	Dishonesty	Do dishonest signals pay for the actor?	No	No	No	Yes	Yes	No

Source: Inspired by Connelly, Certo, Ireland, and Reutzel (2010).

Table 9.2 Added Signal Theory Term Definitions and Constructs, Environment/Results Stage

Environment/Results	Dishonesty	Do dishonest signals pay?
	Instrument	Is the instrument or process open-ended, or does it have definitive controls?
	Institutional Credibility	Does the institution behind this represent high-quality work that others seek to emulate?
	Requires Certified Firm?	Does the assessment require a credible certified firm?

Source: Parker, Indiana University Health (2022).

to attempt false signaling (Connelly, Certo, Ireland, and Reutzel 2010). If this happens, misleading signals will proliferate until receivers learn how to ignore them. Signal costs need to be structured so that dishonest signals do not pay.

We discuss signaling because there is a significant information asymmetry with regard to Information Security. The HIPAA Security Rule requires organizations to maintain an Information Security Program, conduct risk assessments, and demonstrate ongoing compliance and review. Some organizations, such as HITRUST, go above and beyond what is required and provide a strict controls-based framework and review process by a certified assessor. The asymmetry that we observe being presented is that open-ended risk assessments are given the same weight in the marketplace as strict quantitative frameworks such as HITRUST or CSA Star Level 2.

HIPAA compliance is considered a benefit, and there are no controls to prevent someone from falsely signaling they are HIPAA compliant. ISC has a governing body to regulate who is certified by them based on quantitative exams. The American Institute of Certified Public Accountants (AICPA) can regulate who is a CPA based on exams and continuing education. However, we do not have the same controls for HIPAA Compliance or NIST Cybersecurity Framework certification, as they are open-ended frameworks. We have firms that will put their names on assessments demonstrating compliance. We also have firms that will utilize the name of their consulting firm to signal quality; however, they will deliver assessments that are subjective and do not meet the standards for control testing, ISO/IEC 17065:2012, Conformity assessment – Requirements for bodies certifying products, processes and services, and ISO/IEC 17067, Conformity assessment – Fundamentals of product certification and guidelines for product certification schemes, or separation of duties from consulting work as defined in ISO/IEC 17021, Conformity assessment – Requirements for bodies providing audit and certification of management systems – Part 1: Requirements (ISO 2014; ISO 2021; ISO 2016).

There is nothing stopping someone from saying they are HIPAA compliant and no incentive for them to be it because dishonest signals pay. The signals are often non-controls-tested and quantifiable and are based upon opinions. These are not the same levels of criteria we use for aerospace engineering, accounting, or medicine.

There is nothing stopping someone from signaling they are a security professional who can conduct HIPAA risk assessments. There is nothing stopping a large consulting firm that is already doing work for a company from mentioning they are also a security firm. There are multiple steps to someone from signaling that they are a neurosurgeon when they are not, starting with credentialing and verification, and preventing them from harming someone. You need to be licensed to be a teacher, doctor, or tattoo artist. You don't need to be licensed or meaningfully credentialed to conduct information security assessments. The steps needed to do the latter require both technical and procedural means. The same goes for the former, and below we will discuss the procedural means.

Security firms looking to signal quantifiable quality and security need to signal four characteristics:

- A recognized, credible certification
- Definitive controls as part of the evaluation instrument
- High institutional credibility
- Certification requires a recognized impartial certifier

Our assessment of many of the popular frameworks has indicated that only FedRAMP, ISO/IEC 27001/27002/27017/27018, CSA CCM/CAIQ STAR Level 2, a Third-Party PCI-DSS Attestation of Compliance, AICPA SOC 2, or HITRUST Common Security Framework certification can appropriately signal compliance with the HIPAA Security Rule. These are frameworks that due to definitive security controls, certification requirements of accreditors, and central bodies preventing organizations from claiming false compliance, accurately signal the ability to protect data. However, nothing is stopping companies from using the NIST Cybersecurity Framework or others internally to gauge their maturity. The issue is with external players signaling that they have the same level of credibility and rigor that the recognized, impartial certifications require. This is tantamount to claiming one is a doctor or board-certified specialist without possessing proof of either. Organizations need to align themselves with the right certifications to signal quantifiable, measurable, and credible quality and security (Table 9.3).

Our goal of using information security frameworks to demonstrate information asymmetry in signaling of security competencies was to contrast what Information Technology departments signal for competency and credibility as opposed to clinical practices. What most organizations are signaling for this does not meet the P's and C's of the professional cultures of clinical practice. Self-assessments are being passed off as being as credible as third-party certifications or peer reviews. This does not meet the professional standards that our medical practice customers are held to, and makes IT look less credible. In addition, since almost anyone can say they audit against many of these frameworks, that also makes many third-party firms look less credible, especially in the eyes of clinical practices. They have structured their organizations to protect against frauds and fabulists when many security teams have not.

Management Past the Book – What the Thought Leaders Didn't Tell You

One of the significant differences between IT and traditional disciplines such as Accounting, Finance, or Human Resources has been the use of the P's and C's within those professional groups. There is a corpus of written and unwritten communication along with professional organizations such as the Society of Human Resources Management (SHRM), AHIMA, and AICPA that provides leadership development and frameworks to build programs from. While IT does have organizations such as HIMSS and CHIME, a significant portion of the training they offer is not focused on building strategic programming or analytics. It is also not built upon building quantitative analysis. We need to get better at management to be able to execute a strategic plan and do it quickly. As a side note, as a participant in developing the Certified Digital Health Program

Table 9.3 Assessment of Popular Cybersecurity Frameworks

Signal Stages/Constructs			Representative Frameworks							
Stage/Actor	Construct	Definition	HIPAA/HITECH Third-Party Assessment	CSA CCM/CAIQ STAR Level 2	PCI-DSS Self-Assessment	PCI-DSS Third Party	AICPA SOC 2	HITRUST CSF	"HIPAA Certified"	Other Risk Assessment
Signaler	**Honesty (genuineness, veracity)**	Extent to which the signaler actually has the unobservable quality being signaled	Low	Low	Low	High	High	High	Low	Low
	Reliability (credibility)	The combination of a signal's honesty and fit that corresponds with the sought-after quality	Low	Low	Low	High	High	High	Low	Low
Signal	**Signal cost**	Transaction costs associated with implementing a signal	High	High	Low	High	High	High	Low	Low
	Observability (intensity, strength, clarity, visibility)	Signal strength, not accounting for distortions and deception	High	Low	High	High	High	High	Low	Low
	Fit (value, quality)	Extent to which the signal is correlated with unobservable quality	Low	Low	Low	Low	High	High	Low	Low
	Frequency (timing)	Number of times the same signal istransmitted	High	Low	Low	High	High	High	Low	Low
	Consistency	Agreement between signals from one source	High	High	Low	High	High	High	Low	Low

(Continued)

Table 9.3 (Continued)   Assessment of Popular Cybersecurity Frameworks

Signal Stages/Constructs			Representative Frameworks							
Stage/Actor	Construct	Definition	HIPAA/HITECH Third-Party Assessment	CSA CCM/CAIQ STAR Level 2	PCI-DSS Self-Assessment	PCI-DSS Third Party	AICPA SOC 2	HITRUST CSF	"HIPAA Certified"	Other Risk Assessment
Receiver	**Receiver attention**	Extent to which receivers vigilantly scan the signaling environment	High	High	High	High	High	High	Low	Low
	Receiver interpretation (calibration)	Amount of distortion introduced by the receiver, and/or weights applied to signals by the receiver	High	Low	High	Low	Low	Low	High	High
Feedback/ environment	**Counter signals (feedback)**	Responsive signaling from the receiver designed to improve signal interpretation	High	Low	Low	High	High	High	Low	Low
	Distortion	Noise that can be introduced by the signaling environment, external referents, or other signalers	High	High	High	High	Low	Low	High	High
Environment/ results	**Dishonesty**	Do dishonest signals pay?	Yes	No	Yes	No	No	No	Yes	Yes
	Instrument	Is the instrument or process open-ended, or does it have definitive controls?	Open-ended	Definitive	Definitive	Definitive	Definitive	Definitive	Open-ended	Open-ended

(Continued)

Table 9.3 (Continued)   Assessment of Popular Cybersecurity Frameworks

Signal Stages/Constructs			Representative Frameworks							
Stage/Actor	Construct	Definition	HIPAA/HITECH Third-Party Assessment	CSA CCMI CAIQ STAR Level 2	PCI-DSS Self-Assessment	PCI-DSS Third Party	AICPA SOC 2	HITRUST CSF	"HIPAA Certified"	Other Risk Assessment
	Institutional credibility	Does the institution behind this represent high-quality work that others seek to emulate?	Depends on Firm	Yes	No	Yes	Yes	Yes	No	Depends on Firm
	Requires certified firm?	Does the assessment require a credible certified firm?	No	Yes	No	Yes	Yes	Yes	No	No

Source: Inspired by Connelly, Certo, Ireland, and Reutzel (2010).

for Security, our book editor believes that prospective CHIME certification can and will become the source of truth, competence and legitimacy about trustworthy security in healthcare IT. This is while bridging leadership and knowledge gaps between medicine and technology. Ashley Jester and David Finn, leader from the certification and VP education division respectively at CHIME explain well and paraphrased that:

> The purpose behind the CDH Security certification is to bring out leadership, strategy, and change management (that's 67% exam weighting) and how that interacts with technology and services (at a 33% weighting). The CHIME Domain competency areas have stood the test of time. Originally developed in 2008 from the Certified Healthcare CIO (CHCIO) Program, their seven Domain competency areas were critically assessed and identified to capture the essence of building and identifying what makes up true leaders mixed with healthcare IT expertise. Many of the top healthcare CIOs we know hold CHCIO certification. This was a long process of development training. They held a workshop that was led by a third-party certification and testing agency; they helped CHIME identify targets, develop candidate requirements, build consistent metrics that tie to the exam, and a thorough training for the exam build. This third party service helped provide the structure, but ultimately intellectually –it was CIOs developing the content and metrics for other CIOs. Many of those involved in the initial process still serve on those boards or as item writers for continued development and refreshment. The program then evolved as CHIME University which was born in just 2021 and opening CHIME up to all healthcare professionals, not just qualified senior level healthcare IT leaders. With that, they aim to move out to all levels of healthcare professionals, with certification, culminating in the launch of the Certified Digital Health (CDH) Program. Now, CHIME continues to enhance CDH and branch out to other focus areas within the healthcare industry such as security. They are becoming the source of truth and credibility while bridging the gap in this digital healthcare space. There is no other certification designed like the Certified Digital Health, CDH from CHIME.

We believe that this certification program could go a long way towards building trust and respect with stakeholders including physicians when it comes to IT competency.

Further, we need to start by building partnerships across the organization. We need to start with the clinical practices. The goal we have here is to build these with goals of sharing information and better aligning with organizational goals and strategies. We need to have the intent of breaking down the silos. However, we have to be able to leverage quantitative techniques as part of the strategic plan to do so.

We also must directly address toxicity in IT. One of the major concerns in the tech industry is that the workplace has an issue with this. Alison DeNisco Rayome, in her *TechRepublic* article, "52% of tech employees believe their work environment is toxic," discusses a 2018 workplace survey from Blind, an anonymous workplace review service. This survey had 12,549 respondents and indicated that 52% of IT employees surveyed did not consider their workplace to be healthy (DeNisco Rayome 2018). This article further indicated that a toxic workplace culture is one of the top three reasons for burnout.

Toxic team members feel that their hatred and dislike of others is more important than the well-being of others. Some also believe that their need for attention and feeling important is more important than others. Self-interest, toxicity, and bad behavior is the opposite of collaboration. Customers and people will go out of their way to avoid toxic and negative behavior. This means that they will even cause additional information risk if the security people are the toxic ones. Technology is infamous for toxicity, and the indifference of tech CEOs doesn't help. If we see signs of toxicity, we must remove it immediately. We cannot let people, no matter how skilled they are, hold us hostage and hurt others because they possess a certain skill set.

We need to build to engage our team members and use our long-range planning to build development plans to keep them engaged. We can't just throw team members to the wolves. We also just can't expect them to figure it out for themselves. We must lead. The goal of the plan we're building here is to give you a framework to follow, modify, and use to lay out a well-organized long-range plan that addresses your strategic alignment with the practices, strategic well-managed goals, and provide measurable improvement as defined by Healthcare IT and Security frameworks with well-defined controls sets. In other words, we want to give you a framework comparable to ones used in other areas that aligns with, improves, and provides self-determination mechanisms for the stakeholders. It's not about what we can do for them. It's about improving with them.

Where Do We Start?

1. We must see how the organization itself measures success and enterprise risk. We need to understand their metrics in detail. We need to look at:
   - Enterprise risk metrics and the previous enterprise risk assessment, if applicable
   - Financial metrics and forecasting
   - Benchmark goals for practice areas and business units
   - Recommended benchmark metrics for practice areas
   - Recommended benchmark metrics for support services such as revenue cycle, finance, real estate, and home health
   - Ethical and bias concerns, specifically with how P's and C's may inadvertently lead to bias
   - Any other metrics your interviews with leadership uncover

   We need to then take all these metrics and determine which ones are most important to stakeholders and why, not just opinions of one executive. Understand why these metrics are important to them and what they demonstrate. Ask these questions and comprehend why they measure performance.

2. After this, conduct another enterprise risk assessment, focusing on the metrics you gathered information on in the previous step. Understand current numbers and metrics and gather information to understand exactly where the organization is. Data elements you want to gather are:
   a. Assigned Unique ID
   b. Practice Area/Business Unit
   c. Risk Statement
   d. Discussion Notes
   e. Gaps/Risks/Ideal Metrics
   f. Recommendation
   g. Risk Scoring (Negligible = 1, Low = 2, Low/Medium = 3, Medium = 4, Medium/High = 5, High = 6):
      i. Likelihood of Control Failure Causing an Issue
      ii. Velocity (Speed of Risk)
      iii. Impact
      iv. Potential Income Loss
      v. Reputation Loss

    h. Risk Score is Likelihood × Velocity × Impact × Potential Income Loss × Reputation Loss

3. Perform a secondary gap analysis, review the results, and refine the scores.
4. Pick the highest-scored risks to address. Pick the top 20% or so. Your organization will not have the bandwidth to address 100% of risks. Doing so would be the biggest risk as likely you do not have resources to conduct business needed to generate income at the same time.
5. Identify accountable executives and team members who will hold themselves responsible to complete these tasks. No digital transformation will work without accountability starting at the top.
6. Leverage the Digital Transformation Meta Framework (DTMF) we articulate here to build your Framework strategy.
7. Continually follow up and keep this framework current. It's not one and done! After all, Michael Jordan didn't retire after his first championship. Neither did John Elway's Denver Broncos.

The Digital Transformation Meta Framework (DTMF)

The purpose of this framework is to draw from:

- Organizational Strategies
- Practice Strategies
- Practice Goals
- Organization Goals
- Healthcare IT/Benchmarking Frameworks
- Aligned Goal-Setting
- Enterprise Architecture
- Detailed Plan Walkthroughs/Development in alignment with Goal-Setting
- Applicable Security Frameworks
- Targeted Metrics
- Communication Plans

The goal is present a unified picture that aligns from Strategy to Measurable Metrics leveraging multiple existing frameworks to do so cohesively. This is to take the inputs of many, process them in a realistic way, and then demonstrate how reaching these goals provides demonstrable organizational improvement resulting in multiple transformations resulting in organizational benefits. We have six main tenets of what we're trying to accomplish that provide overarching vision:

- Build cross-organizational and cross-stakeholder partnerships
- Identify accountable executives who will see initiatives through
- Align initiatives with mission and values
- Reduce silos of information and resources across the organization
- Reduce overall organizational risk
- Reduce legacy systems and stovepipe processes

We want to list out those key strategies for organizations and practices, their time frames, and which ones we can objectively meet within the given time frames. We then want to list out the

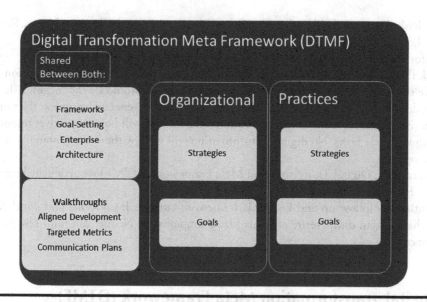

Figure 9.3   Digital Transformation Meta Framework. Source: Parker, Indiana University Health (2022).

Practice and Organization Goals, their time frames, and what we can objectively achieve with them. What we then can do is put those together with shared steps to get a clearer picture of what we are able to do (Figure 9.3).

Next, we want to evaluate what Healthcare IT Benchmarking framework/maturity model to use to demonstrate progress with the system. The two major choices are the HIMSS Electronic Medical Record Adoption Model or the College of Healthcare Information Management Executives' (CHIME) Most Wired. The EMRAM measures clinical outcomes, patient engagement, and clinical use of EMR technology (HIMSS 2022 2). It also serves as an excellent method to optimize multiple organizational characteristics based on evidence-based data, including reducing errors, length of stay, and streamlining data access and use (HIMSS 2022 2). It also has goals of improving patient safety and satisfaction. It has seven levels, which are called stages, shown in Figure 9.4

HIMSS Electronic Medical Records Adoption Model Capability Levels

The goal with large strategic initiatives and goals is to move up to reach or maintain Stage 7 for the organization. The EMRAM is a highly accepted standard, and very few organizations have achieved level 7 status. Level 6 or 7 maturity model certification requires the use of peer and HIMSS reviewers to be certified (Rayner 2015). In this way, this aligns with many of the requirements of clinical practices that must be certified by CAP, ACS, Joint Commission, or some other similar organization.

The CHIME Most Wired survey and certification program is a consistently updated "Digital Health Check-up" for healthcare organizations (CHIME 2022). It covers EMR usage, Telemedicine, Infrastructure, Security, Digital Transformation, Innovation, Analytics, and multiple other areas. It is organized into ten levels within four groupings (CHIME 2022):

■ Levels 9 and 10: In addition to meeting the criteria for levels 1–8, organizations in levels 9 or 10 are often leaders in healthcare technology who actively push the industry forward. Not only have many of them implemented advanced technologies, but they

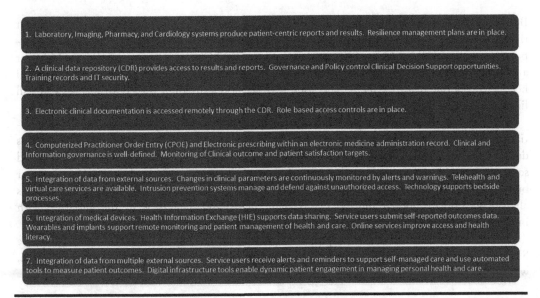

Figure 9.4   EMR adoption model capability levels. Source: HIMSS (2022).

often leverage these technologies in innovative ways and have encouraged deep adoption across their entire organization. As a result, they are realizing meaningful outcomes, including improved quality of care, improved patient experience, reduced costs, and broader patient access to healthcare services. Some of the advanced technologies used to achieve these outcomes include telehealth solutions, price transparency, cost-analysis tools, access to data at the point of care, and tools to engage patients and their families throughout the care process.

- Levels 7 and 8: Organizations in levels 7 and 8 meet the criteria for being designated as Most Wired. These organizations have deployed technologies and strategies (e.g., population health/cost-of-care analytics, HIEs/integration engines, and patient portals) to help them analyze their data and are starting to achieve meaningful clinical and efficiency outcomes. Some of these organizations are experimenting with more advanced technologies, like telehealth, that expands access to care.
- Levels 4–6: Organizations in levels 4–6 have made progress in expanding their core IT infrastructure to support internal strategic initiatives. Often, they have implemented basic technologies to protect patients' health and financial information (e.g., firewalls, spam/phishing filters, endpoint encryption), but they may lack more advanced technologies that would mediate other vulnerabilities. Many are actively collecting patient data electronically; however, they may not effectively leverage the data they collect and may encounter significant barriers in exchanging patient data with external organizations.
- Levels 1–3: Organizations in levels 1–3 are in the early stages of developing their technology infrastructure and may still be transitioning, or may have more recently transitioned, to electronic formats for collecting patient data and performing clinical activities. Some may have deployed technologies that capture data (e.g., EMRs, ERP solutions, revenue cycle management solutions) but may not fully leverage the functionality these technologies offer. Additionally, these organizations may still be working to help end users adopt the technologies that have been implemented.

Both maturity models are commensurate with a high level of digital transformation maturity at the higher levels. HIMSS has multiple other maturity models to assess other organizational components, such as INFRAM for Infrastructure, O-EMRAM for Outpatient EMRs, AMAM for analytics, CCMM for Continuity of Care, and CISOM for Supply Chain (Lyons 2021). The goal here is to pick one maturity model and stick with it. Trying to align with both means, you will likely not achieve either.

The next step is to find an Evaluation Partner that can help your organization identify the initiatives needed as part of the long-range plan. These need to be aligned with the P's and C's. This doesn't have to be an external resource. This can be someone internal; however, we recommend someone external so that you have that impartial view. However, we do not have control over your budgets. This partner needs to be able to develop a strategic plan, organize it along strategic paths aligned with organizational and practice strategic plans, and then associate the initiatives with improvements in the practice P's and C's and associated maturity model.

For each of the initiatives, develop a set of goals that meet the SMARTEST criteria. While SMART (Specific, Meaningful, Achievable, Relevant, and Time-bound) goals have been there, we've added three extra steps on adding empathy, synchronization, and tracking:

1. *Specific* – Are your goals targeted, with numbers or other measurable metrics behind them?
2. *Meaningful* – Do they matter to your stakeholders?
3. *Achievable* – Are these goals realistic, concrete, and able to be completed within a defined time period? Do they allow for self-determination by customers to reach their goals?
4. *Relevant* – Are they relevant to your life, work, and what your customer wants to improve? Do they align with any of the identified areas of the risk assessment? Do they align with the business?
5. *Time-bound* – Are you able to set an exact date for goal achievement? You need to make them measurable.
6. *Empathic* – Are the goals considerate and understanding of the needs of others, or are they focused just on one person? Does it focus on putting you above others because something you know or perceive puts you above them, or understanding them better as equals?
7. *Synchronized* – Are they aligned with the specific qualities you want to improve in the organization, and/or with further improving its positive qualities?
8. *Tracked* – Are they written down, reviewed, and measured on a daily, weekly, and monthly basis toward their goal?

Use a tool such as an A3 problem-solving format methodology to track metrics and goals as part of the overall picture (Healy 2020). This puts all the critical information in one place. This lets executives understand what success looks like and where potential issues with delivery may lie.

Choosing an Architecture

For your initiatives, select an Enterprise Architecture to have a common structure and methodology to develop and manage them. Even though most enterprises have a technical architecture in place to define the technology requirements, Enterprise Architectures has many underlying structural requirements, which in a heterogenous environment like healthcare are not as pervasive as many leaders think. The first of the most prevalent models is the Zachman Framework™, developed by John A. Zachman, which is more of a metamodel that is based on an ontology. It is more of a toolset designed to build architectures on top of it (Zachman 2008). The founding company, Zachman International, owns FEAC Institute, the organization that certifies practitioners as Certified Enterprise Architect (CEA) Black Belt, Associate Certified Enterprise Architect (ACEA) Green Belt, or Winter Certified Enterprise Architect Practitioner (CEAP) Yellow Belt (Zachman 2008). While this framework has certifications signaling competency, it requires significant work to build an Enterprise Architecture that meets customer needs.

The second one, The Open Group Architecture Framework (TOGAF), is an architectural standard developed by The Open Group, currently in its 10th edition (The Open Group 2022). It is divided into nine phases plus a shared Requirements Management task (The Open Group 2022 2):

- Preliminary Work
- Requirements Management
- Phase A: Architecture Vision
- Phase B: Business Architecture
- Phase C: Information Systems Architecture
- Phase D: Technology Architecture
- Phase E: Opportunities and Solutions
- Phase F: Migration Planning
- Phase G: Implementation Governance
- Phase H: Architecture Change Management

The level of detail in these phases, as documented in the TOGAF Standard – 10th Edition Reference Cards, addresses building out the structure needed to architect, develop, manage, and sustain multiple large initiatives. As TOGAF is a supported standard, The Open Group provides educational materials, courses, and certifications for practitioners (The Open Group 2022).

Depending on your organization's needs and technical sophistication, it is important to pick the Enterprise Architecture that meets them. Most healthcare organizations aren't going to have the sophistication to build off the Zachman Framework and would opt for TOGAF as it is an established standard with certifications for implementors available.

Leveraging Use Cases to Identify and Score Potential Failure Points

Leverage Use Cases, hopefully written in the Object Management Group's (OMG) Unified Modeling Language (UML), as part of the Requirements Management phase to model processes and usage. This is a design process used to model software behavior; however, it can also be used to model non-software systems (OMG 2005). It can model three types of diagrams: Structure Diagrams, Behavior Diagrams, including Use Cases, and Interaction Diagrams, which include Sequence Diagrams. UML can associate data elements with Use Cases and track their usage via Sequence Diagrams. This is incredibly useful for having to track data usage in regulated scenarios, such as patient data covered under HIPAA or Personal Data covered by the European Union General Data Protection Regulation (GDPR). System modeling using Use Case Diagrams, Data via Class Diagrams, and Sequence and Communication Diagrams provides an excellent tool to visualize both processes and data usage.

Understanding Use Cases provides a framework for building a testable methodology of seeing where processes can fail. Leverage Failure Mode and Effects Analysis (FMEA), a systemic, proactive method for evaluating processes as part of process walkthroughs to see where and how they may fail and identify their impact, with these use cases to identify potential failure points (IHI 2017). The recommended process is to follow the American Society for Quality's, which is available at https://asq.org/quality-resources/fmea, for each Use Case (ASQ 2022). Leverage the Sequence Diagrams to identify process flows and potential failure points. For each failure, calculate the Severity, which is how serious each effect is, from a scale of 1–10, the Occurrence, which is the probability of failure, also from 1–10 scale, and the Detection Rating, which is how well the current controls can detect either the cause or failure more after they have happened, but before the customer is affected. That is also 1–10. Multiply the three together to get the Risk Priority Number (RPN).

When these have been identified, rank the potential failures, highest RPN first, and develop plans to address them. These can include technical, non-technical, and process-based methods. Address these changes and reevaluate the use cases to determine if the risks have been mitigated.

TOGAF covers monitoring as part of Phase H, Architecture Change Management (The Open Group 2022 2). Develop Operational and Security Management Plans for the initiatives. Include detailed operational plans for monitoring for system exceptions and planning for system failures. Leverage knowledge from the FMEA analysis to develop methods for addressing failures and incident management plans in accordance with your chosen architectural framework.

When you apply the DTMF, which will recontextualize and change the initiative itself, there will be significant other changes. Previous assumptions that were made will no longer apply. This requires a recontextualization, starting with the budgets, cost models, and total cost of ownership (TCO), as they will be affected the most due to the additional visibility. Don't make assumptions about resource availability, or that IT is a bottomless resource that can be cut. It cannot be cut out, and if it is, it is to the detriment of the organization.

As part of both EMRAM and Most Wired, security is a requirement. This must be evaluated and scored to be able to demonstrate improvement. To do so, select a credible Security Management Framework that meets organizational requirements to evaluate maturity against. We recommend leveraging HITRUST, ISO/IEC 27001/27002, FedRAMP, or quantitatively scored versions of the NIST Cybersecurity Framework or CMS Security Risk Assessment Tool (SRA Tool) (HHS ONC 2022) done by a credible certified third party. Associate initiatives and goals with score improvements on this framework. Leveraging these processes will signal higher credibility and align with similar requirements that practices have.

Change is also a significant part of any initiative. Lack of information and Fear of Missing Out (FOMO) can lead to anxiety with team members. When we change processes, even to make them better, if our stakeholders don't know what to do, they're not going to execute if they do not know what they are. Develop Communication Plans to help address organizational and stakeholder communication. These are incredibly important to plan for operational, security, and risk management. While these have been often overlooked in the past, these are critical as part of the operational management plans.

How Do We Track This?

You can use a project tracking tool such as Trello, Project Online, Basecamp, or Airtable to track these. We've also developed a small spreadsheet you can use as part of reporting to map out all of this and provide an easy-to-use mapping (Table 9.4).

Table 9.4   Example Goal Tracking Spreadsheet Table for Tracking SMARTER Goals

Initiative Name	Initiative Description	Expected Duration	Executive Sponsor	Strategy Component	Strategic Pillar	Target Improved Aligned Associations/ Frameworks/ P's and C's	Target SMARTER Goals	Current/ Target IT Framework Levels	Current/ Target Security Framework Levels	Communication Plan?

Source: Parker, Indiana University Health (2022).

What Do We Get from Using the DTMF?

We get a framework that we can use to plan out long-range plans in alignment with the rest of the organization. Instead of just focusing on one framework, we take a holistic approach to aligning with business requirements and needs of practices, establishing goals, and using tested maturity models and frameworks to increase the overall maturity. This is in opposition to pushing down initiatives that do not align.

What Do We Have to Do After?

We must continually follow up. Conduct your annual security risk assessments. Revisit this plan at least annually to update it. Conduct use case modeling and FMEA analysis on major changes. Leverage this plan as the management framework going forward. Don't go back to the old ways just because the project is done. The plan is continual, much like digital transformation is.

Conclusion

This chapter used security as the character foil for the management struggles and challenges that healthcare organizations face in organizational and digital transformation projects and having that strategic seat at the table. It illustrated that the general lack of maturity in IT as compared to the practices they serve leads to disconnects much of the time. These disconnects have manifested in the implementation of IT strategies and systems that often do not serve who they need. Addressing this requires understanding the characteristics of these practices. It also involves understanding the concept of signaling, and how signaling in security does not appropriately communicate assurance in terms that our customers need. From there, we developed an operational plan that organizations can use to effectively implement the frameworks and processes needed to develop and maintain a realistic and holistic secure long-range digital transformation plan using quantitative management measures. These allow organizations to demonstrate increasing maturity along multiple dimensions and improve both the organizations and the associated practices they serve. The goal here isn't to build yet another system to replace the ones the medical staff are disillusioned with. It's to build a framework that incorporates their needs while providing AID, Alignment, Improvement, and self-Determination. Shadow IT isn't shadow IT if it predates the IT department. If we want the buy-in, we must present a better solution than our customers already have with a goal of their improvement in mind. It is our hope that security as that character foil, and the work here, present you with concrete tools and techniques you can use, so you can use what you have learned in this chapter to build more comprehensive digital transformation plans, including security.

References

"Part A – Introduction to AAMI." Association for the Advancement of Medical Instrumentation (AAMI). Accessed May 9, 2022. https://www.aami.org/standards.old/participate-in-standards-development.old/effective-participation-in-standards/resources-for-new-committee-members/part-A---introduction-to-aami#:~:text=The%20Association%20for%20the%20Advancement,safe%20and%20effective%20health%20technology.

"Apa Dictionary of Psychology." American Psychological Association (APA). Accessed May 9, 2022. https://dictionary.apa.org/cultural-norm.

"Asymmetric Information." Corporate Finance Institute (CFI). February 26, 2020. https://corporatefinanc
einstitute.com/resources/knowledge/finance/asymmetric-information/.

"Our History." Who We Are. American Health Information Management Association (AHIMA). Accessed
May 9, 2022. https://www.ahima.org/who-we-are/about-us/history/.

"CHIME DHMW Resource Page 2022." College of Healthcare Information Management Executives
(CHIME). Accessed May 9, 2022. https://info.dhinsights.org/dhmw-2022.

"Credibility." Oxford University Press. Accessed May 9, 2022. https://www.oxfordreference.com/view/10
.1093/oi/authority.20110803095646819.

"Definition of Business Process – Gartner Information Technology Glossary." Gartner, Inc. Accessed May
9, 2022. https://www.gartner.com/en/information-technology/glossary/business-process.

"Electronic Medical Record Adoption Model (EMRAM)." HIMSS. January 3, 2022. https://www.himss
.org/what-we-do-solutions/digital-health-transformation/maturity-models/electronic-medical-record
-adoption-model-emram.

"Facts about the Joint Commission." The Joint Commission (TJC). Accessed May 9, 2022. https://www
.jointcommission.org/about-us/facts-about-the-joint-commission/#35f1e7a473fe442a961daac
a511819de_6c212db320904c41b347a0a3858eaf65.

"Failure Mode and Effects Analysis (FMEA)." What is FMEA? American Society for Quality (ASQ).
Accessed May 9, 2022. https://asq.org/quality-resources/fmea.

"Failure Modes and Effects Analysis (FMEA) Tool: IHI." Institute for Healthcare Improvement
(IHI), 2017. http://www.ihi.org/resources/Pages/Tools/FailureModesandEffectsAnalysisTool
.aspx#:~:text=Failure%20Modes%20and%20Effects%20Analysis%20(FMEA)%20is%20a%20
systematic%2C,most%20in%20need%20of%20change.

"Framework Definition & Meaning." Merriam-Webster. Accessed May 9, 2022. https://www.merriam
-webster.com/dictionary/framework.

"ISO/IEC 17021-1:2015." ISO. December 1, 2016. https://www.iso.org/standard/61651.html.

"ISO/IEC 17065:2012." ISO. June 1, 2014. https://www.iso.org/standard/46568.html.

"ISO/IEC 17067:2013." ISO. September 1, 2021. https://www.iso.org/standard/55087.html.

"Journal Policies." Oxford Academic. Accessed May 9, 2022. https://academic.oup.com/rev/pages/Policies
#Peer%20review%20process.

"Licensure vs. Certification." International Certification & Reciprocity Consortium (IC&RC). Accessed
May 9, 2022. https://internationalcredentialing.org/lic-cert.

"History of Patent Medicine." Hagley Museum. June 21, 2017. https://www.hagley.org/research/digital
-exhibits/history-patent-medicine.

"75th Anniversary." HFMA Turns 75. Healthcare Financial Management Association (HFMA). Accessed
May 9, 2022. https://www.hfma.org/about-hfma/75th-anniversary.html#:~:text=HFMA%20was
%20launched%20in%201946,as%20AAHA%20president%2C%20Frederick%20T.

"HIMSS Celebrates 60 Years with Major Membership Milestone." Healthcare Information and Management
Systems Society (HIMSS). May 6, 2021. https://www.himss.org/news/himss-celebrates-60-years
-major-membership-milestone.

"NCI Dictionary of Cancer Terms." National Cancer Institute. Accessed May 9, 2022. https://www.cancer
.gov/publications/dictionaries/cancer-terms/def/protocol.

"The PROSCI ADKAR® Model." Prosci, Inc. Accessed May 9, 2022. https://www.prosci.com/methodology
/adkar.

"RSNA." Radiological Society of North America – Life Science MRI Facility – Purdue University. Purdue
University. Accessed May 9, 2022. https://www.purdue.edu/hhs/mri/NewsEvents/RSNA.html.

"Security Risk Assessment Tool." US Department of Health and Human Services, Office of the National
Coordinator (HHS ONC). February 15, 2022. https://www.healthit.gov/topic/privacy-security-and
-hipaa/security-risk-assessment-tool.

"Florida Family Indicted for Selling Toxic Bleach as Fake 'Miracle' Cure for Covid-19 and Other Serious
Diseases, and for Violating Court Orders." The United States Department of Justice. USDOJ. April
26, 2021. https://www.justice.gov/usao-sdfl/pr/florida-family-indicted-selling-toxic-bleach-fake-mir-
acle-cure-covid-19-and-other.

"Sputnik, 1957." MILESTONES: 1953–1960. U.S. Department of State, Office of the Historian (US State Department). Accessed May 9, 2022. https://history.state.gov/milestones/1953-1960/sputnik.

"The TOGAF® Standard, a Standard of the Open Group." The Open Group. Accessed May 9, 2022. https://publications.opengroup.org/standards/togaf.

"The TOGAF Standard, 10th Edition Reference Cards." The Open Group, 2022. https://online.anyflip .com/nnfi/kmwp/mobile/index.html.

"What Does it Mean When a Publication is Peer Reviewed?" U.S. Geological Survey. US Geological Survey (USGS). Accessed May 9, 2022. https://www.usgs.gov/faqs/what-does-it-mean-when-publication -peer-reviewed?qt-news_science_products=0.

"What is Shadow it?" Cisco. December 14, 2020. https://www.cisco.com/c/en/us/products/security/what -is-shadow-it.html.

"What is UML." Unified Modeling Language. Object Management Group (OMG). July 2005. https:// www.uml.org/what-is-uml.htm.

Akalp, Nellie. "LLC's: Best Biz Structure to 'House' Multiple Properties." *CorpNet*, April 1, 2022. https:// www.corpnet.com/blog/llcs-best-biz-structure-to-house-multiple-properties/.

Baum, Stephanie. "6 Must-Haves for BYOD to Work: A Chief Information Security Officer's Perspective." *MedCity News*, March 14, 2014. https://medcitynews.com/2014/03/6-things-needed-byod-chief -information-security-officers-perspective/.

Bergengruen, Vera. "'America's Frontline Doctors' Peddle Bogus Covid-19 Treatment." *Time*, August 26, 2021. https://time.com/6092368/americas-frontline-doctors-covid-19-misinformation/.

Bertram, Colin. "Tammy Faye and Jim Bakker: Inside their Relationship and the Scandals that Brought down their Empire." *Biography.com*. A&E Networks Television, September 15, 2021. https://www .biography.com/news/tammy-faye-jim-bakker-relationship-scandals.

Blizzard, Rick. "Nurse Engagement Key to Reducing Medical Errors." *Gallup.com*, April 11, 2021. https:// news.gallup.com/poll/20629/nurse-engagement-key-reducing-medical-errors.aspx.

Brown, Seth. "Alex Jones's Infowars Media Empire is Built to Sell Snake-Oil Diet Supplements." *Intelligencer*, May 4, 2017. https://nymag.com/intelligencer/2017/05/how-does-alex-jones-make-money.html.

Carpenter, Perry, and Kai Roer. *The Security Culture Playbook*. Hoboken, NJ: John Wiley & Sons, Inc., 2022.

Clement, Nicole. "Cybersecurity Assessment Tool – FFIEC Home Page." FFIEC Cybersecurity Assessment Tool – May 2017. FFIEC, June 1, 2017. https://www.ffiec.gov/pdf/cybersecurity/FFIEC_CAT_May _2017.pdf.

Connelly, Brian L., S. Trevis Certo, R. Duane Ireland, and Christopher R. Reutzel. "Signaling Theory: A Review and Assessment." *Journal of Management* 37, no. 1 (December 20, 2010): 39–67. https://doi .org/10.1177/0149206310388419.

Davis, Jessica. "'Voluntary Practices' in Healthcare Insufficient for its Dependence on Legacy Tech." *SC Media*, May 18, 2022. https://www.scmagazine.com/feature/critical-infrastructure/voluntary-prac -tices-in-healthcare-insufficient-for-its-dependence-on-legacy-tech.

DeNisco Rayome, Alison. "52% of Tech Employees Believe their Work Environment is Toxic." *TechRepublic*. TechnologyAdvice, November 29, 2018. https://www.techrepublic.com/article/52-of-tech-employees -believe-their-work-environment-is-toxic/.

Dennedy, Michelle, Jonathan Fox, and Tom Finneran. *The Privacy Engineer's Manifesto: Getting from Policy to Code to QA to Value*. Berkeley, CA: Apress, 2014.

Dubinsky, Lauren. "From Alarm Management to AI, Patient Monitoring Gets a Facelift," February 10, 2020. https://www.dotmed.com/news/story/49813%20Date:%20February%2019,%202020.

Espinosa-Brito, Alfredo D., and José M. Barmúdez-López. "The Clinical Method is the Scientific Method Applied to the Care of a Patient." *The BMJ*, June 27, 2004. https://www.bmj.com/rapid-response/2011 /10/30/clinical-method-scientific-method-applied-care-patient.

Fiore, Kristina. "America's Frontline Doctors' Prescriber Stripped of All State Licenses." *Medical News*. MedpageToday, April 26, 2022. https://www.medpagetoday.com/special-reports/exclusives /98407.

Fornell, Dave. "7 Reasons Clinicians are Leaving Jobs in the Era of the Great Resignation." *Health Exec. Innovate Healthcare*, April 19, 2022. https://healthexec.com/topics/healthcare-management/

healthcare-economics/7-reasons-clinicians-are-leaving-jobs-era-great?utm_source=newsletter&utm
_medium=he_weekly.

Frederik, Jesse. "Blockchain, the Amazing Solution for Almost Nothing." *The Correspondent*, August 21,
2020. https://thecorrespondent.com/655/blockchain-the-amazing-solution-for-almost-nothing.

Gawande, Atul, Siddhartha Mukherjee, and Jerome Groopman. "Why Doctors Hate their Computers."
The New Yorker, November 5, 2018. https://www.newyorker.com/magazine/2018/11/12/why-doctors
-hate-their-computers.

Hannigan, Dave. "Joe Rogan's Snake-Oil Shop the Go-to for the Likes of Aaron Rodgers." *The Irish Times*,
November 10, 2021. https://www.irishtimes.com/sport/other-sports/joe-rogan-s-snake-oil-shop-the
-go-to-for-the-likes-of-aaron-rodgers-1.4724590.

Healy, Brendan. "A3 – a Lean Approach to Problem Solving." *Lean Construction Blog*, December 9, 2020.
https://leanconstructionblog.com/A3-A-Lean-Approach-to-Problem-Solving.html.

Hutchinson, Lee. "How We Learned to Break down Barriers to Machine Learning." *Ars Technica. Conde
Nast, Inc.*, May 19, 2022. https://arstechnica.com/information-technology/2022/05/how-we-learned
-to-break-down-barriers-to-machine-learning/.

Lawsky, Benjamin M. "Banking Division Industry Letter: FS-ISAC Participation Recommended for all
NYS-Chartered Depository Institutions." New York State Department of Financial Services, February
6, 2014. https://www.dfs.ny.gov/system/files/documents/2020/03/il140206.pdf.

Leviss, Jonathan, Richard Kremsdorf, and Mariam F. Mohaideen. "The CMIO–a New Leader for Health
Systems." *Journal of the American Medical Informatics Association* 13, no. 5 (September 1, 2006): 573–
78. https://doi.org/10.1197/jamia.m2097.

Lincoln, Don. "Did 'The Big Bang Theory' Get the Science Right? A Lesson in Supersymmetry and
Economy Class." *LiveScience*. Future US Inc., January 22, 2019. https://www.livescience.com/64557
-the-big-bang-theory-super-asymmetry.html.

Lyons, Jeanne. "Maturity Models." *HIMSS*, August 24, 2021. https://www.himss.org/what-we-do-solutions
/digital-health-transformation/maturity-models.

Medellin, Alejandro. "Here's What You Need to Know about Foil Characters." *The Beat*, December 8,
2020. https://www.premiumbeat.com/blog/foil-characters/.

Monegain, Bernie. "A Look inside Epic's EHR Design and Usability Teams." *Healthcare IT News*. HIMSS
Media, September 8, 2016. https://www.healthcareitnews.com/news/look-inside-epics-ehr-design
-and-usability-teams.

Powell, Alvin. "How Sputnik Changed U.S. Education." *Harvard Gazette*, October 11, 2007. https://news
.harvard.edu/gazette/story/2007/10/how-sputnik-changed-u-s-education/.

Rayner, John. "Electronic Medical Record Adoption Model (EMRAM)ˢᴹ." *HIMSS-UK*, July 8, 2015.
https://na.eventscloud.com/file_uploads/91881bd83f4bafaa97c35daaef89f635_John_Rayner.pdf.

Russell, William, and David Feinberg. "Lessons in Health it from Dr. David Feinberg's 25 Years in
Healthcare." *This Week Health*. Health Lyrics, April 12, 2022. https://www.thisweekhealth.com/vid-
eos/lessons-in-health-it-from-dr-david-feinbergs-25-years-in-healthcare/.

Salas, Erick Burgueño. "Number of Satellites by Country 2021." *Statista*, July 21, 2021. https://www.statista
.com/statistics/264472/number-of-satellites-in-orbit-by-operating-country/.

Salter, Jim. "Jim Bakker to Pay $156K over Covid-19 Cure Claim." *News & Reporting*. Christianity Today,
June 24, 2021. https://www.christianitytoday.com/news/2021/june/jim-bakker-show-covid-cure
-lawsuit-silver-missouri.html.

Shanafelt, Tait D., Edgar Schein, Lloyd B. Minor, Mickey Trockel, Peter Schein, and Darrell Kirch.
"Healing the Professional Culture of Medicine." *Mayo Clinic Proceedings* 94, no. 8 (August, 2019):
1556–66. https://doi.org/10.1016/j.mayocp.2019.03.026.

Sinno, Michael, Snehal Gandhi, and Molly Gamble. "8 Problems Surrounding Meaningful Use." *Becker's
Hospital Review*. Becker's Healthcare, April 28, 2011. https://www.beckershospitalreview.com/
healthcare-information-technology/8-problems-surrounding-meaningful-use.html.

Staff, MasterClass. "Writing 101: What is a Foil Character in Literature? Learn about 2 Types of Literary
Foils and the Differences between Foil and Antagonist –2022." *MasterClass*, September 9, 2021.
https://www.masterclass.com/articles/writing-101-what-is-a-foil-character-in-literature-learn-about-2
-types-of-literary-foils-and-the-differences-between-foil-and-antagonist.

Vyas, Dinesh, and Ahmed Hozain. "Clinical Peer Review in the United States: History, Legal Development and Subsequent Abuse." *World Journal of Gastroenterology* 20, no. 21 (June 7, 2014): 6357. https://doi .org/10.3748/wjg.v20.i21.6357.

Zachman, John A. "The Concise Definition of the Zachman Framework by: John A. Zachman." *Zachman International | Enterprise Architecture,* 2008. https://www.zachman.com/about-the-zachman -framework.

Zadrozny, Brandy. "Parents are Poisoning their Children with Bleach to 'Cure' Autism. These Moms are Trying to Stop it." *NBCNews.com.* NBCUniversal News Group, May 21, 2019. https://www .nbcnews.com/tech/internet/moms-go-undercover-fight-fake-autism-cures-private-facebook-groups -n1007871.

Chapter 10

Managing Third-Party Risk: Framework Details for Risk Management in Medical Technology

Mitch Parker, Brittany Partridge,
Eric Svetcov, and Allison J. Taylor

Contents

DOI: 10.4324/9781003348603-13

An old Jewish story from the teachings of Rabbi Shimon bar Yochai in the Midrash once discussed a group of people traveling in a boat. One of them took a drill and began to drill a hole beneath himself. When he was confronted about why, his response was "What concern is it of yours? Am I not drilling under my own place?" The response from the other occupants of the boat was "But you will flood the boat for us all!" (Chabad-Lubavitch Media Center 2022).

This metaphor from the 5th century is a perfect one that illustrates the vulnerabilities of large multi-party interconnected networks of both local and cloud-based resources containing critical data that support integrated healthcare delivery systems. The systems themselves are the boat, and all of its services rely upon efficient operation to keep the Integrated Delivery Network (IDN) from sinking. The attack on Target, which originated from a vendor who had a Virtual Private Network (VPN) connection to their network, illustrates this. So does the attack on Solarwinds, where a software supply chain attack was utilized to compromise a version of Orion network management software while it was being developed and packaged. Those vulnerable versions were then used to compromise multiple clients.

An organization is only as strong as its weakest link, or in these cases, the insecure code running with Root or Domain Admin permissions on a critical server or services that everyone uses. This also can be extrapolated to the person or vendor that needs local administrator access on every PC to make sure that a critical application is able to remain functional.

No code or system is free from vulnerabilities. No system is perfect and fully protects against or mitigates risk. Yet we acquire goods and services on the false premise that third-party risk is absolute. We do not consider our own resilience or effects of service absence in our environment. We consider the act of writing a check or purchase order to a third party to be appropriate risk transference. Given the Target and Solarwinds attacks, and the insistence by multiple third-party vendors that they are secure, this is not the case. We can do better than this, and by our customers.

Why Are We Here?

Digital Transformation and Mobile Applications have left us in a position where we are reliant upon third parties that have access to the most critical data that belongs to our patients and our organizations. Healthcare organizations, no matter their size, are not capable of digital transformation on their own due to the complexity of the ecosystem they operate in. They need numerous partners to enable this. These partners need to integrate with our networks. They give patients access to data that used to be protected behind multiple levels of security or kept offline. Now this data is online and available on mobile devices used in the patient care process.

We need to know the specificity of services provided to a high level of detail to determine the credibility of stewardship. What this means is that we have to consider all third-party services provided by a vendor or consortium, and how they are managed and delivered, not just the primary product. Sales techniques such as Negative Reverse Selling, which are utilized to put customers on the defensive and use anger to override logic by intimating that they do not care about what matters, or really want what's being sold, are often used to get prospective customers to stop asking detailed questions by short-circuiting decision processes (Nicasio 2020). The sales process in itself is geared toward selling products and services to whoever will buy because it is based on the amount of sales. The chapter on Burnout in Information Security goes into greater detail about how some organizations put extreme pressure on team members to make sales, and how they will sell products that don't even work. The Cycle of Software Disillusionment, as we call it, feeds upon the promises of fixing broken systems and processes to sell products by any means necessary to health systems. Given the emphasis on stock prices, quarterly earnings, and booking sales for funding and higher valuations, these are the fundamental processes by which products are sold into organizations. This is also what we have to watch out for because security is not a consideration. If security is even noted or sold to us in the transaction process, we are ultimately responsible and need to build in checks and balances to validate it, or otherwise negotiate.

To get a better understanding of what healthcare organizations require for security and from their third parties, we'll start this chapter with a basic survey of the security regulations and standards they have to follow. The HIPAA Security Rule, 21st Century CURES Act, Payment Card Industry – Data Security Standards (PCI-DSS), Cyber Insurance, and Credit Ratings are all important and provide the base standards that healthcare organizations have to follow. As many healthcare organizations leverage the bond markets and financial institutions for financing, an excellent credit rating is a must for being able to be funded.

Regulations such as the Health Insurance Portability and Accountability Act (HIPAA) and Joint Commission require organizations to conduct appropriate due diligence on third-party vendors; however, they do not define what appropriate is. 45 CFR 164.308, better known as the HIPAA Security Rule – Administrative Safeguards, requires organizations to implement policies and procedures to prevent, detect, contain, and correct security violations (Administrative Safeguards 2003). This means that they need to conduct a risk analysis of the potential risks and vulnerabilities to the confidentiality, integrity, and availability of electronic protected health information held by the covered entity or business associate (Administrative Safeguards 2003). This also requires them to implement sufficient security measures to reduce risks and vulnerabilities to a reasonable and appropriate level (Administrative Safeguards 2003). It also requires appropriate sanctions against workforce members who fail to comply. Most importantly, it also requires the implementation of procedures to regularly review activity records, including audit logs, access reports, and tracking reports, to ensure that risks are being appropriately managed (Administrative Safeguards 2003). A responsible security official also needs to be identified to develop and implement these policies (Administrative Safeguards 2003).

More recently, Cyber Insurance has become paramount in corporate due diligence. John Reed Stark, in his article "Cyber-security Due Diligence: A New Imperative," discusses how cyber weaknesses imperil the entire operation, and that multidimensional risk analysis is required (Reed Stark 2016). Cyber insurance is now significantly important. However, the growing risks, mainly due to the 125% increase in ransomware attacks, according to Jonathan McGoran in his November 21, 2021, article, "Wondering Why Your Cyber Insurance Rates Are Going Up? Thank the 125% Increase in Attacks from Ransomware Gangs, for Starters," are leading to both a large price increase and a strengthening of cyber risk mitigation. Underwriters are taking a much stronger look at cybersecurity efforts of the companies they serve (McGoran 2021). This article also cites an AGCS report that estimates that 80% of losses from ransomware attacks could have been avoided if best practices were followed. That includes third-party risk management.

The credit rating agencies, particularly Moody's, also closely examine cybersecurity. In her article, "Moody's Warns Cyber Risks Could Impact Credit Ratings," from November 24, 2015, Marianne McGee indicates that Moody's, a major credit rating agency, is making cyber defenses, breach detection, prevention, and response higher priorities in their creditworthiness analysis of companies (McGee 2015). In September 2021, Moody's invested US$250 million in Bitsight, a cyber risk ratings company, and acquired VisibleRisk, a cyber risk ratings joint venture (Kak and Mielenhausen 2021). The purpose of this investment and acquisition is to better understand and address cyber risks as they apply to the organizations they rate.

The Payment Card Industry-Data Security Standards, better known as PCI-DSS, is the set of standards set by the credit card companies to ensure the security of merchants and systems that process credit cards. According to the PCI Security Standards Council (PCI-SSC), maintaining payment security is required for all entities that store, process, or transmit cardholder data (PCI-SSC 2022, 2). Guidance is provided in the PCI-DSS security standards. There are 12 PCI-DSS requirements (https://www.pcisecuritystandards.org/pci_security/maintaining_payment_security):

- Install and maintain a firewall configuration to protect cardholder data.
- Do not use vendor-supplied defaults for system passwords and other security parameters.
- Protect stored cardholder data.
- Encrypt transmission of cardholder data across open, public networks.
- Use and regularly update antivirus software or programs.
- Develop and maintain secure systems and applications.
- Restrict access to cardholder data by business need-to-know.
- Assign a unique ID to each person with computer access.
- Restrict physical access to cardholder data.
- Track and monitor all access to network resources and cardholder data.
- Regularly test security systems and processes.
- Maintain a policy that addresses information security for employees and contractors.

PCI compliance is required. However, it is often not made clear in contracts or language that it is a requirement. Credit card companies and banks require everyone to be compliant, not just the entities who have agreements with them. The software used to process cards is strongly encouraged to be Payment Applications Data Security Standards (PA-DSS) compliant and certified. The credit card machines, even the little Square machines used at the county fairs, have to meet PCI Pin Transaction Security Requirements (https://www.pcisecuritystandards.org/ptsdocs/SquareE MVPolicyProcedures-2-4-40152.pdf).

The 21st Century CURES Act is a series of federal laws designed to ultimately provide patients with their information so that they can gain visibility into the services, quality, and costs of healthcare (ONC 2022). It prohibits information blocking, which is the act of withholding information being made with a reasonable and necessary request (ONC 2022, 2). It also defines reasonable and necessary activities that would not constitute information blocking, including privacy and security, and addressing situations where moving data is not feasible (ONC 2022, 2). It also allows patients their choice of applications that are compliant with the Fast Health Interoperability Resources (FHIR) Application Programming Interface (API) standard version 4.01, US Code Data for Interoperability (USCDI) version 1, and Transport Layer Security version 1.2 or greater (Dynamic Health IT 2020).

Now that we understand the major regulations and standards that drive healthcare IT security, we need to drive toward a process that we can use for evaluating vendors that helps organizations make better decisions that include multiple considerations in the decision-making processes. This helps with the emphasis on cloud-based services for mobile device application deployments and ensuring that adequate resourcing has been considered in the due diligence process. These steps are:

- Evaluating Architectural Fit
  - Assessing what standards are required
  - Applicable standards compliance
  - Security
    - Advanced security – scans aren't enough anymore
    - You do the scans if possible!!!!
  - Integration
  - Actual operational work for business architecture
- The Difference between Monitoring and Participation
- Finding the Right Standards and Guidance
  - Healthcare doesn't map as easily as Finance or Accounting
  - Choose Your Own Adventure – reading the book backward like when you were a kid to find the best outcome. Think of this as doing so professionally.
  - What are the standards and how can we use them?
  - Using FMEA to map out what to do when things fail
  - Advanced API testing and pen testing for risk management
- Policies, Standards, and Procedures
- Phases of the Third-Party Risk Management Process
  - Including how to address it technically
- How to Deal with Nth Party Risk

Our goal with this chapter is to provide healthcare IT executives a good example of a comprehensive process to realistically manage third-party risk to the requirements that they must face. Mobile applications make it more complex. However, what has occurred is that the illusion of risk transference has masked the additional security and operational complexities of operating mobile applications within the healthcare environment. This chapter is meant to address those comprehensively. It's meant to give you the tools you need to bring the partners you need aboard rationally to deliberately enable the digital transformation and mobile applications your organization requires to thrive.

Evaluating Architectural Fit and Integration

When bringing in a new product or service, it's important to understand what standards the applications and services are required to comply with first. The concern here is bringing in cloud-based or business-managed systems, putting critical patient data in them, and then having them sit unprotected on the internet. Numerous data breaches have resulted from unprotected cloud-based systems being exposed to the internet containing critical personal or patient data. Knowing what the applicable standards are, starting with HIPAA and HITECH for patient data, and the numerous federal, state, and local laws, regulations, and industry standards that apply to the environment in question is critical. The first, and most critical step, is to perform that assessment.

When you've performed that assessment, and have determined which standards are applicable, the next step is to map out what steps are needed to assure compliance with them.

- How will you get copies of the actual certifications required?
  - Salesforce has done significantly good work here by making theirs available online at https://compliance.salesforce.com/en/documents

- How often do they need to be updated?
- Do you have someone reviewing them?
- What do you do if there are issues? There are tools you can use to track standards compliance, which are commonly known as Governance, Risk, and Compliance (GRC) tools.

We next need to worry about security. We need to make sure that the security required for the criticality of the system and data handled is commensurate with applicable standards. Due to integration, all systems that interface with these systems we're bringing in all share in the risk, like the passengers on the boat. This means that new systems can significantly lower risk. This also means that we don't just do scans using vulnerability scans and say they're "hacker proof" like old online stores. Security now requires continual monitoring and ongoing follow-up. There is a lot of work to do to provide affirmative assurance of security through evaluation of logs, system behavior, traffic, and understanding and resolving vulnerabilities and anomalies. This means more than vulnerability scans, it means a risk assessment. It means having intentional operations in place to monitor these behaviors and actions, not just an occasional scan to show that years-old exploits can't get through. It's insider behavior and misconfigurations, not the absence of malware or vulnerabilities, that cause many of our security issues.

One of the largest areas of misconfiguration is the integration points. There are many standards for integrating different systems together. Getting two vendors to agree on them and having even fully standards-compliant implementations work together does not work 100% of the time. Many of them require custom code to be written to address translation or integration errors. They also need custom code to adapt and convert between standards that these systems do not support natively. The addition of custom code to the integration steps adds an additional degree of risk and can significantly lower the security of the systems as that code will not undergo the same degree of scrutiny and testing as production code. Production code, especially Application Programming Interface (API) code, will often undergo vulnerability analysis, code review, and testing processes. Integration code used to glue systems together does not often have that, if at all, and usually does not as a rule. The reviews for that code involve testing to see whether data correctly translates between systems. To paraphrase a famous internet meme about cats, if it fits, it ships.

Integration needs to be appropriately budgeted for as well. Too many times the integration code is handled last minute as it's assumed that following the same standards equals being tested. The level of effort needed to integrate the systems continues to be a weakness and is often overlooked in favor of closing a sale. These integration efforts, especially as indicated in the Cycle of Software Disillusionment, often significantly raise overall project costs, both through the level of effort to customize/build and also through additional products and services needed to facilitate promised integration. Most importantly, there needs to be operational resources to address what happens when the integration breaks.

It's possible to have multiple systems that are certified and highly secure in the environment protecting data. However, if the integration points between them are weak, or involve a lot of custom code, there is still significant risk. You can have incredibly secure cloud environments. However, if you have an insecure waypoint between your secure internal systems and theirs, such as insecure file transfer protocols, API implementations, or network protocol vulnerabilities, you can have major issues. A recent network hack for $625 million using a bridge between different blockchain implementations demonstrates this (Thurman 2022). This further illustrates the real weakness of "stablecoins," which are cryptocurrencies used to trade for others. One Bitcoin CEO revealed in an interview that he buys billions in a stablecoin, to trade in multiple cryptocurrencies because banks are nervous to work with cryptocurrency traders (Wu 2021). The hack mentioned demonstrates that the weakness in a stablecoin that isn't discussed is with its use as a trading instrument. People focus on the backing, not integration. With the numerous cryptocurrency exchange hacks that have occurred, this is a significant risk to trading that has not been discussed, and a threat toward its continued use and value. DeFi has the issue of not having centralized monitoring for fraud, waste, abuse, and scams and protections like the Federal Reserve system does. This means that even if you have the most secure endpoints, independently operating exchanges that do not share info and act independently present a significant risk because they cannot work together to piece together the information needed to detect these. We only hear about the big hacks. We don't hear about the attackers under the radar.

We cannot blindly rely upon the word of third-party risk companies as the final word for security of our systems. While they can do great work in due diligence for systems themselves, that work means nothing if there are insecure integration points, which involves operational work.

Evaluating Actual Operational Work Involved

Understanding the actual operational work is another challenge. Many of these products and services are sold under the premise that IT does not need to be involved in the implementation process. As the examples of numerous data breaches show, every third-party system implementation and associated integrations introduce significant planned and unplanned risks. If anyone tells you otherwise, they are lying to make a sale or they have never tried the solution themselves and have no idea of its true impact on real-world operations.

We need to first start out by understanding the underlying business architecture. According to CIO Wiki, a business architecture reveals how an organization is structured and can clearly demonstrate how component elements such as capabilities, processes, organization, and information fit together. The relationships among these demonstrate what the organization does and what it needs to do to meet goals (CIO Wiki 2021). It can also be further defined, according to the same source, as an enterprise blueprint that provides a common organizational understanding for aligning strategic objectives and tactical demands (CIO Wiki 2021). According to Gartner, Enterprise Architecture is a discipline for proactively and holistically leading enterprise responses to disruptive forces by identifying and analyzing the execution of change toward desired business vision and outcomes (Gartner 2022). Business Architecture focuses on understanding the business itself, while Enterprise Architecture targets a transformational subset. For more on enterprise architecture frameworks including TOGAF and Zachman, see Mobile Medicine, Overcoming People, Culture, and Governance (Douville 2022).

Solutions the business purchases will need to augment and further define the business architecture in order to best manage risk. Samuel Holcman from the Enterprise Architecture Center of Excellence and Business Architecture Center of Excellence defines the Business Architecture Framework as having six major components. Solutions need to demonstrate alignment to the following components (BACOE 2022):

1. *Strategies and Goals* – The "Why" component of the model which describes the future and current business drivers.
2. *Processes and Activities* – The "How" that describes the future and current business actions that drive the business.
3. *Materials and Things* – The "What" that documents the tangible and intangible things of interest for the business today and in the future.

4. *Roles and Responsibilities* – The "Who" that describes the organizations, roles, and people that exist now, and need to exist in the future.
5. *Locations and Geography* – The "Where" which describes where the business and its stakeholders operate today and in the future.
6. *Events and Triggers* – The "When" that describes the event categories that cause the business to respond today and in any way in the future.

If the organization cannot articulate how the solution aligns with the business architecture, then it may not be a good fit and no resourcing is going to change that. However, if it does, then it sets the groundwork for further questions to determine how to further align the third-party solution to the business. This involves addressing the following more detailed components:

■ *Architectural Plan* – Statement of how the solution(s) align with the business architecture.
■ *Project Plan* – Plan for implementing the solution(s) in the business.
■ *Operational Management Plan* – Detailed operational plans on how the system will function and support the business.
■ *Integration Plan* – Identification of any dependent systems, how they will integrate, and what data will be interchanged.
■ *Data Flow Diagram* – Document showing data elements stored and interchanged between systems. Required by the HIPAA Security Rule.
■ *User Management Plan* – How to provision and deprovision users, computers, and services from the system(s).
■ *Security Plan* – What security components need to be monitored, how, and what actions to take on anomalous conditions.
■ *Privacy/Fraud Monitoring Plan* – With the privacy requirements in the HIPAA Privacy Rule and increased fraud monitoring requirements in the PCI-DSS 4.0 standard along with other financial standards, organizations need to be able to monitor user and system behavior to spot potential fraudulent behavior (PCI-SSC 2022).
■ *Monitoring Resources* – There needs to be subject matter experts assigned to monitor the behavior of the application. This needs to be integrated into job duties and job descriptions to ensure that resources understand what to do and it's part of their job goals and performance evaluations.
■ *Accountable Business Owner* – An executive-level sponsor and owner needs to be identified to ensure a responsible point of contact, appropriate budgeting, and continued resource allocations.
■ *Accountable Internal Resources* – Appropriate internal resources needed to support the third-party application and alignment to the business architecture need to be identified and documented as part of the project plan.
■ *Use Cases and Workflows* – How will this system work? What are the processes that users/actors will follow to accomplish tasks. What data will be used at each step? Use of Unified Modeling Language (UML) or a similar industry standard helps illustrate this.
■ *Exceptions* – What are the exceptional conditions that can occur during use cases, with integration, or with the Security, Privacy/Fraud Monitoring, or Operational Management plans? Use of Failure Mode Effects Analysis (FMEA), an engineering-based standard for evaluating processes to see where and how they may fail, and assessing their relative impact, helps identify where these may occur (IHI 2017). This needs to be integrated into the project plan itself to test the major steps.

- – Identifying these exceptions has an impact on monitoring resources and the other plans.
- *Vulnerability Management Plan* – When there are identified vulnerabilities, identify who is assigned to address them, how they will be addressed, and what the acceptable timelines are for addressing them based on criticality. As this is both a security and operations plan, it's better to have this separated out as multiple teams jointly own this.
- *Two-factor Authentication/Federation Plan* – Ensure that system users use either a separate two-factor authentication system or federation with enterprise authentication sources. Separate authentication sources without either lead to passwords stored in insecure places and data breaches.
- *Privileged Access Management Plan* – Ensure that users cannot make any changes to system configurations, bulk data, users, user access, or data without separate accounts that have two-factor authentication using an application. Text messaging as a two-factor authentication method is highly vulnerable to SIM Swapping (Kelly 2021).
- *Who to Call at the Vendor* – Know who to call or contact at the vendor and when, and make sure there is an escalation plan. Having Slack or Email as support is not enough with critical data. There needs to be a way to reach humans when something bad happens and critical systems such as Slack or Email cannot be reached. Critical Infrastructure has this on paper and in multiple places.
- *Decommissioning Plan* – There needs to be a plan to safely and securely decommission the system when it has been taken out of daily service for the organization. Due to Accounting of Disclosures requirements under HIPAA (Accounting of disclosures of protected health information 2003), audit log and disclosure records need to be kept for six years. Each state and locality has their own record retention requirements. These records include access and audit logs. If the data is stored offsite or in a third-party's system, there needs to be a plan to store this data someplace for archival and compliance with the rules. We don't want to keep systems online or in a vulnerable state just to retain data. As retention laws vary by state and locality, please consult with qualified legal counsel to make a determination as to how best to account for this in third-party agreements and contracts.

Having a good set of templated plans to deal with third-party vendors is important to managing their risk. There is significant inherent risk from any system changes. However, with the criticality of the data healthcare systems handle, there needs to be an organized and disciplined manner aligned with the business architecture to ensure continued successful operation. Unlike what the vendors promise, these systems don't run themselves, and continued operation of them can put the entire business at risk. Aligning with the business can minimize the risks by providing a better understanding of what the application or service is and developing plans to address them. When systems run outside of central IT management or assistance, such as the dreaded "Shadow IT," this often falls apart. Even if a system isn't run by IT, they need to be involved to align its operations with the business architecture.

The Difference between Monitoring and Participation

One of the most common responses from managers trying to appear like they are on top of situations is to say that they are monitoring it. This really doesn't do anything to assist with determining impact to the business and technical architecture. When a situation happens, you need to be an active participant in working with the third parties to address it and proactively reporting back to senior leadership. Monitoring without an affirmative plan to take action when an event occurs is a substitute for saying you aren't doing anything. When you monitor event logs or alerts and take action when they occur, that's different than just noting an event happened and not responding. When you put a vendor on "watch" when a situation happens, it doesn't answer tough questions that your customers are asking. You need to understand what the exceptional conditions mean. Your organization needs to be prepared to detach and function without third-party services and go to downtime procedures. The Joint Commission and other regulatory agencies ask questions about how organizations will function without critical computer systems in place. This also means that you will need to have operational resources to patch, configure, and maintain your infrastructure in response, especially due to network security. You will also need to have the operational resources to coordinate multiple third-party vendors and resources. Finally, vulnerability management needs to be constant and active. This isn't something you do just in response to the latest warning from Health-ISAC or CISA. It's part of Enterprise Risk Management.

Finding the Right Guidance

The most important part of evaluating a third-party solution is to have the appropriate guidance for applying security criteria. It's not appropriate to have security criteria in place for third-party systems that do not apply. You're just wasting the vendor's time and your own while providing no real security for your clients. We will provide explanations of several common standards and how they apply to healthcare and mobile device security in the context of third-party vendor evaluations.

HIPAA/HITECH

The Health Insurance Portability and Accountability Act, better known as HIPAA, provides guidance on how to secure systems that contain protected health information. It describes the administrative, technical, and physical security and privacy controls that organizations need to implement to properly secure and maintain applications that contain protected health information. The HITECH Act, an amendment to it that was part of the American Reinvestment and Recovery Act of 2009, better known as the Stimulus Act, came in response to the 2008 economic recession, modifies it to increase penalties for breaches, and requires the use of NIST-approved encryption schemes at rest and in transit (Breach Notification for Unsecured Protected Health Information 2009).

One of the biggest misconceptions of encryption and HIPAA/HITECH is that many of the security controls don't have to be implemented because they're addressable. Addressable means that you have to have a plan to address the security controls. If you are not able to address them, you need a plan to have reasonable and appropriate compensating controls (HHS 2021).

HIPAA/HITECH only applies if you're storing and/or processing patient data. There is no centralized agency that certifies systems or companies as being compliant. Any organization that tells you that they are HIPAA Certified or HIPAA Compliant is lying to you to get your money or is not aware that no such certifications exist. Third-party vendors demonstrate compliance with HIPAA by completing an annual risk assessment, developing a risk management plan to address identified risks, and continually following up on open items. An organization can complete their own assessment, or they can hire a third-party to do so. An example of this can be found with the Office of the National Coordinator's Security Risk Assessment tool, available at https://www .healthit.gov/topic/privacy-security-and-hipaa/security-risk-assessment-tool (HHS ONC 2022).

HITRUST

The Health Information Trust Alliance, better known as HITRUST, has a Common Security Framework (CSF) that is a combination of numerous security frameworks that has a goal of unifying together multiple existing regulatory and industry ones (Fruhlinger 2021). It is a constantly evolving framework that adopts controls from over 44 security- and privacy-related standards, regulations, and frameworks, including the HIPAA Privacy and Security Rules, FedRAMP, PCI-DSS, the European Union General Data Protection Regulation (GDPR), IRS Publication 1075, and the Joint Commission Standards (Taule 2020). It is a comprehensive framework with numerous controls. Unlike HIPAA/HITECH, organizations can demonstrate compliance with HITRUST by completing a validated assessment. They can be HITRUST CSF Validated or HITRUST CSF Validated Certified. These have to be completed by an authorized CSF Assessor (Taule 2020).

HITRUST CSF is a comprehensive framework. According to Laurie Leigh of Agio, an IT services and consulting firm, the external cost of achieving a HITRUST CSF Certification is between $60,000 and $70,000 (Leigh 2020). This is not trivial or cheap. However, many third-party vendors adopt HITRUST because it can reduce the cost of responding to security questionnaires. It also provides a level of external assurance to customers that the organization has appropriate security controls. Two of the larger organizations have certified their environments under the HITRUST CSF (Vidich 2022) (Salesforce, Inc. 2022). Certifications are for two years and are required to be renewed for continual compliance.

While this certification is not cheap and will require significant work to implement due to having to comply with multiple controls in the CSF, it is worth it for larger vendors to signal continual compliance with a framework. Not everything a vendor sells, however, will be HITRUST certified. The certification letters document the services in scope. Make sure the services you are purchasing are covered by the CSF certification if you are in healthcare. This certification is excellent for providing third-party assurances of continual controls of environments that store and process patient and personal data. Banks are still going to require a PCI-DSS Attestation of Compliance, however.

PCI-DSS

The Payment Card Industry Data Security Standards (PCI-DSS) also has the emphasis on continual compliance. Merchant Organizations, based on the number of transactions they process annually, are required to comply with one of four levels (Table 10.1).

However, given the positive assurances that cyber insurance providers now require, any third-party vendor your organization utilizes for credit card processing will be considered a Level 1 merchant, and your policies will require an annual ROC from a QSA. Based upon discussions with merchant banks and insurance brokers, requiring all vendors that store and process credit cards is good business practice. This also provides affirmative assurances to third parties that they have done appropriate due diligence and maintain an appropriate degree of security to safeguard cardholder data.

If your organization processes credit cards, your insurance provider will likely require you to complete Level 1 requirements given the increased cybersecurity risks. You need to require your vendors to do the same. Merchant banks, specifically Wells Fargo, require third-party vendors to maintain this level of compliance. Cyber insurance providers also require the same as a condition of acquiring a policy. Mandating this reduces your risks by demonstrating externally verified compliance.

Table 10.1   PCI Compliance Levels. Source: Adapted from https://www.itgovernance.eu/ blog/en/a-guide-to-the-4-pci-dss-compliance-levels and https://www.pcisecuritystandards .org/pci_security/ (Irwin 2022) (PCI-SSC 2022 2).

Level	Criteria	Validation Requirements
1	• Merchants processing more than 1 million JCB, 2.5 million American Express, or 6 million Visa, Mastercard, or Discover transactions annually • Merchants that have suffered a data breach or cyberattack that resulted in cardholder data compromise • Merchants identified by another card issuer as Level 1	• Annual Report on Compliance (ROC) by a Qualified Security Assessor (QSA) or Internal Security Assessor (ISA) • Quarterly network scan by an Approved Scan Vendor (ASV) • Completed Attestation of Compliance (AOC) form
2	• Merchants processing less than 1 million JCB, between 50,000 and 2.5 million American Express transactions, or between 1 and 6 million Visa, Mastercard, or Discover transactions annually	• Annual Self-Assessment Questionnaire (SAQ) • Quarterly network scan by an Approved Scan Vendor • Completed Attestation of Compliance form
3	• Merchants processing less than 50,000 American Express transactions, between 20,000 and 1 million Discover card-not-present only transactions, or between 20,000 and 1 million Visa e-commerce transactions annually • Merchants processing 20,000 Mastercard e-commerce transactions but less than or equal to 1 million total Mastercard transactions annually	• Self-Assessment Questionnaire • Quarterly network scan by an Approved Scan Vendor • Completed Attestation of Compliance form
4	• Merchants processing less than 20,000 Visa or Mastercard e-commerce transactions annually • All other merchants	• Dependent upon merchant's acquiring bank requirements • These can include the SAQ and ASV

ISO Information Security Management Standards 27001, 27002, 27017, 27018

The most accepted Information Security Management Systems (ISMS) standard is ISO/IEC 27001:2013, Information Security Management System. This is a standard that provides the basic framework and controls for information security management programs. It has six main security areas (EC-Council Global Services 2022):

- Company Security Policy
- Asset Management
- Physical and Environmental Security
- Access Control
- Incident Management
- Regulatory Compliance

These areas are further broken down into 14 domains (IT governance UK list):

- Information Security Policies
- Human Resource Security
- Physical and Environmental Security
- Operation Security
- Supplier Relationships
- Information Security Aspects of Business Continuity Management
- Organization of Information Security
- Asset Management
- Cryptography
- Operations Security
- System Acquisition, Development, and Maintenance
- Information Security Asset Management
- Compliance
- Access Control

There are also ten management system clauses, which are present to provide the frameworks for implementation, management, and continual improvement (IT governance USA link):

- Scope
- Normative References
- Terms and Definitions
- Context of the Organization and Stakeholders
- Leadership
- Planning, Including Risk Assessment and Management Plans
- Support
- Operation
- Performance Evaluation
- Improvement

ISO/IEC 27001:2013 provides the management guidance to implement an ISMS. However, it does not provide the guidance on security techniques, control objectives, or controls needed to implement a system. It will have an upcoming revision in ISO/IEC 27001:2021, which will update and modernize it for the current risk management landscape.

ISO/IEC 27002:2022, Information Security, Cybersecurity and Privacy Protection – Information Security Controls, provides these now. According to ISMS online, this recent 2022 revision provides significant changes over the 2013 version, and greatly increases scope (ISMS .online 2022). It also now aligns with the NIST Cybersecurity Framework version 1.1. The 2022 revision encapsulates the NIST framework and provides a superset of requirements by providing actual control sets, not just a framework to manage organization-supplied controls.

This standard is organized into four categories of controls:

- Organizational
- People
- Physical
- Technological

There are 11 new controls in the 2022 revision of ISO/IEC 27002:

- Threat Intelligence
- Information Security for the Use of Cloud Services
- ICT Readiness for Business Continuity
- Physical Security Monitoring
- Configuration Management
- Information Deletion
- Data Masking
- Data Leakage Prevention
- Monitoring Activities
- Web Filtering
- Secure Coding

The 93 controls in the 2022 revision all now have five categories of attributes (Verry 2021):

- Control Type – Detective, Preventative, Corrective
- Cybersecurity Concept – Identify, Protect, Detect, Respond, Recover (these align with NIST Cybersecurity Framework v1.1)
- Information Security Properties – Confidentiality, Integrity, and Availability
- Operational Capabilities – These include (https://www.isms.online/iso-27002/):
  - Governance asset management
  - Information protection
  - Human resource security
  - Physical security
  - System and network security
  - Application security
  - Secure configuration
  - Identity and access management
  - Threat and vulnerability management
  - Continuity
  - Supplier relationships security
  - Legal and compliance
  - Information security event management
  - Information security assurance
- Security Domains – Protection, Defense, Resilience

ISO/IEC 27002:2022, along with ISO/IEC 27001, provides for a very strong Information Security Management program that addresses current cybersecurity needs. The 2022 updates to ISO 27022 also provide clarity via the encapsulation of the NIST Cybersecurity Framework within the Cybersecurity Concept attributes in the 93 controls of ISO/IEC 27002:2022. However, this is for environments that the organization has complete control over, not the cloud.

ISO/IEC 27018 – Public Cloud PII Protection

ISO/IEC 27018:2019, Information technology – Security techniques – Code of practice for protection of personally identifiable information (PII) in public clouds acting as PII processors, extends ISO/IEC 27001. It does so by providing a guide of best practices for the protection of PII in the cloud by processors (GlobalSuite Solutions 2020). It establishes additional requirements on 15 controls in the following clauses:

- Domain 5: Information Security Policies
- Domain 6: Information Security Organization
- Domain 7: Human Resources Security
- Domain 9: Access Control
- Domain 10: Cryptography
- Domain 11: Physical and Environmental Safety
- Domain 12: Operations Security
- Domain 13: Communications Security
- Domain 16: Incident Management
- Domain 18: Compliance

These controls are based on the following eight core principles:

- Consent and choice
- Purpose of legitimacy and specification
- Data minimization
- Limit of use, retention and disclosure
- Opening, transparency and notification
- Responsibility
- Information security
- Privacy compliance

ISO/IEC 27018 can only be certified jointly with ISO/IEC 27001 (GlobalSuite Solutions 2022). A number of cloud providers have adopted this in 2015, and it has been adopted by multiple providers since then. It provides that cloud-based complement to ISO/IEC 27001.

ISO/IEC 27017 – Cloud Services Security Controls

ISO/IEC 27017:2015, Information Technology – Security techniques – Code of practice for information security controls based on ISO/IEC 27002 for cloud services, extends ISO/IEC 27002 for the cloud, much like ISO 27018. It provides cloud-based guidance on 37 of the controls in ISO/IEC 27002, and also has 7 new controls that address the following (BSI Group 2022):

- Who is responsible for what between the cloud service provider and the cloud customer
- The removal/return of assets when a contract is terminated
- Protection and separation of the customer's virtual environment
- Virtual machine configuration
- Administrative operations and procedures associated with the cloud environment
- Cloud customer monitoring of activity within the cloud
- Virtual and cloud network environment alignment

Certified Vendors ISO 17021, and Separation of Duties

If you represent a healthcare organization that is looking to use the ISO 27001/27002 and 27017/27018 certifications as selection criteria, you need to stipulate that the certification be done by an accredited certification body. Non-accredited certification bodies offer services that include consultancy and certification, according to Camden Woollven from IT Governance USA (Woollven 2020). This is considered a conflict of interest when a single organization offers assessment and consultancy services. They also may not be providing services that are monitored for performance, quality, and competence (Woollven 2020). ISO/IEC 17021:2015, Conformity assessment – Requirements for bodies providing audit and certification of management systems, defines the principles and requirements for the competence, consistency, and impartiality of organizations providing audit and certification services for management systems (ISO 2016). In the United States, the ANSI-ASQ National Accreditation Board (ANAB) provides the most current directory of accredited certification bodies at http://anabdirectory.remoteauditor.com/ (ANSI 2022).

If you are considering attempting to certify your vendors against the NIST Cybersecurity Framework, you may consider working with them to certify them against ISO/IEC 27001 and 27002:2022 instead if there is a need for international certification and accreditation. The five concepts (Identify, Protect, Detect, Respond, and Recover) on each of the 93 controls in ISO/IEC 27002:2022 map to the five corresponding Functions in the NIST Cybersecurity Framework (Keller 2021).

If you are considering attempting to certify your vendors against the HITRUST CSF, have them work with a firm that follows ISO 17021. Many firms that offer HITRUST assessment services offer consultancy services. However, there are also many that just perform assessments and certifications across ISO, PCI-DSS, and HITRUST. It is important to demonstrate that you are working to avoid potential conflicts of interest.

American Institute of Certified Public Accountants/ Service Organizational Controls Reports

If you are entrusting financial data to a third-party, you need to request audit reports from a certified public accountancy that attest to the security and controls in the system. While there are multiple audits and criteria that can be utilized, the SOC 2 – Service Organizational Controls for Service Organizations: Trust Services Criteria – Report on Controls at a Service Organization Relevant to Security, Availability, Processing Integrity, Confidentiality, or Privacy, from the American Institute of Certified Public Accountants (AICPA) is the standard method used to gauge this (AICPA 2022). This report on its own does not immediately indicate that the organization and its controls are compliant and that you can trust them, as much as vendors would like you to believe it. A SOC 1 report only covers controls relevant to financial reporting (Marcum LLP 2022), and a SOC 3 is a public report of the internal controls that isn't going to cover the details decision-makers need to make an educated decision (Google, Inc. 2022). The SOC 2 report is the one that covers the controls to a level of an ISO 27001/27002 report.

ISO/IEC 17021:2015, Conformity assessment – Requirements for bodies providing audit and certification of management systems, defines the principles and requirements for the competence, consistency, and impartiality of organizations providing audit and certification services for management systems (https://www.iso.org/standard/61651.html). In the United States, the ANSI-ASQ

National Accreditation Board (ANAB) provides the most current directory of accredited certification bodies at http://anabdirectory.remoteauditor.com/.

Here are some caveats to look out for when examining a SOC 2 report:

- **Is it in scope?** Is this for the application or service being provided, or for the hosting facility? A SOC 2 report for a cloud infrastructure provider does not immediately transfer security to the applications hosted in the same datacenter. For example, cloud giant provides a SOC 2 report for their cloud services that is in scope for their ERP applications.
- **Does it cover separation of duties?** Does the SOC 2 report cover adequate controls for separation of duties with roles within the system to prevent collusion or unauthorized actions? Does it cover appropriate detective mechanisms?
- **Are there audit exceptions?** Does the report discuss any exceptions, how they occurred, and whether they were:
  - **Misstatement**: An error or omission in the description of the service organization's system or services (Foresite Cybersecurity 2022)
  - **Deficiency in control design**: A missing control or one that is not properly designed to achieve the control objective or criteria (https://foresite.com/soc-opinions-and-exceptions/).
  - **Deficiency in the operating effectiveness of a control**: When a properly designed control does not operate as designed or when the person performing it does not do so effectively (Foresite Cybersecurity 2022).
  - **Having an exception which in itself is not bad**: Having them indicates that the auditor is doing their work.
- **What kinds of opinions exist?** (Foresite Cybersecurity 2022):
  - **Unqualified opinion**: The controls are described in a fair and accurate manner and operate effectively.
  - **Modified opinion**: A modified opinion will be issued if the controls fail to meet standards, or if the auditor cannot get sufficient and appropriate evidence.
  - **Qualified opinion**: Controls are mostly effective, however they do fall short. Specific instances of non-effectiveness are documented in detail. However, the controls environment is otherwise robust.
  - **Adverse opinion**: The service organization materially failed one or more of the standards. This will be documented descriptively.
  - **Disclaimer of opinion**: This is when an auditor declines to issue an opinion. If there is not enough evidence to support one, a Disclaimer of Opinion will be issued.
- **Is it a Type 1 or Type 2 report?** (Foresite Cybersecurity 2022)
  - **Type 1**: The report says that on the date the report was performed, and only on that date, the target system was in the state described in the report.
  - **Type 2**: The report utilized evidence to verify the system state over a particular defined time period.

The presence of a report does not signify whether or not a system is actually compliant. What matters is the content of the report. What we've published here is a guide that organizations can actually use to evaluate SOC 2 reports to determine if they are relevant, and more importantly, if the controls are effective. Systems cannot be looked at in isolation. The whole compliance and integration picture relies upon detailed examination of audit reports to get a complete understanding of risks.

NIST Cybersecurity Framework (NIST CSF)

The US government has been known for an approach to security that can be described as Department of Defense first and then everyone else last. This led to numerous security frameworks, and multiple variances, even within DOD and the uniformed services. The issue was that firms that worked with multiple service branches were not able to consistently apply security to the numerous systems they had. To achieve certification and accreditation, vendors had to rewrite the documentation for each branch of DOD or government to meet their standards. This was highly inefficient, and many of those costs ended up being absorbed into project costs, ultimately not benefiting taxpayers. It also led to the Cybersecurity Enhancement Act of 2014. This Act extended the National Institute of Standards and Technology's role. They were now empowered to develop cybersecurity risk frameworks for critical infrastructure (NIST 2018). Version 1.1, released on April 16, 2018, provides that prioritized, flexible, repeatable, performance-based, and cost-effective approach (NIST 2018).

The NIST CSF can be divided into five key areas to better understand it:

■ **Recommended Implementation Program.** This is a seven-step program to implement the Framework itself. It roughly parallels the numerous DOD-specific frameworks that preceded it, specifically DITSCAP, the Defense Information Technology Security Certification and Accreditation Program. The steps are:
  – *Step 1: Prioritize and Scope.* The organization identifies its business and mission objectives and high-level organizational priorities. They determine the scope of supporting systems and assets. The scope may be reflected in risk tolerances.
  – *Step 2: Orient.* The organization identifies:
    • Related Systems and Assets
    • Regulatory Requirements
    • Overall Risk Approach
    • Applicable Threats and Vulnerabilities
  – *Step 3: Create a Current Profile.* The organization develops a Current Profile by determining whether the appropriate subcategories/controls are being fully or partially achieved.
  – *Step 4: Conduct a Risk Assessment.* The organization analyzes the operational environment in scope against the applicable subcategories/controls to determine likelihood and impact of a cybersecurity event. Organizations need to identify emerging risks and leverage cyber threat info from internal and external sources to help determine this. Previous risk assessments can be used in this step to assist in building the current one. If this risk assessment involves protected health information, conduct a risk assessment against the HIPAA Security Rule here. You can use the Administrative, Technical, and Physical controls from the ONC Security Risk Assessment tool, available at https://www.healthit.gov/topic/privacy-security-and-hipaa/security-risk-assessment-tool (HHS ONC 2022). This will also bring healthcare organizations into alignment with the Security Rule.
  – *Step 5: Create a Target Profile.* The organization creates a Target Profile based on the risk assessment from Step 4 that is based on their desired state. They may also add in their own controls/subcategories based on their organizational needs. As part of this Target Profile, consideration of external stakeholders needs to be included, especially customers, suppliers, and business partners.
  – *Step 6: Determine, Analyze, and Prioritize Gaps.* The organization compares the Current and Target Profiles. They develop a prioritized risk management plan to address the

identified gaps between the two. The plan needs to reflect mission drivers, costs, benefits, risks, and desired outcomes (NIST 2018).
  - *Step 7: Implement Action Plan.* The organization makes the appropriate changes, takes appropriate actions, and adjusts current practices to achieve the Target Profile goals.
■ **Core Functions**. The five key functions are Identify, Protect, Detect, Respond, and Recover. ISO/IEC 27002:2022 based its Cybersecurity Concepts on the Core Functions in NIST CSF version 1.1. Explanations are:
  - *Identify.* Develop an organizational understanding to manage cybersecurity risk to systems, people, assets, data, and capabilities.
  - *Protect.* Develop and implement appropriate safeguards to ensure delivery of critical services.
  - *Detect.* Develop and implement appropriate activities to identify occurrences of cybersecurity events.
  - *Respond.* Develop and implement appropriate activities to take action regarding a detected cybersecurity incident.
  - *Recover.* Develop and implement appropriate activities to maintain plans for resilience and to restore any capabilities or services that were impaired due to a cybersecurity incident.
■ **Outcome Categories and Subcategories.** Each key function is subdivided into business categories/functions associated with needs, and further subdivided into specific subcategories/controls/operational activities. Each comes with Informative References that reference other controls and objectives from other frameworks as a crosswalk to them. The Categories under each Core Function are:
  - *Identify.* Asset Management, Understanding the Business Environment, Governance, Risk Assessment, Risk Management Strategy
  - *Protect.* Identity Management and Access Control, Awareness and Training, Data Security, Information Protection Processes and Procedures, Maintenance, Protective Technology
  - *Detect.* Understanding Anomalies and Events, Security Continuous Monitoring, Detection Processes
  - *Respond.* Response Planning, Communication, Analysis, Mitigation, Improvements
  - *Recover.* Recovery Planning, Improvements, Communications
■ **Implementation Tiers**. These are a representation of context on how an organization views its cybersecurity risk for the target area(s) and the processes in place for managing it. These are each subdivided into the contexts of Risk Management Process, Integrated Risk Management Program, and External Participation. The tiers are shown in Figure 10.1.
■ **Target Profiles**. When an organization assesses the area(s) in scope against the Functions, Categories, and associated subcategories/controls, the results constitute a profile of the area(s). The area(s) can represent a system, application, unit or area, or the entire organization. There are two types of these:
  - *Current Profile.* The cybersecurity outcomes currently being achieved.
  - *Target Profile.* Outcomes needed to meet the targeted cybersecurity risk management goals.

If an organization wishes to demonstrate compliance with the NIST Cybersecurity Framework as part of the third-party vendor risk process, they need to do the following:

  ■ Demonstrate they have conducted a risk assessment of each of the subcategories/controls with a representative Tier score and explanations for each.

FOCUS AREA	TIER 1 PARTIAL	TIER 2 RISK INFORMED	TIER 3 REPEATABLE	TIER 4 ADAPTIVE
People	• Cybersecurity professionals (staff) and the general employee population have had little to no cybersecurity-related training. • The staff has a limited or nonexistent training pipeline. • Security awareness is limited. • Employees have little or no awareness of company security resources and escalation paths.	• The staff and employees have received cybersecurity-related training. • The staff has a training pipeline. • There is an awareness of cybersecurity risk at the organizational level. • Employees have a general awareness of security and company security resources and escalation paths.	• The staff possesses the knowledge and skills to perform their appointed roles and responsibilities. • Employees should receive regular cybersecurity-related training and briefings. • The staff has a robust training pipeline, including internal and external security conferences or training opportunities. • Organization and business units have a security champion or dedicated security staff.	• The staff's knowledge and skills are regularly reviewed for currency and applicability and new skills, and knowledge needs are identified and addressed. • Employees receive regular cybersecurity-related training and briefings on relevant and emerging security topics. • The staff has a robust training pipeline and routinely attend internal and external security conferences or training opportunities.

Table 1. Customized Tier Definitions

Figure 10.1   NIST Cybersecurity Framework Implementation Tiers.

- Add additional categories/subcategories as appropriate for their systems and/or scope.
- Discuss Current and Target profiles.
- Produce evidence of a risk management plan that addresses identified gaps.
- Contractually obligate vendors to address these needs by a specific target date.

It is not possible to currently certify against this framework because NIST provides it in an advisory capacity, and as a government agency it is not in a position to provide these services to private industry. The structure that ISO/IEC has for certifying the certifiers is not there. However, a consulting firm that has practices compliant with ISO/IEC 17021:2015 would be able to provide the impartiality needed to evaluate organizations and their in-scope subset against the NIST CSF. We understand that many organizations may not be able to hire an external firm to assess themselves. However, assessing and addressing risk is still possible.

CSA CCM, CAIQ, and STAR Registry

The cloud has been a significant challenge for organizations to comprehend or secure. The controls and applications present significant controls testing challenges that on-premises systems do not. The Cloud Security Alliance (CSA), which is a research, education, and certification organization that also produces events and products, has developed a multi-stage process for addressing how to assess and address cloud security environments (CSA 2022). CSA is organized like standards organizations such as IEEE and ISO by significant research in multiple working areas such as Healthcare and Finance to enhance their standards, advice, publications, instruments, and certifications. These feed back to their instruments and certifications. The use of the Cloud Controls Matrix (CCM), Consensus Assessment Initiative Questionnaire (CAIQ), and the Security, Trust, Assurance, and Risk (STAR) Registry provide multiple ways for organizations to verify and validate the security of cloud-based implementations. The current version of the CCM and CAIQ, version 4, is the result of the feedback from this research. They also offer the Certificate of Cloud Security Knowledge (CCSK) and Certificate of Cloud Auditing Knowledge (CCAK) for individuals (https://cloudsecurityalliance.org/education/ccsk/).

The CCM, which is now offered in parallel with the CAIQ, is a cybersecurity control framework for cloud computing (CSA 2022 2). It has 197 control objectives across 17 domains that cover the depth and breadth of cloud technology. It is designed to be used to assess cloud implementations. It also provides guidance on what controls need to be implemented at different spots in the cloud supply chain (CSA 2022 2). The domains it covers are:

- Audit and Assurance (A&A)
- Application and Interface Security (AIS)
- Business Continuity Management and Operational Resilience (BCR)
- Change Control and Configuration Management (CCC)
- Cryptography, Encryption, and Key Management (CEK)
- Datacenter Security (DCS)
- Data Security and Privacy (DSP)
- Governance, Risk Management, and Compliance (GRC)
- Human Resources Security (HRS)
- Identity and Access Management (IAM)
- Interoperability and Portability (IPY)
- Infrastructure & Virtualization Security (IVS)
- Logging and Monitoring (LOG)
- Security Incident Management, E-Discovery, and Cloud Forensics (SEF)
- Supply Chain Management, Transparency, and Accountability (STA)
- Threat and Vulnerability Management (TVM)
- Universal Endpoint Management (UEM)

The CCM currently maps to the following control sets:

- ISO/IEC 27001/27002/27017/27018
- Cloud Controls Matrix v3.0.1
- American Institute of Certified Public Accountants (AICPA) Trust Services Criteria (TSC) 2017 version
- Center for Internet Security (CIS) Critical Security Controls version 8
- NIST Special Publication 800-53 revision 5, Security and Privacy Controls for Information Systems and Organizations
- Payment Card Industry-Data Security Standards (PCI-DSS) version 3.2.1

The purpose of the CCM is to provide required controls for the specific components of cloud implementations. It maps across a broad set of standards to provide assurance. The use of ISO 27001/27002 is also integral to the STAR certification. However, the CCM is only the controls. Successfully completing the CAIQ questionnaire for cloud components in scope demonstrates a reasonable and appropriate degree of compliance.

CAIQ v4 has 261 yes/no questions that map to the 197 controls of the CCM, and has four response columns that organizations need to complete (Catteddu 2021):

- Cloud Service Provider (CSP) CAIQ Answer, which is yes or no
- Shared Services Responsibility Model (SSRM) Control Ownership, which requires specific identification of who is responsible for control implementation and maintenance
- CSP Implementation Description, which provides a short narrative of control implementation

■ Cloud Service Customer (CSC) Responsibilities, which is a short narrative of customer responsibilities

Successful completion of the CAIQ is a good indicator of good security practices. However, anyone can fill out a questionnaire and not tell the truth about the answers. This is why CSA also has the Security, Trust, Assurance, and Risk (STAR) Registry. Much like how the PCI Security Standards Council maintains public registries of assessors on the PCI Security Standards website at https://www.pcisecuritystandards.org/, or Visa's public Global Registry of Service Providers at https://usa.visa.com/splisting/splistingindex.html, CSA maintains a public one of certified cloud providers at https://cloudsecurityalliance.org/star/registry/ (PCI-SSC 2022; Visa 2022; CSA 2022 3).

The STAR registry has two levels. Level 1 is a Security Self-Assessment that uses a version of the CAIQ v4 that organizations can voluntarily submit to the registry. This information is publicly available in the interests of transparency and visibility into security practices (CSA 2022 4). These self-assessments also need to be updated annually. There is also a European Union General Data Protection Regulation (GDPR) Self-Assessment that addresses and demonstrates compliance to GDPR as part of Level 1. It is designed for low-risk environments that do not handle protected data. It also does not have associated costs.

Level 2, however, is for those environments that do handle protected data or operate in a medium to high-risk environment. It has associated certificate and attestation fees based on the number of effective employees in the organization (CSA 2022 4). These need to be completed by approved assessment firms (CSA 2022 5). The steps required to complete a Level 2 certification are as follows (Chaudhary, Di Maria, and Williams 2021):

■ Complete and submit a Level 1 CAIQ Self-Assessment to the STAR Registry.
■ Prepare for an ISO/IEC 27001 audit against the Cloud Controls Matrix. It is recommended that your organization hire an external resource to assist with this.
■ Choose a STAR Certified Auditor to conduct the audit.
■ Have them conduct the audit and make the submission.
■ Upon successful completion, your organization will be notified.
■ The registry entry will be published, and you can then notify customers.

There are three additional variations of a Level 2 certification (CSA 2022 5):

■ *STAR Attestation for SOC 2.* This is a collaboration between CSA and the AICPA to conduct both the CCM/CAIQ and SOC 2 engagement. This expires after one year.
■ *STAR Certification for ISO/IEC 27001:2013.* This follows the standard ISO/IEC 27001 protocol and audits both the ISO/IEC 27001:2013 management system standard and CCM/CAIQ.
■ *CSA C-Star Assessment.* This is a third-party assessment aimed at the Greater China market that assesses the CCM/CAIQ along with the following Chinese standards. It expires after three years unless updated:
  – GB/T 22080-2008: Information technology – Security techniques – Information security management systems – Requirements
  – Selected controls from GB/T 22239-2008: Information security technology – Baseline for classified protection of information system security

- Selected controls from GB/T 28828-2012: Information security technology – Guideline for personal information protection within information system for public and commercial services

The Level 2 STAR certification provides third-party assurance of the CSA CCM/CAIQ and related standards. It aligns closely with ISO/IEC 27001/27002 and can be jointly certified with it. Future versions of the Level 2 STAR certification will likely include the ISO/IEC 27001/27002 2022 revisions.

If your organization wants to accept a CSA Certification, it's recommended that you accept a Level 2 STAR Registry entry with either the base certification, SOC 2, or ISO/IEC 27001:2013 variations. Anyone can complete a Level 1 certification by filling out a CAIQ, and there will doubtless be many vendors that will and claim that they are certified. A Level 2 requires the same level of scrutiny as a ISO/IEC 27001 assessment. In addition, make sure that the vendor used for the assessment does not provide any other services to the organization and maintains a degree of independence commensurate with Institute of Internal Auditors (IIA) standards.

CURES Act and HITECH Certification Requirements

This is a certification program that can be described as the most misunderstood one in Healthcare IT outside of the HIPAA Security Rule and HITECH Act. The language used in the 21st Century CURES Act Final Rule updates did not help. The purpose of this section is to provide a comprehensive guide for healthcare executives and those evaluating vendors to be able to evaluate certified products and cut through the nebulous messaging.

FHIR is not just another interface you can pay someone to write code for and forget it exists. Unlike the code that links your Enterprise Resource Planning system to downstream systems, or the code that links your portal to a weather application, this is higher risk. This requires deliberate security evaluation and planning. You cannot cheap out by having a junior developer, intern, or consultant write this code.

We're going to lead off and indicate that the scope of these certifications is not going to include all the security you need to properly evaluate it. Security is going to require third-party assurances, either through the use of HITRUST, ISO/IEC 27001/27002/27017/27018, the AICPA SOC 2, CSA CCM/CAIQ/STAR, or a third-party certification your organization finds appropriate. Evidence of this was with the recommendation that the HL7 Patient Empowerment Working Group made to the HL7 Policy Advisory Committee (PAC) for the Office of the National Coordinator (ONC) regarding Alissa Knight's research and her discovery that not even basic security controls protected FHIR APIs (HL7 2021).

ONC certifications are for conformance testing and basic security standards. An ONC-certified application covers functionality and interoperability. While the CURES Act mandates Transport Layer Security 1.2 or greater, FHIR 4.0.1 or greater, and USCDI version 1, it does not mandate other security controls that may be required (Dynamic Health IT 2020). We're going to start with some history, discuss the program itself, and then discuss how healthcare organizations can address validating Healthcare IT for security and conformance. We're then going to finish up by discussing what healthcare organizations can do to protect and defend themselves against insecure applications that consume FHIR data.

The Office of the National Coordinator for Health Information Technology (ONC) developed and operates the ONC Health IT Certification Program (HHS ONC 2022 3). This was granted under section 3001(c)(5) of the Public Health Service Act (PHSA), defined in the Health Information Technology for Economic and Clinical Health (HITECH) Act. This is in itself an outgrowth of the American Reinvestment and Recovery Act of 2009, better known as the Stimulus after the Great Recession of 2008. This is a voluntary third-party conformity assessment program for health IT. It is based on the ISO/IEC framework. ONC, as a government agency, does not perform these assessments. These are completed by third parties. ONC defines the requirements and processes for evaluation, testing, certification, and maintenance (HHS ONC 2022 3). This program covers conformance and interoperability testing, not security, as primary goals. While 45CFR § 170.210, which is part of the ONC Health Information Standards, covers some security components, it does not cover overall security. These standards cover (Health Information Technology: Initial Set of Standards, Implementation Specifications, and Certification Criteria for Electronic Health Record Technology 2010):

■ Encryption and decryption of electronic health information (EHI) using NIST-approved security functions in FIPS 140-2. This will need to be updated for FIPS 140-3 and the corresponding ISO standards.
■ Hashing of electronic health information using NIST-approved hashing algorithms that use Secure Hash Algorithm (SHA) version 2 or greater.
■ Recording treatment, payment, and healthcare operations disclosures.
■ Recording actions related to electronic health information, audit log status, and encryption of end-user devices.
■ Requiring encryption and hashing of electronic health information.
■ Requiring clock synchronization using Network Time Protocol (NTP) version 4, RFC 5905 (Martin, Burbank, Kasch, and Mills 2020).
■ Requiring conformance of audit log content to ASTM E2147-18, Standard Specification for Audit and Disclosure Logs for Use in Health Information Systems (ANSI 2018).

45 CFR § 170.315 – 2015 Edition Health IT Certification criteria amends this in subpart (d) to include the following (2015 Edition Health Information Technology (Health IT) Certification Criteria, 2015 Edition Base Electronic Health Record (EHR) Definition, and ONC Health IT Certification Program Modifications; Corrections and Clarifications 2015):

■ Verifying against a unique identifier (username or number) that a user seeking access is the one claimed.
■ Establishing the type of access and actions user is allowed to perform given the unique ID.
■ Auditing actions in relation to 45CFR § 170.210(e)(1).
■ Recording the audit log status (enabled/disabled) in accordance with 45CFR § 170.210(e)(2) unless it cannot be disabled.
■ Recording the encryption status of EHI /PHI locally stored on end-user devices by technology in accordance with 45CFR § 170.210(e)(3) unless the technology prevents storage of EHI/ PHI locally.
■ Technology must be set to record audit actions, audit log status, and encryption status by default.
■ Only a limited set of users can disable logging and auditing.
■ Protecting the audit log from being changed, overwritten, or deleted.
■ Detecting alteration of the audit log.

- Allowing users to create an audit report for a given time period and to sort entries by date and the criteria in ASTM E2147-18.
- Allowing users to select record(s) affected by a patient's request for amendment and:
  - If accepted: Append the amendment or link to the amendment itself.
  - If denied: Append the request and denial to the record or link to the denial itself.
- Automatic access time-out after a predetermined period of inactivity and require user authentication to resume or regain access.
- Permit an identified set of users to have emergency access to EHI.
- Require end-user devices to encrypt EHI using NIST-approved algorithms and standards or prevent its storage altogether.
- Create message digests of EHI using NIST-approved hashing algorithms and secure protocols at the message and transport level. This means the use of Transport Layer Security 1.2 or greater, and SHA-2 or greater hashing algorithms.
- Auditing actions on EHI.
- Provide for Accounting of Disclosures.
- Attest that the application encrypts stored authentication credentials at rest or not.
  - If it does, document the use cases supported.
  - If not, explain why not. *This does not mean* denial.
- Attest that the application does or does not support multi-factor authentication.
  - If it does, document the use cases supported.
  - If not, explain why not. *This does not mean* denial.

45CFR § 170.215 Application Programming Interface Standards documents the following:

- Use of HL7 Fast Health Interoperability Resources (FHIR) Release 4.0.1
- Use of HL7 FHIR US Core Implementation Guide STU 3.1.1
- Use of HL7 SMART Application Launch Framework Implementation Guide Release 1.0.0 including mandatory support for the "SMART Core Capabilities"
- Use of FHIR Bulk Data Access v1.0.0 including mandatory support for "group-export" "OperationDefinition"
- Use of OpenID Connect Core 1.0 incorporating errata set 1
- Maintain a refresh token for applications that can maintain a client secret for no less than three months (HHS ONC 2020)

The graphic and explanation in Figure 10.2 illustrate the process.

1. ONC develops product certifications and guidelines in accordance with ISO/IEC 17067:2013, Conformity assessment – Fundamentals of product certification and guidelines for product certification schemes (https://www.iso.org/standard/55087.html). This is also done in accordance with 45CFR 170(A)(D), which as discussed earlier, only covers encryption, hashing, audit log, audit log formats, and time synchronization.
2. ONC's partner, the National Voluntary Laboratory Accreditation Program (NVLAP), a division of the National Institute of Standards and Technology (NIST), operates as an unbiased third party to accredit testing and calibration laboratories under ISO/IEC 17011:2017, Conformity assessment – Requirements for accreditation bodies accrediting conformity assessment bodies and ISO/IEC 17025:2017, General requirements for the competence of

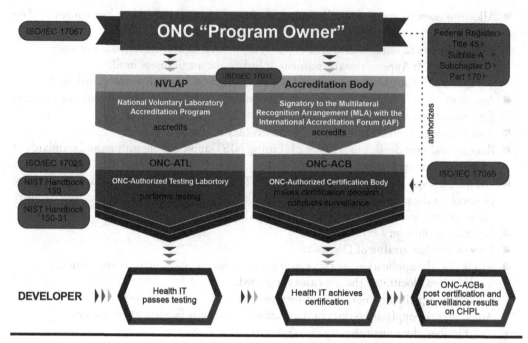

Figure 10.2 ONC Health IT Certification process. Source: https://www.healthit.gov/topic/certification-ehrs/about-onc-health-it-certification-program.

testing and calibration laboratories. NVLAP develops the following documentation and guidance to certify and accredit the labs:

   a. NIST Handbook 150 – National Voluntary Laboratory Accreditation Program (NVLAP) Procedures and General Requirements (Leaman and Hackett 2020)

   b. NIST Handbook 150-31 – NVLAP Health Information Technology Testing (Moore and Clark 2021)

3. A testing laboratory that seeks to become certified by ONC must do the following:

   a. Apply for NVLAP Accreditation (Levey 2021)

   b. Demonstrate conformance with NIST Handbooks 150 and 150-31

   c. Complete the required application steps through the NVLAP Interactive Web System (NIWS)

   d. Submit affirmative proof of accreditation to ONC to become a certified lab (ONC-ATL) defined under 45CFR §170.520 Application. This certification is good for three years. Requirements are listed under:

     i. 45CFR §170.511 Authorization scope for ONC-ATL status.

     ii. 45CFR §170.560 Good standing as an ONC-ACB or ONC-ATL.

     iii. 45CFR §170.565 Revocation of ONC-ACB or ONC-ATL status.

     iv. 45CFR §170.570 Effect of revocation on the certifications issued to Complete EHRs and EHR Module(s).

   e. An ONC-ATL performs the required conformance/certification testing according to NIST Handbooks 150 and 150-31, ISO/IEC 17011:2017, and ISO/IEC 17025:2017 of applications.

4. Accreditation bodies that wish to certify healthcare IT applications and be approved by ONC to become ONC-Authorized Certification Bodies (ONC-ACB) must do the following:

    a. They must become signatories to the International Accreditation Forum's Multilateral Recognition Agreement (IAF-MLA). This requires evaluation of their operations by a peer evaluation team (IAF 2021). The majority of these are done by one of six groups:

        i. Inter-American Accreditation Cooperation (IAAC). This organization is the association of accreditation bodies that accredits ones in the Americas (IAAC 2022).

        ii. Asia-Pacific Accreditation Cooperation (APAC) (APAC 2022).

        iii. European Accreditation (EA), the European Co-operation for Accreditation, formally appointed by the European Commission in Regulation (EC) No. 765/2008 (EA 2022).

        iv. The African Accreditation Cooperation (AFRAC) (AFRAC 2022).

        v. The Arab Accreditation Cooperation (ARAC), established by the Ministerial Decree of the Arab Industrial Development and Mining Organization (AIDMO) in June 2008 (ARAC 2022).

        vi. IAF-MLA themselves if one of these organizations is not capable of doing so.

    b. Becoming a signatory requires evaluation and acceptance of processes and outcomes under the following main ISO/IEC standards and related sub-scopes by the appropriate peer evaluation team operating in their region (IAF 2021):

        i. ISO/IEC 17021-1:2015 – Conformity assessment – Requirements for bodies providing audit and certification of management systems – Part 1: Requirements

        ii. ISO/IEC 17065:2012 – Conformity assessment – Requirements for bodies certifying products, processes and services

    c. When an Accrediting Body achieves signatory status, they must follow the below guidance:

        i. Follow 45CFR §170.520 Application and complete an application identifying:

            1. Type of authorization sought under 45CFR §170.510

            2. General identifying information

            3. Documentation of ISO/IEC 17065:2012 accreditation, which comes with being a IAF-MLA signatory.

            4. An agreement to adhere to the Principle of Proper Conduct for ONC-ACBs, defined in 45CFR §170.523, Principles of Proper Conduct for ONC-ACBs.

        ii. Adhering to 45CFR §170.560 Good standing as an ONC-ACB or ONC-ATL.

        iii. Adhering to 45CFR §170.565 Revocation of ONC-ACB or ONC-ATL status.

        iv. Adhering to 45CFR §170.570 Effect of revocation on the certifications issued to Complete EHRs and EHR Module(s).

        v. Adhering to the Principles of Proper Conduct for ONC-ACBs, available at 45CFR §170.523 (Health Information Technology: Initial Set of Standards, Implementation Specifications, and Certification Criteria for Electronic Health Record Technology 2010).

5. Healthcare IT vendors that wish to certify their applications must have them:

    a. Tested by an ONC-ATL that is able to test them within their scope of authorization criteria. The criteria are available at https://www.healthit.gov/topic/certification-ehrs/2015-edition-test-method (HHS ONC 2021). The list of authorized ONC-ATLs is at https://www.healthit.gov/topic/certification-ehrs/onc-authorized-testing-laboratories-onc-atls (HHS ONC 2022 4).

    b. Certified by an ONC-ACB that is able to certify within their scope of authorization criteria. The list of accredited ONC-ACBs is at https://www.healthit.gov/topic/certification-ehrs/onc-authorized-certification-bodies-onc-acbs (HHC ONC 2022 5).

   c. Follow post-certification guidance from the ONC-ACB.
   d. The ONC-ACBs will post certification and surveillance reports on the Certified Health IT Product List (CHPL), available at https://chpl.healthit.gov/#/search (HHS ONC 2022 6).

What Does This All Mean?

This means that ONC-certified applications, even those for FHIR or consumer apps, are required to encrypt data to NIST standards, properly authenticate, and follow standards. ONC, in their Standards-based Application Programming Interface (API) Certification Criterion, describes how applications have to conform to USCDI v1, FHIR 4.0.1, and TLS 1.2 or higher (HHS ONC 2020). However, these standards do not translate to the ongoing management of healthcare information security, which would be covered by an ISO/IEC 27001/27002, CSA CCM/CAIQ/STAR, or HITRUST certification.

In addition, there are many in healthcare who believe that we have to let any application connect and download healthcare data, or else it is considered information blocking. Nothing could be further from the truth. AHIMA, in their article Understanding the Eight Exceptions to Information Blocking, has the following about the Security Exception. The Security Exception allows organizations to not fulfill a request if the following is met (Slivochka and Warner 2021):

> This exception covers risks to the integrity and security of the information and EHI systems. However, this exception is not to be used as a broad brush for request denials. To trigger this exception, healthcare organizations need to demonstrate that the denial is "directly related to safeguarding the confidentiality, integrity, and availability of EHI; tailored to specific security risks; and implemented in a consistent and non-discriminatory manner." Provider organizations should update relevant privacy and security policies or implement new policies to mitigate practices that prohibit or delay data sharing.

What this means is that if an application is not able to or is:

- Properly authenticate users
- Properly register with the authorization server (HHS ONC 2020)
- Communicate using Transport Layer Security 1.2 or greater
- Conform to FHIR 4.0.1, USCDIv1, or OpenID Core Connect standards
- Encrypt data at rest to NIST standards
- Identified as presenting a specific security risk
- Identified by the organization's security infrastructure as presenting a specific security risk
- Is from a Banned Developer on the Certified Health IT Product List (HHS ONC 2022 6)

You are capable of denying the application access.

How do healthcare organizations actually verify and validate FHIR API services and applications? We have been building to this throughout this chapter. We recommend the following approach for delivering FHIR APIs and services:

- Leverage ONC-ACB certified services to deliver FHIR or other API-based services to clients. They have the conformance testing behind them. Anyone else who claims otherwise does not understand engineering-level testing and will put your organization at risk.

- Leverage HITRUST, ISO/IEC 27001/27002/27017/27018, AICPA SOC 2, or CSA STAR Level 2 certified services to deploy APIs and services.
- If you leverage credit cards in any way, make sure PCI-DSS services are all certified
- Conduct API-level penetration testing on a regular basis using:
  - OWASP Top Ten Web Application Security Risks (https://owasp.org/www-project-top-ten/) (OWASP 2022)
  - Penetration testers with Offensive Security Certified Professional (OSCP) (https://www.offensive-security.com/pwk-oscp/) or Offensive Security Experienced Penetration Tester (OSEP) certifications (https://www.offensive-security.com/pen300-osep/) (Offensive Security 2022; Offensive Security 2022 2)
  - Penetration testers with GIAC Penetration Tester (GPEN) certifications (https://www.giac.org/certifications/penetration-tester-gpen/) (GIAC 2022)
  - Experienced API penetration testers
- Ensure that application and service environments remain certified and approved.
- Protect APIs using API Threat Protection like Approov, Akamai, or Salt Security (UK Government 2022; Akamai, Inc. 2022; Salt Security 2022).
- If you use the NIST Cybersecurity Framework 1.1, make sure that you extend it to include penetration testing.
- Ensure that all environments hosting PHI have appropriate network-level security capable of protecting internal and external assets.
- Ensure that all environments have Endpoint Detection and Response Software.
- Ensure you have operational capabilities for monitoring and reviewing alerts and logs.
- Ensure you are capable of blocking applications that do not meet conformance or security requirements.
- Follow the Certified Health IT Product List and make sure your applications and services remain certified. Make plans to move away from decertified products (HHS ONC 2022 6).
- Block applications from Banned Developers on the CHPL. They present a security risk.

For delivering FHIR applications and mobile apps, we recommend:

- Leverage ONC-ACB certified applications or application frameworks.
- Conduct application-level penetration testing on the applications and mobile applications using tools such as Corellium (https://www.corellium.com/) to test API calls and security at lower levels and on mobile operating systems themselves (Corellium, Inc. 2022).
- Conduct API-level penetration testing on a regular basis using:
  - OWASP Top Ten Web Application Security Risks (https://owasp.org/www-project-top-ten/) (OWASP 2022)
  - Penetration testers with Offensive Security Certified Professional (OSCP) (www.offensive-security.com/pwk-oscp/) or Offensive Security Experienced Penetration Tester (OSEP) certifications (https://www.offensive-security.com/pen300-osep/) (Offensive Security 2022; Offensive Security 2022 2)
  - Penetration testers with GIAC Penetration Tester (GPEN) certifications (https://www.giac.org/certifications/penetration-tester-gpen/) (GIAC 2022)
  - Experienced API penetration testers
- Do not use products or services from Banned Developers on the CHPL (https://chpl.healthit.gov/#/search) (HHS ONC 2022 6).

■ Conduct regular code reviews and conformance testing using the ONC Certification Criterion (https://www.healthit.gov/topic/certification-ehrs/2015-edition-test-method) (HHS ONC 2021).

The CURES Act has information spread across multiple federal laws. It is confusing. It does not provide actionable advice to healthcare providers on what to do. More importantly, it appears to give the impression that it covers security standards required under a standard such as CCM/CAIQ or ISO/IEC 27001/27002. Utilizing several of the other frameworks, we aim to have broken apart the CURES Act and provided what organizations need to do to comply with this.

The 405(d) Program and Healthcare and Public Health Sector Coordinating Council (HSCC) Cybersecurity Working Group (CWG)

The 405(d) Health Industry Cybersecurity Practices program is a collaboration between the numerous industry components and the federal government. It designs and develops guidelines, practices, and methodologies to strengthen the overall industry posture against cyber threats (https://405d.hhs.gov/public/navigation/aboutUs) (405(d) Program 2022). Much of the guidance provided aligns with the NIST Cybersecurity Framework, and is practical advice aimed at organizations. They provide cybersecurity best practices, resources and templates, training, and collateral for organizations to utilize (https://405d.hhs.gov/public/navigation/resources) (405(d) Program 2022 2). Since their work aligns with the NIST Cybersecurity Framework, it's an excellent complement if you choose to use that.

The HSCC CWG is recognized by HHS as the critical infrastructure industry partner for coordinating strategic, policy, and operational approaches for preparation, response, and recovery from significant cyber and physical threats (https://healthsectorcouncil.org/#) (HSCC CWG 2022). These threats would prevent the ability of the sector to deliver critical assets and services to the public (https://healthsectorcouncil.org/#) (HSCC CWG 2022). The HSCC CWG provides best practices, recommendations, and policy comments (https://healthsectorcouncil.org/hscc-recommendations/) (HSCC CWG 2022 2). These include Model Contract Language for Medtech Cybersecurity, Securing Telehealth and Telemedicine, and Supply Chain Risk Management (https://healthsectorcouncil.org/hscc-recommendations/) (HSCC CWG 2022 2). They also have numerous working task groups working on deliverables and initiatives. The 2021 working groups are:

■ 405(d) Health Industry Cybersecurity Practices
■ Health Technology Risk Analysis
  – Future Gazing Sub-Group
■ Intellectual Property Data Protection
■ International Engagement
■ Medical Technology
  – Medtech Legacy Devices
  – Medtech Model Contracts
  – Medtech Vulnerability Communications
■ Metrics for Cybersecurity Adoption

- Policy
- Risk Assessment
- Supply Chain Cyber Risk Management
- Telemedicine
- Workforce Development

Both the 405(d) and HSCC groups work together to deliver messaging and information to healthcare organizations that need it the most. Their websites are at:

- 405(d): https://405d.hhs.gov
- HSCC: https://healthsectorcouncil.org

This section went over significant guidance regarding HIPAA and HITECH, HITRUST, the ISO 27001 series, the AICPA Service Organization Control (SOC) reports, the NIST Cybersecurity Framework, the CSA CAIQ/CCM/STAR Registry, Payment Card Industry – Data Security Standards (PCI-DSS), the 21st Century CURES Act Final Rule and what it really means, and the 405(d) and Health Sector Coordinating Council Programs. There was a lot of ground to cover that ended in how to practically protect organizations and leveraging guidance from HSCC and 405(d). In the next section, we're going to cover what your organization needs to do to wrap these up in policies, other standards, and procedures.

Policies, Standards, and Procedures

Organizations need policies and procedures that cover the administrative, technical, and physical security components. We're not going to tell you exactly which of those to have. We will tell you that you need to cover all of those in the HIPAA Security Rule. ONC has put the documents that describe the controls in the HIPAA Security Rule in their Security Risk Assessment Tool, which is freely available to download from their website at https://www.healthit.gov/topic/privacy-security -and-hipaa/security-risk-assessment-tool (HHS ONC 2022). This has all of the required policies and procedures you need already documented.

As part of your organizational risk assessment, do a gap analysis of the policies, standards, and procedures you already have against the ones in the tool. Document what you need to add as part of it and develop a plan to do so. Most importantly, if you already have policies and procedures, modify those. Don't throw out what people know, or what people have contributed to. One of the big challenges with security has been that people arbitrarily throw out years of organizational knowledge in the pursuit of better security. This causes a larger issue because you throw out years of organizational knowledge, and likely cause discord with people who have built years of processes and methods. Instead, work with them to make the changes needed, and make sure that you also cover outside vendors as part of the process. As we outsource significantly more due to mobile applications and the cloud, we need to realize that they now manage large chunks of our business, and address third parties in our policies, procedures, and processes with that revelation. If you take away policies and procedures someone has spent years building, and then give them something you bought with a credit card off of an internet search, they're more likely to print what you give them out and use it for training their animals where to go to the bathroom than follow it. So will your vendors. These policies also need to be realistic.

Speaking of what happens when things from the bathroom collide with air circulation equipment, insurance is also a requirement, with at least $1M set aside for cyber insurance from vendors and appropriate liability given the organizational risk and criticality of the services you are purchasing, if not more. This is an item where we cannot recommend an exact number for you because we are not your cyber team or legal counsel. Even if your organization does not have dedicated cyber personnel, it is worth your time and money to speak with legal counsel to determine the risk exposure of your organization to products and services you purchase in the contract negotiation process.

Outsourcing is also a double-edged sword. There are two dangers with it. The first is with outsourcing to countries with weak data protection laws like India, which at the time of writing this only has section 43A of the IT Act 2000/8, which has weak regulations, does not specify any governmental agency, and does not lay out penalties for violations (Goswani and Haram 2017). In addition, sending personal data to countries such as China, which has engaged in mass data exfiltration of data on US citizens according to Kevin Collier of NBC news, will only help build their databases (Collier 2020). At minimum, personal data needs to reside in the country of origin, with access only granted through remote desktop technology over an encrypted connection. Given the risks of this data ending up in the hands of a foreign government, or in a jurisdiction where there is little to no protection in case of a data breach, it is important to have the right countermeasures.

The same can be said for countries that are concerned about their data residing in the United States. Specifically, because of the surveillance programs that Edward Snowden disclosed, the European Court of Justice ruled that the US-EU Safe Harbor principles did not provide sufficient data protection guarantees for EU citizens (UC Berkeley School of Law 2015). Privacy Shield was created to replace this, however it was invalidated in 2020 (Bracy 2022). In 2022, a new trans-Atlantic data flow agreement was announced in principle (Bracy 2022). However, the concern still remains that outside jurisdictions may not consider the United States secure enough, and that legislation will likely continue to be challenged. In this case, organizations need to make sure that they are able to process personal data in the country the data originates from given the history of trans-Atlantic data agreements failing.

Don't give your vendors lengthy questionnaires and expect them to be nice to you. Too many of them are taking security engineers off of product design and development efforts to fill out your version of a questionnaire that covers something you didn't like about HITRUST, HIPAA/HITECH, NIST, PCI-DSS, CCM/CAIQ, or ISO/IEC 27001/27002. Questionnaires are the instrument of last resort. Make sure you accept numerous industry certifications first before requiring a questionnaire from a vendor.

Also, don't steer your vendor toward paid certifications that don't follow ISO 17021 standards for separation of audit and consulting work. You are putting your organization at risk, especially if you work for a non-profit, as you are steering work toward a for-profit firm in hopes of getting your business. You are also asking smaller companies to spend significant amounts of money in hopes of getting more business. They may take an initial loss just to do business with you; however, if the certification comes with additional consulting costs, you may cause your vendors, who may include smaller businesses, to lose a lot of money.

Conduct your own risk assessment of your organization, Leverage the Actual Operational Work section of this chapter as a guide to write your own plan to address monitoring, alerting, enforcement, and alignment through policies, procedures, and processes. Leverage what your organization has to make it part of everyday work. These are only as effective as the buy-in from your organization, and how they align with the work already in process. They are not effective if you buy something or have a consultant write something up that is not considerate of stakeholders or organizational history.

Most importantly, realize you cannot test in a vacuum. For your new systems, run a Failure Mode Effects Analysis (FMEA) to understand the use cases in detail, where the highest risks are, and how to address them before they become issues. A group of accredited systems operating together can be at higher risk than one system due to integration issues and concerns. Make sure to do testing in your environment, with integration, and do full security and FMEA testing along with it.

Third-Party Contracting and Risk Management

Third-Party Risk Management needs to be implemented in phases. You can't just drop in a program and have it work, much to the consternation of vendors who will gladly collect six figures for the privilege of telling you that you can. There's several areas that need to be prepared before you can develop and execute a contract:

- Insurance/Cyber Insurance Requirements
- Merchant Bank Requirements for PCI-DSS
- Other Applicable Regulatory Requirements
- Metrics and Measurements of Success
- Operational Management Requirements/Being an Active Participant
- Integration Requirements
- Nth Party Risk
- Ongoing Risk Management
- Decommissioning
- Renewals
- Terminations and Enforcement
- Example Health Sector Coordinating Council Language

Cyber insurance, now more than ever, is required to protect yourself and the organization that you work for. Data breaches are expensive and present significant reputational risk. You do not want to be in a situation where you are wholly responsible for the costs past the fees you paid a third-party vendor to remediate your customers and make them whole. Companies that do business electronically, especially in this time of ransomware and malware, need to carry cyber insurance and maintain it in good standing. Given that the insurance companies have significantly increased their requirements for it, as we discussed earlier, this also serves to demonstrate that the vendors are utilizing reasonable and appropriate protection to be able to have a policy underwritten in the first place.

Likewise, merchant banks are very strict about PCI-DSS because the brands themselves (Visa, Mastercard, Discover, American Express, JCB) enforce compliance (Sysnet and Viking Cloud 2017). They do not want to be in a position where they themselves are not able to process credit cards. As everyone is required to be PCI compliant, it's important to require the appropriate PCI-DSS evidence, and it is not too much to ask for a ROC from a QSA as insurance companies do ask policyholders for these. Requiring PA-DSS certification for payment applications is a basic requirement that the brands and your insurance company will ask for.

Risk assessments are very important. One of the reasons is to uncover the regulatory and accreditation requirements outside of HIPAA that your organization may be subject to, with the largest being the Joint Commission. For example, if you run a laboratory environment on-site, you are subject to review by the College of American Pathologists (CAP), who stringently examines those environments to ensure effectiveness and safety. If your organization has systems that process data for cellular therapy, you may require accreditation from the Foundation for the Accreditation of Cellular Therapy (FACT). You need to ask these questions as part of the risk assessment process so that you can appropriately include the regulatory and accreditation requirements in the contracts.

Following your organization's policies, specifically on Identity Management, Information Security, Vulnerability Management, and Risk, is also critical. A major reason we do the work on augmenting policies and ensuring that they will actually be effective and used is because third parties also have to abide by them and be included in contracts. Contract clauses that don't align with your policies and based on risk management plans deriving from risk assessments aren't worth the paper they're written on.

Determining how to measure success is important, and how your organization will be able to gauge contract effectiveness. If you're contracting for a product or service, you need to understand the metrics for these, which include uptime for remotely hosted or cloud-based systems. This is different than the security metrics themselves, in that you're measuring the individual product effectiveness and availability. These need to be defined in the project requests and Return on Investment (ROI) documents your leaders have prepared. What good is putting a product or system in if you can't define what successful operation and metrics are that define it? These also need appropriate service levels and penalties for noncompliance.

Once you have metrics, you have the operational management processes and requirements. One area in which the CSA CAIQ excels is at delineating responsibilities. Document the operational management processes and requirements, clearly identify ownership and responsibilities, and put those in the contract. This is important to have because when something adverse happens you want it in writing who is responsible, because if it isn't, it will be you. The four-column layout of the CAIQ also can apply to these processes at the contract level. This also means that you and your organizations have to be active participants. You can't think that a contract obligates your vendors to do everything regarding these systems as well, much like Dr. Sheldon Cooper's roommate agreement obligated Dr. Leonard Hofstader to drive him everywhere.

Delineation is good; however, understanding interfaces and responsibilities in the same way as the CSA CAIQ also works. Interfaces are two ways, and we often don't know who is responsible. Putting that language in the contract helps identify who is responsible, which is important, especially given the complexity of these systems and the data they handle. Interfaces need to be documented, down to the protocols used, data types, data formats, and security protocols. The CURES Act requirements and conformance testing requirements illustrate the importance of documenting this in the contracts.

Nth party risk is also important. You need to require that all your contractors and subcontractors follow the same policies and procedures, and ensure minimum levels of security. All it takes is one call center with privileged access to have a compromise and multiple customers can be disrupted. Requiring this is critical, especially given increased cyber insurance requirements and the numerous breaches caused by third parties.

Ongoing risk management is critical. Too many times there are vendors that will only get a certification or give a risk assessment to get a deal, and then slack off afterward. The requirements from HIPAA/HITECH and other applicable healthcare certifications such as the CURES Act require continual compliance. Clearly outline the certification and regulatory requirements, how they will be measured and evaluated, and how often you will ask for evidence for the defined time period. For example, for a third-party data center, you will ask for a SOC 2 report annually with no modified, qualified, or adverse opinions, and no disclaimer of opinions for the applicable defined time period.

When the relationship ends, there needs to be a mutually agreed-upon process by which organizations can dispose of or transfer data, and what methods will be used to decommission data. One of the most common references is to NIST Special Publication 800-88 Revision 1, Guidelines for Media Sanitization, for physical media (https://csrc.nist.gov/publications/detail/sp/800-88/rev-1/final) (Kissel, Regenscheid, Scholl, and Stine 2014). This document describes the NIST-approved methods, which is critical to erasing personal data, and gives an example certificate to use. In addition to this, data and audit logs will likely need to be transferred over to your organization or another designated third party due to Accounting of Disclosures and to remain in compliance with applicable data retention laws for your locality. This needs to be included and disclosed within the contract.

You also need to have specific clauses for termination and penalties. Healthcare organizations are required to have a Sanctions Policy according to the HIPAA Security Rule (https://www.hhs.gov/hipaa/for-professionals/security/laws-regulations/index.html). We do not know your organization, and cannot tell you what to have outside what the Security Rule requires. However, it is important to have one, especially for Service Level Agreement and availability concerns. We recommend that you consult a contracts lawyer to discuss how to effectively implement one of these in a contract. Don't be Dr. Sheldon Cooper and try to put unenforceable penalty clauses in. The Big Bang Theory is a sitcom, not a model to try and flex knowledge from, and he's a fictional character. You're not and trying to be like him will backfire on you.

When it's time to renew the contract, make sure your organization has good processes for renewing contracts. Don't wait until the last minute. Make sure you track start and end dates and start renegotiation processes early. If you wait until the last minute, you will burn out your contracting staff.

The Health Sector Coordinating Council (HSCC), as a coda to this, has provided excellent contract template language for medical technology solutions that has been pre-negotiated with vendors. It's available at https://healthsectorcouncil.org/model-contract-language-for-medtech-cybersecurity-mc2/ (HSCC CWG 2022 3). As the legal teams for many of your vendors have already negotiated this language, leveraging and using this where applicable would save all involved parties time rather than debating minutiae.

Conclusion

Third-party risk is not nearly as easy as some make it out to be. It's not easy to absolve your organization of responsibility by outsourcing functions to others. Much like the story from the Midrash, a hole in the boat sinks everyone. Digital Transformation and Mobile Applications have left all businesses, especially Healthcare Delivery Organizations, reliant upon multiple third parties to deliver solutions leveraging technology to customers and patients. This requires organizations to build and acquire specialist knowledge, specifically in the security and interoperability requirements of systems, to better deliver solutions. The goal of this chapter was to provide healthcare executives with the knowledge needed to understand the regulations and requirements, and the processes by which they can apply them to implement proper risk management. This includes internal and external risk management and developing internal skills and resilience to help address the integration points that third-party vendor risk just doesn't accommodate for currently. Many of the standards that exist are nebulous and not accessible to the people who need to make the decisions. Our goal here was to put this all in one place so that the people who have to understand and make them have an easier time of doing so.

Our goal was also to get organizations like the College of Healthcare Information Management Executives (CHIME) and their CISO organization, the Association for Executives in Healthcare Information Security (AEHIS), to leverage this knowledge with their members. One of the most significant challenges we've observed has been that often healthcare CIOs and CISOs are thrown into these jobs with little knowledge. CHIME and AEHIS have done great work with their bootcamps. However, we feel that the knowledge in this chapter would be an excellent net addition to educate their members on.

References

"2015 Edition Health Information Technology (Health IT) Certification Criteria, 2015 Edition Base Electronic Health Record (EHR) Definition, and ONC Health IT Certification Program Modifications; Corrections and Clarifications." *80 Federal Register* 238 (December 11, 2015): 76868–76872.

"2015 Edition Test Method." HealthIT.gov. US Department of Health and Human Services Office of the National Coordinator (HHS ONC), November 12, 2021. https://www.healthit.gov/topic/certification-ehrs/2015-edition-test-method.

"About APAC." Asia-Pacific Accreditation Cooperation (APAC). Accessed May 10, 2022. https://www.apac-accreditation.org/about/.

"About IAAC – Introduction." InterAmerican Accreditation Cooperation (IAAC). Accessed May 10, 2022. https://www.iaac.org.mx/index.php/en/about-iaac/introduction-en.

"About ONC's Cures Act Final Rule." US Department of Health and Human Services Office of the National Coordinator (ONC). Accessed May 10, 2022. https://www.healthit.gov/curesrule/overview/about-oncs-cures-act-final-rule.

"Accounting of Disclosures of Protected Health Information." *Code of Federal Regulations* 45 (2003): 755–757. https://www.govinfo.gov/content/pkg/CFR-2003-title45-vol1/pdf/CFR-2003-title45-vol1-part164.pdf.

"About – Overview." Cloud Security Alliance (CSA). Accessed May 10, 2022. https://cloudsecurityalliance.org/about/.

"About the IAF MLA." International Accreditation Forum (IAF), December 16, 2021. https://iaf.nu/en/about-iaf-mla/.

"About Us." 405(d). 405(d) Program, Office of Information Security (OIS), US Department of Health and Human Services (HHS). Accessed May 10, 2022. https://405d.hhs.gov/about.

"About Us – ARAC." Arab Accreditation Cooperation (ARAC). Accessed May 10, 2022. https://arab-accreditation.org/about/.

"Administrative Safeguards." *Code of Federal Regulations* 45 (2003): 737–739. https://www.govinfo.gov/content/pkg/CFR-2007-title45-vol1/pdf/CFR-2007-title45-vol1-sec164-308.pdf.

"ANAB CB Directory." American National Standards Institute (ANSI). Accessed May 10, 2022. http://anabdirectory.remoteauditor.com/.

"App and API Security." API Security | Data Security | Akamai DDoS Protection. Akamai, Inc. Accessed May 10, 2022. https://www.akamai.com/solutions/security/app-and-api-security.

"Approov API Threat Protection." Approov API Threat Protection – Digital Marketplace. UK Government. Accessed May 10, 2022. https://www.digitalmarketplace.service.gov.uk/g-cloud/services/753375422325655.

"ASTM E2147-18 Standard Specification for Audit and Disclosure Logs for use in Health Information Systems." ANSI Webstore. American National Standards Institute (ANSI), 2018. https://webstore.ansi.org/standards/astm/astme214718.

"Breach Notification for Unsecured Protected Health Information; Interim Final Rule with Request for Comments." *74 Federal Register* 172 (August 24, 2009): 42740–42770.

"Business Architecture." CIO Wiki, February 6, 2021. https://cio-wiki.org/wiki/Business_Architecture #Definition_of_Business_Architecture.3F.

"Business Architecture Framework." The Business Architecture Center of Excellence (BACOE). Accessed May 10, 2022. https://www.bacoe.org/framework.

"Certified Health IT Product List (CHPL)." US Department of Health and Human Services Office of the National Coordinator (HHS ONC). Accessed May 10, 2022. https://chpl.healthit.gov/#/search.

"Certified STAR Auditors." Cloud Security Alliance (CSA). Accessed May 10, 2022. https://cloudsecurityal liance.org/star/certified-star-auditors/.

"Cloud Controls Matrix (CCM)." Cloud Security Alliance (CSA). Accessed May 10, 2022. https://cloudse curityalliance.org/research/cloud-controls-matrix/.

"CSA STAR Registry." Cloud Security Alliance. Accessed May 10, 2022. https://cloudsecurityalliance.org /star/registry/.

"CURES ACT FINAL RULE Standards-based Application Programming Interface (API) Certification Criterion." US Department of Health and Human Services Office of the National Coordinator (HHS ONC), March 7, 2020. https://www.healthit.gov/cures/sites/default/files/cures/2020-03/APICertific ationCriterion.pdf.

"Definition of Enterprise Architecture (EA) – Gartner Information Technology Glossary." Gartner, Inc. Accessed May 15, 2022. https://www.gartner.com/en/information-technology/glossary/enterprise -architecture-ea.

"Demystifying SOC Opinions and Exceptions." Foresite Cybersecurity. Accessed May 10, 2022. https:// foresite.com/soc-opinions-and-exceptions/.

"Evasion Techniques and Breaching Defenses (PEN-300)." Offensive Security. Accessed May 10, 2022. https://www.offensive-security.com/pen300-osep/.

"Failure Modes and Effects Analysis (FMEA) Tool: IHI." Institute for Healthcare Improvement, 2017. http://www.ihi.org/resources/Pages/Tools/FailureModesandEffectsAnalysisTool.aspx.

"GIAC Penetration Tester (GPEN)." GIAC Penetration Tester Certification | Cybersecurity Certification. GIAC Certifications. Accessed May 10, 2022. https://www.giac.org/certifications/penetration-tester -gpen/.

"Health Information Technology: Initial Set of Standards, Implementation Specifications, and Certification Criteria for Electronic Health Record Technology" *75 Federal Register* 8 (13 January 2010): 2013–2047.

"HSCC Recommendations." Health Sector Council. Healthcare and Public Health Sector Coordinating Council Cyber Security Working Group (HSCC CWG). Accessed May 10, 2022. https://healthsec torcouncil.org/hscc-recommendations/.

"ISO/IEC 17021-1:2015." ISO, December 1, 2016. https://www.iso.org/standard/61651.html.

"ISO 27002 Ultimate Guide." ISMS.online. Accessed May 10, 2022. https://www.isms.online/iso-27002/.

"ISO 27018. Security and Protection of Personal Information in the Cloud." GlobalSUITE Solutions, May 4, 2020. https://www.globalsuitesolutions.com/iso-27018-security-and-protection-of-personal-infor mation-in-the-cloud/.

"ISO/IEC 27017." Security Controls for Cloud Services ISO/IEC 27017 | India. BSI Group. Accessed May 10, 2022. https://www.bsigroup.com/en-IN/Security-controls-for-cloud-services-ISO-IEC27017/.

"Model Contract-Language for Medtech Cybersecurity (MC2)." Health Sector Council. Healthcare and Public Health Sector Coordinating Council Cyber Security Working Group (HSCC CWG). Accessed May 10, 2022. https://healthsectorcouncil.org/model-contract-language-for-medtech-cybersecurity -mc2/.

"ONC-Authorized Certification Bodies (ONC-ACBs)." HealthIT.gov. US Department of Health and Human Services Office of the National Coordinator (HHS ONC), January 5, 2022. https://www .healthit.gov/topic/certification-ehrs/onc-authorized-certification-bodies-onc-acbs.

"ONC – Authorized Testing Laboratories (ONC-ATLs)." HealthIT.gov. US Department of Health and Human Services Office of the National Coordinator (HHS ONC), April 29, 2022. https://www .healthit.gov/topic/certification-ehrs/onc-authorized-testing-laboratories-onc-atls.

"ONC Health IT Certification Program Overview." HHS Office of the National Coordinator (HHS ONC), March 31, 2022. https://www.healthit.gov/sites/default/files/PUBLICHealthITCertification ProgramOverview.pdf.

"OWASP Top Ten." OWASP. Open Web Application Security Project (OWASP). Accessed May 10, 2022. https://owasp.org/www-project-top-ten/.

"Payment Card Industry Data Security Standard Requirements and Testing Procedures Version 4.0." PCI Security Standards Council (PCI-SSC), March 2022. https://www.pcisecuritystandards.org/documents/PCI-DSS-v4_0.pdf.

"PCI Compliance Guide Frequently Asked Questions: PCI DSS Faqs." PCI Compliance Guide. Sysnet and Viking Cloud, September 5, 2017. https://www.pcicomplianceguide.org/faq/.

"PCI Security." PCI Security Standards Council® (PCI-SSC). PCI Security Standards Council® (PCI-SSC). Accessed May 9, 2022. https://www.pcisecuritystandards.org/pci_security/.

"PEN-200 – The Official OSCP Certification Course." Offensive Security. Accessed May 10, 2022. https://www.offensive-security.com/pwk-oscp/.

"Resources." 405d. 405(d) Program, Office of Information Security (OIS), US Department of Health and Human Services (HHS). Accessed May 10, 2022. https://405d.hhs.gov/resources.

"Salt Security: API Security across Build, Deploy, Runtime." Salt Security. Accessed May 10, 2022. https://salt.security/.

"Security Risk Assessment Tool." US Department of Health and Human Services, Office of the National Coordinator (HHS ONC), February 15, 2022. https://www.healthit.gov/topic/privacy-security-and-hipaa/security-risk-assessment-tool.

"Snowden's Legacy: Ruling from the Court of Justice of the European Union." Berkeley Technology Law Journal. UC Berkeley School of Law, November 23, 2015. https://btlj.org/2015/11/snowdens-legacy-ruling-from-the-court-of-justice-for-the-european-union/.

"SOC 1 Report." The SSAE 18 Reporting Standard – SOC 1 – SOC 2 – SOC 3 (Formerly SSAE 16). Marcum LLP. Accessed May 10, 2022. https://www.ssae-16.com/soc-1/.

"Soc 2® – SOC for Service Organizations: Trust Services Criteria." Assurance and Advisory Services. AICPA. Accessed May 10, 2022. https://us.aicpa.org/interestareas/frc/assuranceadvisoryservices/aicpasoc2report.html.

"SOC 3." Google, Inc. Accessed May 10, 2022. https://cloud.google.com/security/compliance/soc-3.

"Security, Trust, Assurance and Risk (STAR)." Cloud Security Alliance (CSA). Accessed May 10, 2022. https://cloudsecurityalliance.org/star/.

"Statement on Enhancing Patient Privacy and Security without Compromising the Patient's Right of Access." Health Level 7 (HL7), October 25, 2021. https://confluence.hl7.org/download/attachments /81012315/v6-2021-10-25-HL7-Patient-Empowerment-WG-meeting-proposed-response-to-Playing -with-FHIR%20%281%29.docx?version=1&modificationDate=1638509137602&api=v2.

"The 21st Century Cures Act Final Rule & What It Means for Developers." Dynamic Health IT, April 23, 2020. https://www.dynamichealthit.com/post/the-21st-century-cures-act-final-rule-and-what-it -means-for-software-developers.

"The Boat – From the Teachings of Rabbi Shimon Bar Yochai." Chabad-Lubavitch Media Center. Accessed May 9, 2022. https://www.chabad.org/library/article_cdo/aid/386812/jewish/The-Boat.htm.

"Trust: Compliance." HITRUST | Salesforce Compliance. Salesforce, Inc., February 28, 2022. https://compliance.salesforce.com/en/hitrust.

"Virtual Devices with Real-World Accuracy." Corellium, Inc. Accessed May 10, 2022. https://www.corellium.com/.

"Visa Global Registry of Service Providers." Visa. Accessed May 10, 2022. https://usa.visa.com/splisting/splistingindex.html.

"Welcome to AFRAC." Pages – Home. African Accreditation Cooperation (AFRAC). Accessed May 10, 2022. https://www.intra-afrac.com/Pages/Home.aspx.

"Welcome to the Healthcare and Public Health Sector Coordinating Council (HSCC) Cybersecurity Working Group (CWG)." Health Sector Council. Healthcare and Public Health Sector Coordinating Council Cyber Security Working Group (HSCC CWG). Accessed May 10, 2022. https://healthsectorcouncil.org/#.

"What Do You Know About ISO 27001?" EC-Council Global Services. EC-Council. Accessed May 10, 2022. https://egs.eccouncil.org/what-do-you-know-about-iso-27001/.

"What is the Difference between Addressable and Required Implementation Specifications in the Security Rule?" US Department of Health and Human Services (HHS), June 28, 2021. https://www.hhs.gov /hipaa/for-professionals/faq/2020/what-is-the-difference-between-addressable-and-required-imple-mentation-specifications/index.html.

"What ONC's Cures Act Final Rule Means for Clinicians." US Department of Health and Human Services Office of the National Coordinator (ONC). Accessed May 10, 2022. https://www.healthit.gov/cures-rule/what-it-means-for-me/clinicians.

"Who Are We?" European Accreditation (EA), January 20, 2022. https://european-accreditation.org/about -ea/who-are-we/.

Bracy, Jedidiah. "EU, US agree 'in principle' to new trans-Atlantic data agreement." International Association of Privacy Professionals (IAPP), March 25, 2022. https://iapp.org/news/a/eu-us-agree-in -principle-to-new-transatlantic-data-agreement/.

Catteddu, Daniele. "CAIQ v4 Released – Changes from v3.1 to v4." Cloud Security Alliance (CSA), August 16, 2021. https://cloudsecurityalliance.org/blog/2021/06/07/caiq-v4-released-changes-from -v3-1-to-v4/.

Chaudhary, Ashwin, John Di Maria, and Walter Williams. "Code of Practice for Implementing STAR Level 2." Cloud Security Alliance (CSA), June 23, 2021. https://cloudsecurityalliance.org/artifacts/ code-of-practice-for-implementing-star-level-2/.

Collier, Kevin. "China Spent Years Collecting Americans' Personal Information. The U.S. Just Called It out." NBCNews.com. NBCUniversal News Group, February 10, 2020. https://www.nbcnews.com/ tech/security/china-spent-years-collecting-americans-personal-information-u-s-just-n1134411.

Douville, Sherri. *Mobile Medicine: Overcoming People, Culture, and Governance.* New York: Routledge, 2022.

Fruhlinger, Josh. "HITRUST Explained: One Framework to Rule Them All." CSO Online, May 31, 2021. https://www.csoonline.com/article/3619534/hitrust-explained-one-framework-to-rule-them-all.html.

Goswami, Suparna, and Varun Haran. "Analysis: Data Protection in India – Getting It Right." Bank Information Security. Information Security Media Group (ISMG), April 26, 2017. https://www .bankinfosecurity.asia/analysis-data-protection-in-india-getting-right-a-9866.

Irwin, Luke. "A Guide to the 4 PCI DSS Compliance Levels." IT Governance European Blog, January 18, 2022. https://www.itgovernance.eu/blog/en/a-guide-to-the-4-pci-dss-compliance-levels.

Kak, Shivani, and Joe Mielenhausen. "Moody's and BitSight Partner to Create Integrated Cybersecurity Risk Platform." Moodys, September 13, 2021. https://ir.moodys.com/press-releases/news-details/2021 /Moodys-and-BitSight-Partner-to-Create-Integrated-Cybersecurity-Risk-Platform/default.aspx.

Keller, Nicole. "An Introduction to the Components of the Framework." NIST Cybersecurity Framework. NIST, May 14, 2021. https://www.nist.gov/cyberframework/online-learning/components-framework.

Kelly, M. J. "What Is a Sim Swapping Scam and How Can You Protect Yourself?" The Mozilla Blog, April 7, 2021. https://blog.mozilla.org/en/internet-culture/mozilla-explains/mozilla-explains-sim-swapping/.

Kissel, Richard, Andrew Regenscheid, Matthew Scholl, and Kevin Stine. "Guidelines for Media Sanitization." National Institute of Standards and Technology (NIST), December 15, 2014. https:// nvlpubs.nist.gov/nistpubs/SpecialPublications/NIST.SP.800-88r1.pdf.

Leaman, Dana S., and Bethany Hackett. "NIST Handbook 150 2020 Edition." National Institute of Standards and Technology (NIST), August 18, 2020. https://nvlpubs.nist.gov/nistpubs/hb/2020/ NIST.HB.150-2020.pdf.

Leigh, Laurie. "The Hidden ROI of HITRUST Certification." Agio Healthcare, July 21, 2020. https:// healthcare.agio.com/newsroom/the-hidden-roi-of-hitrust-certification/.

Levey, Cheryl. "Apply for NVLAP Accreditation." National Voluntary Laboratory Accreditation Program. National Institute of Standards and Technology (NIST), December 20, 2021. https://www.nist.gov/ nvlap/apply-nvlap-accreditation.

Martin, Jim, Jack Burbank, William Kasch, and David L Mills. "Network Time Protocol Version 4: Protocol and Algorithms Specification RFC 5905." RFC 5905 – Network Time Protocol Version 4: Protocol and Algorithms Specification, January 21, 2020. https://datatracker.ietf.org/doc/rfc5905/.

McGee, Marianne Kolbasuk. "Moody's Warns Cyber Risks Could Impact Credit Ratings." Bank Information Security. Information Security Media Group (ISMG), November 24, 2015. https://www.bankinfosecurity.com/moodys-warns-cyber-risks-could-impact-credit-ratings-a-8702.

McGoran, Jonathan. "Wondering Why Your Cyber Insurance Rates Are Going up? Thank the 125% Increase in Attacks from Ransomware Gangs, for Starters." Risk & Insurance, December 1, 2021. https://riskandinsurance.com/wondering-why-your-cyber-insurance-rates-are-going-up-thank-the-125-increase-in-attacks-from-ransomware-gangs-for-starters/.

Moore, Bradley, and Asara Clark. "NVLAP Health Information Technology Testing." National Institute of Standards and Technology (NIST), November 30, 2021. https://www.nist.gov/publications/nvlap-health-information-technology-testing.

National Institute of Standards and Technology (NIST). "Framework for Improving Critical Infrastructure Cybersecurity, Version 1.1," April 16, 2018. https://doi.org/10.6028/nist.cswp.04162018.

Nicasio, Francesca. "Negative Reverse Selling – Tips, Power Phrases and Examples." Userlike Live Chat, July 24, 2020. https://www.userlike.com/en/blog/negative-reverse-selling.

Reed Stark, John. "Cyber-Security Due Diligence: A New Imperative." Compliance Week, June 7, 2016. https://www.complianceweek.com/cyber-security-due-diligence-a-new-imperative/10728.article.

Slivochka, Sharon, and Diana Warner. "Understanding the Eight Exceptions to Information Blocking." Journal of AHIMA. American Health Information Management Association (AHIMA), February 25, 2021. https://journal.ahima.org/page/understanding-the-eight-exceptions-to-information-blocking.

Taule, Jason. "Introduction to the HITRUST CSF." HITRUST Alliance, December 4, 2020. https://hitrustalliance.net/content/uploads/CSFv9.4_Introduction.pdf.

Thurman, Andrew. "Axie Infinity's Ronin Network Suffers $625M Exploit." CoinDesk, March 29, 2022. https://www.coindesk.com/tech/2022/03/29/axie-infinitys-ronin-network-suffers-625m-exploit/.

Verry, John. "What the New ISO 27001:2021 Release Will Mean to You." Pivot Point Security, October 19, 2021. https://www.pivotpointsecurity.com/blog/what-the-new-iso-270012021-release-will-mean-to-you/.

Vidich, Stevan. "HITRUST – Azure Compliance." Azure Compliance | Microsoft Docs, April 6, 2022. https://docs.microsoft.com/en-us/azure/compliance/offerings/offering-hitrust.

Woollven, Camden. "List of US Accredited Certification Bodies for ISO 27001." IT Governance USA Blog, July 15, 2020. https://www.itgovernanceusa.com/blog/list-of-us-accredited-certification-bodies-for-iso-27001.

Wu, Ethan. "Crypto Mega-Billionaire Sam Bankman-Fried Says He's Bought Billions of Tethers in Order to Trade Other Coins." Business Insider, October 10, 2021. https://markets.businessinsider.com/news/currencies/sam-bankman-fried-tether-crypto-ftx-bitcoin-stablecoin-miami-fraud-2021-10.

Chapter 11

Hospital at Home: Managing the Risks of Delivering Acute Care in the Patient's Home

Mitch Parker and Peter McLaughlin

Contents

Medical devices, patient monitoring tools, and remote access to patient files are significantly more complicated to use outside the familiar healthcare facility because the patient data does not benefit from the layers of security within the traditional environment. A wireless device depends on cellular or Wi-Fi networks available, as opposed to the hospital IT system with access controls, firewalls, and myriad other defenses. And the application of HIPAA is not limited to traditional healthcare scenarios. Thus, patient data on devices that themselves are not robustly secure or that transmit patient data without encryption place the patient and the provider at unnecessary risk. This chapter reviews the rules and risks and delivers recommendations for a better, more secure patient experience at home.

The Trend and Importance of Acute Care Delivery at Home

In November 2020, the HHS Center for Medicare and Medicaid Services announced the "Acute Hospital at Home" program. While the idea of providing healthcare at home is not new – remember the quaint concept of house calls? – just one challenge has been the availability of reimbursement. For all the tragedy of the COVID pandemic, one repercussion was the necessity of managing capacity in hospital beds and emergency departments. Additionally, given the highly transmissible nature of COVID, healthcare providers wanted to protect acute care patients who might otherwise catch COVID while waiting to be seen (Hospital at Home: A Shift in Thinking for Acute Care, 2022).

There are many anecdotal reasons for stakeholders to like the Hospital at Home model. Patients likely prefer the familiar environment of home, and providers manage the capacity of space within more traditional healthcare settings. A 2019 randomized controlled trial at Brigham and Women's Hospital found that hospital-level care at home for acute patients yielded 38% lower cost than the traditional setting; fewer imaging studies; a 50% increase in patient mobility; and a 70% reduction in readmissions (Id.)

The benefits should not lead a hospital leader to think that implementing a healthcare at home program (whether for acute or chronic conditions) is without both the familiar and unfamiliar challenges. Change is difficult, and older patients may have greater faith in the traditional healthcare setting. Clinicians are able to see fewer patients per day, and those patients may not always live in relatively safe neighborhoods. Both patients and clinicians likely find reassuring the supporting infrastructure in the traditional environment, and with this infection control, sanitation, and imperfect patient monitoring add to the hurdles, both technical and personal (Chandrashekar, Moodley, and Jain, 2019).

There remain bases for optimism, however, as a McKinsey & Company survey concludes:

[b]ased on a survey of physicians who serve predominantly Medicare fee-for-service (FFS) and Medicare Advantage (MA) patients, we estimate that up to $265 billion worth of care services (representing up to 25 percent of the total cost of care) for Medicare FFS and MA beneficiaries could shift from traditional facilities to the home by 2025 without a reduction in quality or access.

(Bestsennyy, Chmielewski, Koffel, and Shah, 2022)

The Security Is the Tricky Part

Most security frameworks, whether the HIPAA Security Rule or any number of publications from the Commerce Department's National Institute of Standards and Technology (NIST), incorporate concepts of layered defenses and compensating controls. When understanding a layered defense, often referenced as defense in depth, it's easy to imagine a castle on a hill. Being elevated on the hill means that it is more difficult for adversaries to launch projectiles from below. In so many fables the castle has a moat, which presents another obstacle after the attackers have made it up the hill. A deep moat housing any number of carnivorous species increases the difficulty of crossing. The castle has high walls, of course, and the defenders of yore might have had several weapons to throw, drop, or pour upon the attackers. The relatively simple idea is that multiple layers of defense make whatever assets the castle contains less vulnerable.

The security of a healthcare facility is not that different in concept. While welcoming patients, staff, and visitors physically into the facility, leadership has applied defense in depth to protect patient data and information assets in accordance with HIPAA's mandates. In this way, in an acute care context, any patient data and a medical device within a clinical environment benefits from several layers of security. Physical security protects against the risk that the device is simply removed from the premises improperly. Administrative safeguards require staff-handling devices to wear proper identification and to be alert for misuses of data. And technical security measures typically involve physically present security measures such as network firewalls and antivirus protection, secure Wi-Fi protocols, and network segmentation so that improper access to one part does not necessarily result in access laterally across the entire network.

Just one reason why layered defense is so important within the healthcare environment is that medical devices are rarely designed with security as a core functional element. Medical devices are designed to perform their primary diagnostic or whatever diagnostic or therapeutic function is targeted and then to transmit the associated data through the network to the designated recipient. Over the years, this network transmission has increasingly been wireless, rather than wired. This is not to say that any individual or category of medical devices does not contain security measures to protect the confidentiality of whatever patient data is within. Though rarely are these devices designed to be used outside these castle walls. In other words, a networked medical device – even one with relatively good security characteristics – will likely not be as protective of the information it contains when used outside the traditional hospital setting and without the benefit of that associated protection from the application of related defense in depth tactics.

So what is a health system or hospital to do when dispatching clinicians and other providers to patient homes? How does one anticipate and manage the different vulnerabilities and risks when deploying mobile medical devices that will typically collect individually though protected identifiable health information and will often transmit that information wirelessly to the main facility? At the risk of stating the obvious, HIPAA and the HIPAA Security Rule direct covered entities and business associates to protect patient information regardless of where the care is provided, be it in the Emergency Department with its layered defenses or a patient's home with little, if any of that.

The Do's and Don'ts

A fundamental prerequisite to a Hospital at Home program is a new risk analysis, as required by 45 CFR 164.308(a)(1) of the Security Rule. It is impossible to protect information assets effectively if one does not understand what data will be collected in the remote acute care setting and how it differs from that captured in the traditional setting. Similarly, an assessment of the vulnerabilities to each device holding and transmitting patient data will often produce more red flags, given the absence of robust safeguards within the device itself. While visiting clinicians and relevant medical devices in a patient's home will confront myriad scenarios that are starkly different from the consistency of the traditional setting, it is important to complete as thorough a risk analysis as possible. The results of the risk analysis will then help inform leadership on how to tackle some of the related issues we discuss in the following section.

Hospital Grade versus Consumer Grade

There are major differences between devices meant to be used in a hospital and consumer-level devices. Consumer-level devices don't have to follow the same guidelines for protecting data as Hospital Medical Grade, mainly because their intended users are not Covered Entities which are regulated or Business Associates using them to store, process, or interchange patient data. They are mainly used to transfer personal data for the purpose of self-analysis. Therefore, the Confidentiality, Integrity, and Availability spoken of in the HIPAA Security Rule, and the logging/auditing required as part of the Security and Privacy Rules do not apply to these devices.

This is not just for the devices that directly interface with patients. Any devices that support the connectivity for purposes of storing or processing Protected Health Information need to meet these requirements. This drives the development of medical-grade networks to be able to effectively use these devices reliably. These networks are normally centrally managed. Hospital Medical Grade infers several characteristics:

- Centrally Managed Networks, including use of enterprise authentication techniques using identities as opposed to shared passwords like home networks
- Centrally Managed Devices leveraging constant connectivity back to central management servers
- Compliance and security often enforced end-to-end with technology frameworks like Zero Trust
- Centrally Managed Identity using Active Directory, or similar directory services
- Centrally Managed Authentication using RADIUS, LDAP, Oauth2, or SAML authentication
- Some degree of redundancy for power and critical services due to Joint Commission, the hospital accreditation body's requirements

Consumer grade does not have any of these characteristics. It is an unreliable island in a sea of lack of management. Putting critical devices to monitor patients elsewhere without the supporting infrastructure needed to support them is a challenge that needs to be overcome to address these risks.

The first item to think of is device connectivity. 5G network slicing, while spoken about glowingly by pundits, is still a nascent technology that is being deployed in limited proof-of-concept campus deployments (Weissberger, 2021; NTT, 2022). In addition, the nature of the kind of multiple input–output, MIMO systems is that they require several antennas to handle cellular traffic, thus contributing to interference. This negatively impacts both device and application reliability. Further, these multiple antenna systems require the implementation of several not yet mature though necessary enabling technologies such as base stations to accommodate these systems.

Thus, deployment of these networks over multiple carriers and areas has not been finalized yet, including because of the inconsistent security implementations. The best example has been that it has been successfully deployed at AT&T Stadium for Dallas Cowboys; however, it's not fully compatible with another telecom company's Wireless which is deployed at Lincoln Financial Field for Philadelphia Eagles games and has slight implementation differences that can make an impact (AT&T, 2020) (Verizon Wireless, 2019) (Pegoraro, 2021). In other words, if you want to take the game on the road, it's going to be a more hostile environment than originally planned for, or that the visionary pundits have imagined. Beyond security, there is also widespread ignorance to the actual status and progress for enabling and required technologies for 5G.

Devices have a maintenance schedule and usually a Clinical Engineering department or tech department to ensure acceptable tolerances and tests. Consumer devices and home environments don't have this. You're going to have to plan for a hostile world, much like when the Eagles play at the Cowboys' home field. What are those steps that organizations need to take?

- Define how these devices will be monitored and maintained over a less secure environment. The managed networks expected in hospitals will not exist. 5G network slicing, at the time of this writing, is restricted to campus environments.
- Undertake Failure Mode Effects Analysis (FMEA) processes to determine workflows per case type and what to do when they fail. While some conditions can be monitored remotely without network or device connectivity, others will require alternatives to protect patient

safety. You cannot rely upon the compensating controls of a healthcare facility to protect patients when they are at home. You need to develop them per use case and environment. Who will be alerted? What will they do?

■ What is your organization's Command Center strategy? While Command Centers are making inroads because of how they can manage workflows and intake within hospitals, they also play an important role in managing Hospital at Home patients, especially when there are issues. Hospital at Home has to be part of any successful Command Center strategy.

■ Define how devices will be virtually managed. Mobile Device and Application Management is paramount to ensuring prevention of configuration changes that can have adverse effects.

■ Define physical management processes. Medical devices in acute and outpatient settings need periodic maintenance and monitoring to be kept in optimal operating condition. Processes need to be defined to ensure these devices are physically reviewed and maintained by qualified personnel, and repaired if need be.

■ Understand network connectivity. Don't rely upon home routers, which have numerous security flaws. Michael Horowitz, on his website, routersecurity.org, catalogs the numerous vulnerabilities in these devices (Horowitz, 2022). Leverage managed cell phones or LTE/5G hotspots configured by your organization to provide your own connectivity, and bypass the network connectivity in the location you're placing the devices in. With the numerous security holes in home routers, plus the sharing of bandwidth with YouTube, gaming, and numerous other uses, critical monitoring data can be crowded out. Cell networks, especially with 4G LTE and 5G, are reliable enough that T-Mobile now offers 5G as a replacement for wired connectivity (T-Mobile, 2022). While network slicing is not available, it's possible to use 4G or 5G networking with Virtual Private Networking (VPN) to provide appropriate security back to the hospital.

■ Understand the target area. If it's a truly rural area, check to see if high-speed service is even possible. Starlink offers 43 ms latency, as compared to 4G LTE offering 47.2–54 ms (Vaughan-Nichols, 2021; OpenSignal, 2021). If it's not possible to offer Starlink, 4G, or 5G with a good backup plan if they fail, then offering safe services is not possible. Often wireless coverage maps are inaccurate so don't rely on them. Have a plan to test coverage just like the telecom analyst at one large health system does for every inch and corner of every campus building.

■ Use devices with excellent security that meet standards such as UL 2900-1, ISO 11073, the MDS² 2019 standard, or the upcoming IEEE/UL P2933 TIPPSS standards. Devices used in a hostile environment need to be even more resilient than those used within hospitals because they do not have the protections of more secure hospital networks.

While Hospital at Home is an excellent way to conceptually aspire to reduce bed count in hospitals, shift resources to the most critical patients, and help patients heal in a familiar environment, there are still multiple risks that need to be addressed. Understanding these risks and crafting a strategy based upon providing your own connectivity over cell networks to hardened devices is a good place to start. Understanding use cases and workflows and planning for failure to know who to call in case of an adverse event is also critical. Integrating it into a Hospital Command Center strategy is also required to efficiently manage Hospital at Home alongside the inpatient cases and through a hostile environment. Like the Philadelphia Eagles playing at AT&T Stadium, it takes a lot more effort to pull out a win outside of the supportive "home" or hospital environment and fans.

Bibliography

"5G Built Right: Verizon 5G Ultra Wideband Service Live in 13 NFL Stadiums." About Verizon. Verizon Wireless, December 12, 2019. https://www.verizon.com/about/news/verizon-5g-ultra-wideband -service-live-13-nfl-stadiums.

"AT&T and the Dallas Cowboys Upgrade Fan Experiences." AT&T News, Wireless and Network Information. AT&T, November 18, 2020. https://about.att.com/story/2020/att_dallas_cowboys_att _stadium.html.

"High-Speed 5G Home Internet Service Plans: T-Mobile 5G Home Internet." T-Mobile, Inc. Accessed June 21, 2022. https://www.t-mobile.com/home-internet.

"Hospital at Home: A Shift in Thinking for Acute Care." Modern Healthcare, April 1, 2022. https://www .modernhealthcare.com/technology/hospital-home-shift-thinking-acute-care

"Private 5G." NTT. Accessed June 21, 2022. https://services.global.ntt/en-us/services-and-products/ networks/mobile-and-wireless-networks/private-5g.

Health Insurance Reform; Security Standards; Final Rule, 45 CFR Parts 160, 162, and 164, (2003) as amended by Modifications to the HIPAA Privacy, Security, Enforcement, and Breach Notification Rules Under the Health Information Technology for Economic and Clinical Health Act and the Genetic Information Nondiscrimination Act; Other Modifications to the HIPAA Rules; Final Rule, 45 CFR Parts 160 and 164, (2013) ("Security Rule").

Horowitz, Michael. "Router Bugs Flaws Hacks and Vulnerabilities." Router Security. Accessed June 21, 2022. https://routersecurity.org/bugs.php.

Oleg Bestsennyy, Michelle Chmielewski, Anne Koffel, and Amit Shah, "From Facility to Home: How Healthcare Could Shift by 2025." McKinsey & Company, February 1, 2022. https://www.mckinsey .com/industries/healthcare-systems-and-services/our-insights/from-facility-to-home-how-healthcare -could-shift-by-2025.

OpenSignal. "Mobile Provider Latency in the US 2019." Statista, December 6, 2021. https://www.statista .com/statistics/818205/4g-and-3g-network-latency-in-the-united-states-2017-by-provider/.

Pegoraro, Rob. "Here's How AT&T, Verizon and T-Mobile Slice and Dice 5G Plans and Pricing." Light Reading. Informa Tech, April 14, 2021. https://www.lightreading.com/ossbsscx/heres-how-atandt -verizon-and-t-mobile-slice-and-dice-5g-plans-and-pricing/d/d-id/768746.

Pooja Chandrashekar, Sashi Moodley, and Sachin H. Jain, "5 Obstacles to Home-Based Health Care, and How to Overcome Them." Harvard Business Review, October 17, 2019. https://hbr.org/2019/10/5 -obstacles-to-home-based-health-care-and-how-to-overcome-them.

Vaughan-Nichols, Steven. "Starlink Is Better than Its Satellite Competition but Not as Fast as Landline Internet." ZDNet. Ziff Davis, August 5, 2021. https://www.zdnet.com/article/starlink-is-better-than -its-satellite-competition-but-not-as-fast-as-landline-internet/.

Weissberger, Alan. "AT&T EXEC: 5G Private Networks Are Coming Soon + 5G Security Conundrum?" Technology Blog. Institute for Electrical and Electronics Engineers (IEEE), April 15, 2021. https:// techblog.comsoc.org/2021/04/15/att-exec-5g-private-networks-are-coming-soon-5g-security/.

THE PRACTICAL TECHNICAL, LEGAL, MANAGEMENT, AND LEADERSHIP STEPS TO A MORE INTERACTIVE HEALTH SYSTEM

IV

Chapter 12

Mitigating Value Risk in Innovation: A Physician Product Design Expert's Lessons Learned Driving Clinical Product Development

Joshua Tamayo-Sarver

Contents

DOI: 10.4324/9781003348603-16

Introduction

As an emergency physician, I have spent my clinical career managing high-risk uncertainty across multiple patients simultaneously. Like most emergency physicians who have been practicing for more than ten years, I developed a process to handle my anxiety and ensure that I delivered good patient care (Carrière et al., 2009). I would walk into an exam room where the patient was not breathing well and formulate a quick working hypothesis that the patient had congestive heart failure with high blood pressure. I would ask the nurse to start nitroglycerin and BiPAP for the breathing and order diagnostic testing to ensure I had the correct diagnosis. I would then ask the nurse to inform me if the patient's breathing did not improve in 15 minutes or if the blood pressure dropped below 110 systolic (Allen and O'Connor, 2007). I would then walk into the next room, where a different patient appeared to be septic with altered mental status and low blood pressure. I would order a sepsis bundle and ask the nurse to inform me if the blood pressure was not responding within 20 minutes so I could place a central line and start pressers after the appropriate fluid bolus (Nguyen et al., 2011). I could manage many critical patients simultaneously and have a clear plan to address the uncertainty for each. Although each nurse's skills, attitude, and approach were highly variable, I trusted that the bedside nurse and I were aligned and that the patient would receive the proper care.

When I began managing multiple innovation projects, I was an anxious mess, even though no lives were on the line. I discovered that while innovation was exciting, it also had a lot of uncertainty as innovation requires flexibility and constant changing movements. We were trying to create numerous new products in various domains, and each initiative had its own leadership, vision, and way of doing things. Nonetheless, I needed to balance resources and investment across this portfolio riddled with Gestalt, worry, endless counsel from many others, and coffee to guide resource allocation and anticipate potential returns on investment and unseen pitfalls. Projects that seemed promising failed to take off because there had been a fundamental flaw in the concept after significant investment and development. Other projects that appeared to be small and almost trivial turned out to be big winners with an enormous impact. To make matters even worse, I often carried a backlog of 150–200 proposed projects where at least one person felt they were addressing the most critical thing in the universe. Every time one of the active projects started to hit a wall, someone in leadership would inevitably point to something in that backlog as the thing we should have been working on. Those were challenging times.

While innovating is imagining a future that does not currently exist, managing an innovation portfolio with multiple individuals attempting innovations is essentially managing a portfolio of uncertainty to ensure a positive return. It seemed that if I could navigate multiple high-stakes uncertainties clinically, I should be able to figure out how to navigate an innovation portfolio full of uncertainties. When I reflected on what I was doing in the emergency department, I concluded that I had a framework of the hierarchy of the most important things I needed to ensure were working: airway, breathing, circulation, etc. At each stage of the hierarchy, I also had a consistent context of threats: infectious, vascular, cardiac, pulmonary, etc. I realized that I was mitigating the risk by first identifying my most significant assumption about the patient and testing it until it was satisfied. I could continue to move through the framework with the next assumption down the line. Over time, my colleagues and I have developed a method for building and overseeing a complex and diverse innovation portfolio to mitigate risk and maximize value return. We have taken the clinical context of pathophysiological processes and put that into the innovation context of stakeholders and shareholders. Similarly, we took the hierarchy of physiological assumptions with the most critical assumption first and put that into the innovation framework with the most critical innovation assumption first. In this chapter, I share our experience and the structure that allows less coffee consumption.

Current State of Innovation in Many Organizations

Doing innovation work in healthcare is hard. Having been involved in creating and implementing health technology innovation solutions since 1991, I can attest that every day is a new opportunity for humility. At its core, innovation is nonstandard work where creativity and unbridled passions can create a future only imagined by a very small few. Although leading an innovation project presents significant challenges and a high failure rate, being the executive overseeing an innovation portfolio introduces a new set of difficulties and complications.

You Can Only Rely on Uncertainty

When managing a portfolio of innovations, the uncertainties are multiplicative rather than additive. The primary uncertainty is whether or not the innovation will succeed and ultimately return value on the investment. Especially in the healthcare space, the long time horizon between starting the project and learning that it will not work means that uncertainty is a prominent part of the planning process and the day-to-day operations. This can make innovation team members hesitant, question themselves within the process, and take on different directions, which may be more detrimental to innovating. Thus, the goal for managing uncertainty is to develop a method to identify if a project is likely to succeed or fail as quickly and inexpensively as possible. No one wants to abandon a project that would have been successful with more remarkable persistence or continue to invest in a project that will not succeed. The great challenge in innovation is simply knowing when to stop and when to keep going, which is also known as strategic quitting (https://www.amazon.com/Dip-Little-Book-Teaches-Stick/dp/1591841666/ref=tmm_hrd_swatch_0?_encoding=UTF8&qid=1651023286&sr=1-1).

There Are as Many Approaches to Innovation as There Are Innovations

Managing a portfolio of innovation projects with constrained resources and competing demands is even more challenging because most innovators have their way of doing things and are as much artists as they are scientists (The Discipline of Innovation n.d.). As a leader, we do not want to stifle the artistic creativity of innovators, given its strong association with business performance (Briganti and Samson, 2019). Still, we need to be able to quickly know whether the project is progressing, succeeding, or failing in a way that allows us to compare it to other projects and prioritize resources appropriately (Arrow, 1962). While there is evidence that higher risk tolerance is associated with greater innovation (Tian and Wang, 2014), we assert that a structured framework allows you to embrace high-risk ideas and mitigate investment risk (Kim and Mauborgne, 2005).

Our Solution

We propose a method to ensure the potential to create and realize real value, coupled with a framework we have developed over many years and countless projects in multiple contexts. The framework is a stepwise approach to evaluating and nurturing innovation that builds on the previous step. The framework's purpose is to keep each stage focused on answering a specific question. Then resources and teams can be aligned with the minimum necessary to answer that question as quickly as possible. Only once that step is a success can the project be moved to the next step with additional resource allocation.

At each step of the framework, a specific hypothesis is being tested. Still, the particular hypothesis will need to be looked at from multiple stakeholder perspectives with two main questions for each stakeholder: what value does this create for me, and how do I capture that value?

The framework is broadly applicable, and most innovations do not originate within your organization. However, before implementing someone else's solution in your organization, it is worth ensuring that the solution has externally demonstrated that it will internally answer each of the hypotheses at all the steps preceding where it is being integrated into your organization.

To make the framework more tangible, I will use a case study with a palpation device we are creating. In addition to the case study, throughout the process, I will draw on firsthand experience where I have gotten the element right or wrong and describe how the element was applied to provide a better illustration. For context, we saw well over 1 million patients virtually every year. Many patients complained of abdominal pain, but it is difficult to evaluate abdominal pain without palpating the abdomen. Ultimately, as a physician evaluating a patient with abdominal pain, I palpate the abdomen to determine whether or not they need advanced imaging (CT scan or ultrasound). The goal was to create an inexpensive device that patients could use on themselves or on their children that would reliably determine if the patient would get advanced imaging if seen in person.

Before diving into the framework, let's focus on the method to ensure that there is a potential to create and realize value. This method involves understanding (1) which entities will be involved or affected by the innovation, (2) what value is created by the innovation and how it does so, and (3) how that created value is distributed to the different entities.

Methodology to Ensure Innovation Creates Value

Finding Landmines with Your Eyes Instead of Your Feet: Navigating the Landscape

As a physician, I believe myself to be a very rational thinker. What's worse is that I expect everyone else to be a rational thinker, but I am often surprised when arguments that I think are well reasoned and persuasive seem to fall on deaf ears. The framework we lay out is rational – it starts with an understanding of the entities, what value is created, and how it is distributed. I have had countless projects fail that made things better for all parties. Frustrated that I could not get buy-in for the idea that one plus one equals two, I searched for answers. I have discovered a dramatic increase in success from adopting the perspective that Daniel Shapiro discusses in *Negotiating the Nonnegotiable*. Each person or group must have both their core and relational identities addressed before they can engage the brain's rational part that allows them to change in a way where they can receive value (Shapiro, 2017). A quick summary of Shapiro's conceptualization of core and relational identity can be invaluable in mitigating value risk in innovation.

Core Identity

The core identity is a spectrum of characteristics that define an individual or group. It is the structure that allows us to synthesize experiences into a coherent narrative. When a new technology or innovation reinforces one of these elements, it will likely be embraced and adopted. On the other hand, if new technology or innovation threatens or weakens any of these elements, it will face stiff opposition and likely failure. When contemplating an innovation or technology, ask if, for each entity, the new technology or innovation reinforces, creates, or threatens each of the following core identities for the entity:

Beliefs

Beliefs that the entity holds as truth. In the case of the palpation device, we can anticipate that physicians believe that their hands and physical exam are reliable and are unlikely to be willing to rely on a device. We will need to figure out how to change that belief in a way that resonates with the physician. As another example, if someone believes that psychotherapy requires a sacred space, then a telehealth solution for therapy will not be successful for that person. By contrast, an innovation that resonates with core values, like helping vulnerable populations, will be more readily accepted. While the alignment of a given innovation may be intuitively clear to some, one may take a structured approach to manage the values in a diverse portfolio of innovations (Stilgoe et al., 2013; Hansen, Grosse-Dunker, and Reichwald, 2009).

Rituals

Rituals are activities that are personally meaningful for the entity. Part of the ritual of a patient encounter is the relationship-building of the physical exam. For the palpation device, we will need to ensure that the palpation device is used in a way that respects the physical exam ritual. As another example, if caring for their elderly parent's wounds every morning is part of a ritual connection between the parent and the adult child, then providing a home nursing service to improve the quality of the wound care will threaten the core identity of the adult child, the parent, or both.

Allegiances

Allegiances are felt loyalties to an individual or a group, such as family members, friends, authority figures, nation, or ancestors. In the case of the palpation device, a patient may have loyalty to their primary care physician. If the primary care physician provides them with the device, their allegiance may help motivate their utilization. In contrast, if the patient does not feel loyalty to their health insurer, then they may not feel motivated to use the device if it is provided to them by the insurer. As another example, a self-registration and check-in system may provide a better patient experience. Still, if that means firing the registration clerk who has worked at a facility for several years, there may be significant resistance to its implementation and adoption.

Values

Values are guiding principles and overarching ideals. For the palpation device, providing quality care is a value that is closely held by the overwhelming majority of physicians. The palpation device must align with this value and ensure the physician feels they are providing quality care by using the device for palpation of the abdomen. As another example, if patients with a particular type of insurance are eligible for an attractive service but others are not, this may threaten the value of equity, which those responsible for enrollment may feel.

Emotionally Meaningful Experiences or Memories

Emotionally meaningful experiences or memories are intense events, positive or negative, that define a part of you. For the palpation device, a patient may feel emotionally cared for and understood when they have had abdominal pain in the past and had their physician lay hands on their abdomen. For this patient to adopt and use the palpation device, we will need to make sure that we acknowledge and address it so that the encounter is still emotionally meaningful. As another example, if an entity comes out strongly and publicly on a particular side of a debate, but new information surfaces, which makes the other side clearly correct, then changing the position may recast their public stand to undermine their core identity.

Relational Identity

There are five basic emotional needs, which Shapiro labels "core concerns," that must be met in a relationship before someone can engage the rational part of the brain. I have found it invaluable to ensure that each entity involved in a new health technology or innovation meets these five emotional needs (Shapiro, 2017). When contemplating an innovation or a technology, ask if, for each entity, the new technology or innovation reinforces, creates, or threatens each of these five things:

Appreciation

Appreciation is feeling recognized for the positive aspects of perspective or effort. For the palpation device, there is a clear threat to the physician feeling appreciated for doing a physical exam. Therefore, we will need to make sure that the physician feels appreciated for enabling more convenient care for the patient with a virtual abdominal exam. As another example, a clerk who previously called patients to remind them of their appointment now facilitates that appointment reminder with the digital system. Previously, the clerk could speak with patients, establish a human connection, and was thanked for the effort. In the new workflow, how does one ensure that the clerk feels appreciated so that they will adopt and embrace the new system?

Affiliation

Affiliation is feeling aligned and part of the in-group, with a sense of belonging. For the palpation device, utilization will depend on both patients and physicians feeling that their peers are using the device. For example, a new telehealth service enables patients to treat minor ailments asynchronously and essentially anonymously. This creates significant convenience for treating urinary tract infections or sexual dysfunction. However, it erodes the sense of affiliation and belonging that the patient has with his or her primary care physician, which will be critical for the far more significant and expensive when the physician recommends a specific course of treatment, or there are discussions around end-of-life care.

Autonomy

Autonomy is feeling one has the freedom to choose and decide to act as one believes is best. For the palpation device, neither the physician nor the patient will say "yes" to use the device unless they are comfortable that they can say "no." The implementation needs to empower patients and physicians not to use the device so that, paradoxically, there will be good utilization. As another example, a clinical decision support window forces a clinician to justify a particular action that may be poorly supported in the literature. Feeling that their autonomy is being threatened, the clinician may assume a passive-aggressive posture and always select a nonsensical response to get past the stop. Thus, by ignoring this core concern and implementation, the decision support does not improve decisions but increases provider burnout.

Status

Status is feeling that one's knowledge, expertise, value, and ability to contribute are recognized. While appreciation is recognition for what one contributes, status is deference for what one can do. For the palpation device, a successful implementation will need to acknowledge and reinforce the status of the physician as the medical expert deciding medical care. For example, it may seem reasonable to implement a computerized billing and coding solution based on a deep analysis of millions of medical claims. Suppose the implementation does not recognize the billing and coding teams' expertise status by incorporating it into the solution. In that case, the likelihood of a successful deployment is remarkably close to zero.

Role

Ambiguity around one's role is exceptionally frustrating, so it is essential that one has the feeling that one's role is clearly defined and desirable. For the palpation device, the patient may not want the role of examining themselves or their child. A successful implementation must ensure that the patient is comfortable with their role in self-examination. For example, a paging tool in the emergency department allows the physician to directly page consultants without needing the emergency department clerk to place the page for the emergency physician. This seemed like a good idea to reduce the burden on the emergency department clerk, but its failure was actually because the clerk's role was no longer well defined. Is the clerk responsible for keeping track of pages they do not know about? Are they still responsible for ensuring the consultant returns the page? How do they connect the return call with the requesting physician? While any of these things could have been addressed during product development, they were not, and the product failed from lack of utilization.

Consider All the Possible Users, Stakeholders, and Shareholders

Healthcare can be a particularly complicated ecosystem because so many different individuals, ideas, and institutions are involved with complex and changing interdependencies (Omachonu, 2010). We often think that the entities involved are users, stakeholders, and shareholders of any specific innovation. We conceptualize the user as the individuals who interact with the innovation in one form or another. Stakeholders are once removed from this and may be affected by the innovation but do not directly interact with it. Shareholders are those whose financial interest is affected by the innovation but do not directly or indirectly interact with it. For example, a clinician in a hospital would be a user of an electronic health record, corporate counsel would be a stakeholder as it affects the medical-legal risk profile, and the hospital CEO would be a shareholder. These different roles change depending on the innovation in question and the context. A quick review of the various entities may help provide a framework for evaluating an innovation to ensure a holistic view of all involved (Alexander, Miesing and Parsons, 2005).

When we look at the different entities involved in innovation and the healthcare ecosystem, it becomes apparent that complexity is part of why it is such a complicated ecosystem to innovate (Conklin, 2002). Furthermore, it becomes mind-boggling to realize that with this number of potential users, stakeholders, and shareholders, the potential for different incentive systems, interests, and desires changes from one local setting to another local setting (Chaudoir et al., 2013). This is one of the main reasons successful technologies often fail to translate to another healthcare environment and should offer a cautionary tale for the importance of evaluating any potential new technology in each local environment to which it will be deployed. One size doesn't fit all.

Beneficiaries of Service

First and foremost, the beneficiaries of healthcare are an essential entity. Generally speaking, these entities directly benefit from having a health concern effectively addressed.

Patient

The patient is the individual receiving healthcare interventions, such as medications, education, medical devices, etc. If we think that the healthcare service, broadly speaking, is one of diagnosis and intervention, the patient is the one who is being diagnosed and then intervened upon (Epstein and Street, 2011). Typically, the patient is trying to address a health concern. This could range from concerns about knee pain to end-of-life care, wanting to lose weight, and chronic disease management. While this would seem a reasonably straightforward persona to understand, wanting effective, high-quality, convenient, low-cost, satisfying service, the reality can sometimes be different. Health concerns can offer other psychosocial benefits. Most practicing providers have seen numerous situations where the patient's ailment is the focal point around which the family's psychosocial structure is built. Sensitivity to what it would mean for the patient to address their health concerns ensures successful technology implementation (Smith, 2002).

Patient Family

The patient's family benefits indirectly from the patient receiving healthcare interventions. As those healthcare interventions enable the patient to resume their social function, it may relieve the burden on other family members, whether financial, logistical, or emotional. More directly, the patient's health concern may concern the patient's family. One of the things that we have observed through our customer discovery efforts is that aging parents' health is often more distressing to their adult children than it is to the patients themselves. While this would seem to be another straightforward desire for health concerns to be addressed, we have all encountered situations where a patient's health concern creates a need for a family member to help the patient. A meaningful relationship is created and maintained because of that need. Any health technology or innovation that would resolve that particular health concern and threaten the bond maintained by that need is likely to be quickly sabotaged by the patient family. In such a case, a successful implementation must acknowledge a method for establishing a new relationship dynamic that is also rewarding.

Patient Caregivers

Those who provide care for the patient, whether family or unrelated, paid or unpaid, professional or unprofessional, have a vested interest in the patient's health and functioning. Depending on the nature of the caregiver and the caregiver arrangement, that vested interest may or may not align with the patient's desires. However, given the patient caregiver's proximity to the patient, they are critical in ensuring either utilization or failure of a particular intervention.

Employer

When a patient is quickly diagnosed, treated, and returned to normal function in zero time and with zero cost, the employer benefits. Although most entities can have complex relationships that make what they're trying to accomplish from benefiting from healthcare services challenging, the employer's interest tends to be straightforward and consistent: happy and healthy employees without cost in terms of time or dollars.

Providers of Service

The healthcare system exists to address a patient's health concerns through diagnosis and intervention. Therefore, most health technologies involve the providers of that service in one way or another. Most service providers want to accomplish their care tasks as quickly and efficiently as possible. Most have pride in their job, so high quality tends to resonate. Conversely, any deviation from the current state must address any anxieties about potential failures, misses, or mistakes. While a 95% accurate model is statistically impressive and works excellent in a consumer goods or retail setting, a 5% malpractice rate would quickly lead to death and loss of licensure in a medical environment. This zero-error expectation for healthcare providers of service creates a change (or risk)-averse culture.

Clinician

The clinician is the licensed medical professional who is ultimately responsible for the patient's care. While there are undoubtedly bad clinicians, just as there are certainly thieves and murderers, the overwhelming majority of clinicians take the responsibility for another human life very seriously. To address this momentous responsibility, medical training teaches clinicians to develop standardized processes and scripts (Schaik et al., 2005). Clinicians can be notoriously challenging to manage. Much of that challenge can often be traced back to each clinician's rigid process developed to address the weight of the patient responsibility and the Hippocratic oath. Innovations need to be developed and deployed to address clinicians' concerns sincerely (Scott et al., 2008). At the same time, the financial structure of the service has a profound influence on the clinician. If the clinician is fee for service, then an innovation that decreases "unnecessary" utilization is a way to reduce income.

On the other hand, if the clinician is in an employee or capitated arrangement, then an innovation that decreases demand for their services is typically welcomed. In many practices, there is a heterogeneous population with various financial structures that differ from one patient to the next. A clear understanding of the clinician's interests is needed to understand the potential value of an innovation for which the clinician is a user, stakeholder, or shareholder.

Ancillary Service

The clinician may be responsible for the ultimate legal medical order (such as a prescription) that directs the patient's care journey. Still, numerous ancillary service providers do most of the care work: nurses, medical technicians, therapists, etc. (Needleman and Hassmiller, 2009). Like clinicians, ancillary service providers want to provide quality and rewarding care with minimal effort. Much of the ancillary services are the direct delivery of care, but a substantial amount of bureaucratic work is also involved. Given that this work tends to be more task-oriented and less symbolic knowledge management, many of these tasks are frequent targets of technology innovation. An acute sensitivity to whose job or status may be threatened is often required to navigate an implementation successfully.

Facility

The facility plays a vital role in delivering care. Although there is a strong push to increase the capacity to provide care virtually or in a patient's home, most clinical encounters and care are still delivered in outpatient and inpatient facilities. The number of diverse incentives, interests, and arrangements for the different facilities is staggering, often with a perplexing diversity of interests and incentives within the same facility and even for the same patient with different conditions. We have made the mistake before and noted that others have made the same mistake. We assumed that one facility's web of incentives and interests are generalizable broadly without closely examining how those things play out in each location we intended to target.

Laboratory

From a general perspective, the laboratory's role is to analyze some patient-derived specimens to provide information that aids in diagnosing or monitoring disease or treatment. Similar to facilities, the incentives and interests for each laboratory and particular use cases related to innovation need to be carefully explored as we are always surprised at the heterogeneity in this landscape.

Imaging

From a general perspective, the role of imaging is to provide an enhanced visualization of some aspects derived from the patient to provide information that aids in diagnosing or monitoring disease or treatment. Similar to facilities, the incentives and interests for each imaging entity in particular use cases related to innovation need to be carefully explored. We are always surprised at the heterogeneity in this landscape.

Medications/Pharmacy

While much of the medical apparatus revolves around diagnosing and monitoring a disease state, the most crucial question for the patient is often "so what now?" Frequently, healthcare's answer to this critical question of what we will do about symptoms or disease is medication. As a provider who wants to do something to help the patient feel better, it is difficult not to give the patient a prescription for something tangible to feel like they walk away with more than just empathy and moral support. Indeed, even medications that cannot withstand a randomized trial frequently have significant placebo benefits to that individual patient. Innovations that address and improve prescribing practices from a scientific perspective need to be sensitive to clinician-patient interaction and account for positive placebo effects that the patient receives from not scientifically indicated medications. Additionally, the entity providing such medications has its interests and incentives that need to be addressed when disrupting this aspect of medical care.

Logistics

There is a lot to healthcare logistics, and it is easy to despair the number of resources spent on healthcare that do not directly benefit a patient's health. Given the amount of non-value-added resources spent in this arena, innovations from technology to structure to contractual arrangements have significant potential (Cutler, Wikler and Basch, 2012). However, one person's wasted resources are another person's steak dinner. Therefore, it is hard to have innovation in a space that is not displacing an entrenched interest.

Coding

Coding is the transformation of medical documentation into a structured bill for reimbursement or payment of those services (Beck and Margolin, 2007). It has been a focus for computer-assisted technology and is ripe for disruption. The process and goal of coding are pretty straightforward. The industry currently employs many people to do the manual coding, and disrupting that industry is a threat to those jobs. Part of the coding and claims process is the continual arms race between providers of care who get paid for providing services to patients and the payers of care who often can make money by not paying for those services. This results in an endless back-and-forth of mechanisms to either increase reimbursement by gaming the coding system, such as is prominent in Medicare advantage providers, or finding ways to deny payment, such as is prominent in payment denial systems masquerading as quality metrics (Zarabozo and Harrison, 2008; Claffey et al., 2012).

Billing

Collecting the dollars associated with the service from those responsible for paying for that service to those who provided the service is remarkably complicated. It has taken the form of an arduous process that looks like a structure set up between two Cold War adversaries, much more than two complementary components in a single system with a shared goal of patient care. This led to multiple intermediary innovations to take advantage of the substantial friction between the payers and the providers. While this is a space that is ripe for disruption, understanding what creates the structure that leads to that value proposition is essential to understanding the long-term value of that innovation.

Transportation

Traditionally, resources needed to provide medical care were fixed in one location, such as a hospital, clinic, or outpatient laboratory. The patient needed to be physically moved to where those locations could provide that resource. Currently, an exciting area of innovation is reimagining how to get a patient and a given resource physically together, whether by delivering the diagnostic tool as an application on the patient's smartphone, home phlebotomy, or rideshare to transport the patient.

Legal, Compliance, and Security

Healthcare is a very highly regulated environment that includes many high-stakes interactions (Field, 2008). There is little standardization in much of the healthcare system to make matters worse without clear evidence to guide process standardization and optimization (Altman and

Morgan, 1983). In such a hostile environment, a healthcare entity's legal, compliance, and security apparatus has a naturally conservative and risk-averse stance.

Legal

It is hard to find many examples where a new innovative approach to healthcare has resulted in positive publicity for the legal department that figured out how to make it possible. However, the lead story of many news outlets has frequently been about one healthcare entity or another doing something innovative, creative, novel, and illegal. While this is sometimes a clear breach of ethical and moral principles, more often than not, it is simply a violation of the Kafkaesque legal and regulatory environment in healthcare. Therefore, the legal perspective on new healthcare innovations tends to be skeptical, and more often than not, a boxer waiting for the other person to throw the first punch. Medical malpractice and medical-legal liability are high costs in our healthcare system, and technologies and innovations are, by definition, doing something nonstandard. An appreciation for the increased potential legal liability exposure associated with something nonstandard is important in figuring out how to collaboratively address and mitigate that risk. Given the critical function of the legal department to keep the healthcare entity afloat, finding ways to add value from a legal perspective is paramount.

Compliance

Healthcare is dangerous, with many high-stakes decisions complicated by a confusing web of interests and incentives. With the best intentions to protect all parties, volumes of regulations have been instituted at local, county, state, and federal levels. Navigating these regulations to ensure compliance with new health technology innovations makes me feel like the Sean Connery character in "Hunt for Red October." Yes, we have experts and some maps, but the maps weren't drawn for what we were trying to do and are not entirely applicable, so we were steering blind, and if we get it wrong, we sink everything. From this perspective, it is easy to have compassion for those in compliance trying to enable innovative and improved care in outdated but commanding guidelines and regulations. Having participated in government advisory groups, I can say that this is one area that I find most frustrating as an operator and disappointing because I don't think there is anyone to blame. It is hard to write regulatory and compliance laws to protect people at their most vulnerable when you are trying to anticipate a future that does not yet exist and is changing much more rapidly than you can legislate.

Patient Safety

As noted above, healthcare is making many high-stakes decisions with lifesaving but potentially fatal interventions. To help mitigate the risk of undesired outcomes, a patient safety apparatus attempts to take a systems approach to improve processes for everybody's benefit. Healthcare innovations, especially when they are first starting, may not have much evidence to demonstrate the risks, benefits, and pitfalls of implementing them.

Information Security

Healthcare entities are under daily information breach attacks, and there is a significant monetary value to selling stolen healthcare data. Therefore, the information security entity within healthcare plays a vital role in enabling the rest of the healthcare system to function. Given that those functions happen in a zero failure tolerance environment, any new healthcare innovation or technology that introduces greater risk or has not been proven secure poses a threat to information security. Information security's critical role in enabling health system functioning must be acknowledged. Methods to address and mitigate concerns around new technologies and innovations need to be built early in the design process for successful implementation and deployment.

Information Privacy

Few things are more personal than health information, and who can access that information is of utmost importance to patients. Much of the healthcare apparatus is information management: managing information with expertise to diagnose and identify an optimal treatment. Therefore, it is no surprise that most health technology innovations seek to leverage information in new ways. However, most of the time, this information is derived or somehow associated with a human being who may have concerns about the privacy of the information derived from them. The information privacy entity within the system must ensure that the patient's privacy rights are respected, even if that decreases the development or capability of innovation and technology.

Payer

The payer is the one paying for services. Multiple different entities pay for the healthcare services and many other arrangements. Generally, one can think of payment arrangements as managing (1) the required capital to pay for services and (2) the risk of paying for services.

Patient

The patient may be financially responsible for no aspect of the care or all of the care. The patient may also encounter a moral hazard after paying for health insurance and then wanting to take advantage of services that the patient feels they have already paid for. Conversely, literature has demonstrated that patient shared responsibility (co-payments) decreases utilization, but patients may make poor purchase decisions between helpful treatments (analgesics) and critical treatments (insulin). Understanding the financial motivations of the patient helps to figure out the value proposition for a particular innovation.

Employer

The employer is often financially responsible for the patient's care through insurance premiums or self-insurance. Similar to the patient, the employer's interest may vary depending on how much financial responsibility they have for the patient's health.

Third-Party Payer

Commercial, government, and other third-party payers only make or save money by not paying, or paying less, for services. The third-party payer wants to decrease overall costs as much as possible, which does mean that early care, which results in an overall lower cost burden, and usually better quality for the patient, is aligned with the interests of the third-party payer (Feldman, Novack and Gracely, 1998). When the third-party payer is the government, as it often is, the interests can be diverse and conflicting (Tang et al., 2004).

Implementation/Deployment Team

Although we covered the various entities involved in the healthcare ecosystem, the individuals involved in implementing and deploying a new technology or innovation are critical to understand at the individual level.

Executive Sponsor

The executive sponsor is responsible for representing the project at the executive level and ensuring that there are adequate organizational resources. Ensuring that the value that the executive sponsor wants from the innovation aligns with the value that the specific deployment implementation will create is critical to project success and long-term adoption. Understanding this individual's perspective is well worth that investment in time. For example, we had a tool that helped anticipate staffing needs and decrease labor costs. We believed that helping the executive in charge of this particular labor force decrease their labor costs would be a significant victory for them. However, we found that the loss of political clout from having a substantially reduced budget was often more important than the short-term success of decreasing labor costs. This effective technology proceeded to go nowhere.

System-Level Leadership

Many of the users and stakeholders of technologies happen at the point of the patient or the patient's encounter with the healthcare system. But it is at the system level, not the local delivery level, where many decisions are frequently made regarding new technologies or innovations that can be implemented. Given the separation between the frontline and the system-level leadership, it is often the case that the system-level leadership does not feel or share the frustrations of the local leadership, who must hear the venting from the frontline workers and patients. Figuring out what value the innovation or technology delivers to the system-level leadership can be challenging but necessary for project success. In our experience, we find a frequent misalignment between those who work on the venture side and those on the front lines. The result is that the local leadership and frontline workers would like to try a particular technology or innovation. Still, they are blocked from doing so by the system-level leadership, which has already invested in an alternative technology that the local leadership or frontline workers do not desire.

Project-Level Leadership

Where the boots are on the ground, there is project-level leadership in determining what the local project-level leadership is getting out of the innovation or technology is essential to provide that continued motivation and drive necessary to complete change management and overcome the inevitable setbacks. There are many different ways that it may be giving value, as reviewed below. Still, clarity around the value proposition to project-level leadership can be the difference between a successful implementation and one that never gets off the ground.

Project-Level Stakeholders

The individuals involved in implementing and deploying their particular innovation or technology where their boots are on the ground and their hands are on the keyboard or on the patient can be a heterogeneous group. The personalities can range from the ambitious who are trying to build a career, the passionate trying to make a difference, individuals volunteered by their manager for the project, and even those concerned that the innovation or technology may threaten their job and are involved to ensure project failure. For each individual, the value they can receive from the innovation or technology needs to be clear to create both the motivation for change and the necessary solidarity to communicate back to all the different stakeholder groups for successful buy-in.

Provider of Innovative Technology or Service

The innovation technology or service provider must also receive value from the deployment. This is one area where both the customer and the vendor often think too narrowly about potential value from the relationship. Usually, the vendor needs assistance with development, market fit, market traction, commitment to future purchases, clarity on market demands, or more significant subject matter expertise, especially in a complicated healthcare ecosystem. We can often develop mutually beneficial contracts in ways that a simple arrangement for cash is not.

Value Creation

Value is an intuitively important concept that is both essential and ambiguous and seems to have lost any power once it gained widespread status as a buzzword. Nonetheless, it is impossible to discuss return on investment without first diving into the value of that return and to whom that return is valuable. A quick review of the types of value we have seen from different innovation projects within healthcare may be helpful when evaluating a potential project (Porter, 2010).

Financial

Direct financial value is always relatively easier to quantify. Still, we have found multiple different financial returns on projects depending on the mechanism for capturing the economic benefits.

Revenue

An increase in revenue is relatively straightforward, depending on which stakeholder has the revenue increased. We tend to think of an increase in revenue as an increase in the unit compensation for a particular good or service. For example, technologies that allow clinicians to bill for bedside ultrasounds that they were performing in their clinical care when they were not previously able to increase the unit compensation for that care.

Cost Avoidance

Like an increase in revenue, cost avoidance provides financial value by decreasing the unit cost for a particular good or service. This is often much more challenging to quantify as it can be challenging to measure what doesn't happen or attribute something not happening to a particular cause (Kb and Ds, 1999; Krueger et al., 2012). For example, care navigation for frequent emergency department utilizers with non-emergent complaints decreases the cost for the third-party payer (Gardner et al., 2014). Still, it can be difficult to attribute the change in patient behavior to the care navigation program.

Increase Market Share

Increasing market share or increasing the volume of goods or services delivered is another mechanism that some innovations provide (Devers, Brewster and Casalino, 2003). For example, technologies that help manage social media reputation for an outpatient clinic or health system may help drive greater volume and demand to that clinic.

Decreased Loss of Market Share (Attrition or Leakage)

Patients are behaving increasingly like consumers with more minor barriers to changing their venue for healthcare. Mechanisms to keep patients from seeking care elsewhere can help decrease a loss of market share (Forrest et al., 2006). For example, health system or physician-sponsored virtual care options, a comprehensive patient portal with medical records, and patient engagement technologies generally have a value proposition to decrease the loss of market share.

Efficiency

While quantifying a financial ROI can be somewhat challenging, the rules of the game are relatively straightforward as it comes down to dollars. However, efficiency is often of greater value to many stakeholders in the healthcare system, especially patients and their families (Glied, 2008). Quantifying the value of that efficiency can be more challenging (Hussey et al., 2009). It may be done directly by looking at behavior, labor costs, and productivity, or indirectly by looking at willingness to pay and NPS scores.

Time

Time is a finite resource and everyone can agree is valuable. Innovations that decrease the amount of time something takes provide value through that decrease in time. This may be realized as increased productivity, reduced delay of care, or greater satisfaction with the experience.

Decrease the Number of Steps or Processes

Simplification, especially from the user perspective, often provides significant value. Amazon's famous 1-Click purchase simplifies the online purchasing experience to the delight of purchasers and sellers everywhere. Healthcare is a notoriously inefficient and process-heavy industry with many opportunities to create value by decreasing the number of steps or required processes for the user or stakeholder. Assessing the financial value associated with simplifying a job can be challenging.

Quality

It is hard to imagine hearing someone say that they would like poor-quality healthcare. Whether or not there is an active investment in achieving quality healthcare, everyone endorses quality as a goal (Batalden and Davidoff, 2007; Becher and Chassin, 2001). Indeed, many of the health technologies and innovations that garnered the most attention claim to improve some aspect of the quality of care delivered. Measuring the quality can be challenging, regardless of whether one looks at process measurements or outcome measurements, and quantifying the value of that quality can be even more difficult. Nonetheless, when we are the patient, and it is our lives that we are talking about, it would seem that few things have as much value as quality.

Experience

The experience in the healthcare system, regardless of the role, is generally not a favorable or enjoyable one. Whether one is a patient, family member, payer, or provider in our healthcare system, the experience is often most notable for frustration, confusion, and despair. As many direct primary care startups have demonstrated, there is significant value to a rewarding and positive experience; consumers appear to be willing to spend substantial sums out-of-pocket for this improved experience.

Satisfaction

One component of the experience satisfaction and patient satisfaction scores has become a key focus as an indicator of how well we meet patients' needs (Rivers and Glover, 2008). While there are undoubtedly many potential opportunities and challenges focusing on patient satisfaction, improving patient satisfaction can provide a clear metric of value. Given that many stakeholders within the healthcare ecosystem have financial or other incentives tied to their performance on patient satisfaction measures, this often becomes one of the surprisingly more straightforward metrics to quantify the value and get alignment.

Sense of Accomplishment

As anyone who has raised a child can attest, there is a significant value when someone has a sense of accomplishment that seems different than a measurement of satisfaction with an experience. The poorly cooked and barely edible breakfast served on Mother's and Father's Day is often accompanied by the child's feeling of elation from that accomplishment. The value we get from accomplishing something creates a problematic paradox for innovations that do things for us and may take away that sense of accomplishment. Although difficult to quantify, the ever-elusive engagement may be closely related to ensuring there is a sense of accomplishment.

Sense of Mastery

More than a mere sense of accomplishment, a sense of mastery provides significant value to the individual. Although this is very difficult to capture as a return on investment, it is a critical concept when looking to maximize utilization and decrease the risk of product failure. The joy of a teenager who can now drive a car themselves is easy to recognize. With a bit of distance, one can recognize that the teenager values that sense of mastery so much that they would prefer to drive their uncomfortable, beat-up old car of questionable reliability rather than ride as a passenger in the passive luxury of their parents' expensive German car.

External Recognition

It is only human to enjoy and thus value recognition. The apparent value associated with external recognition leads us to question celebrities' motives for philanthropic acts. At a more individualistic level, the importance of external recognition in the form of comments, likes, and emojis on social media has created societal change at levels completely unfathomable 15 years ago. Although this value proposition is difficult to quantify, many health technologies are attempting to leverage external social recognition for engagement and behavior change. Specifically for successful health technology innovations, external recognition for those implementing the new technology is often an important motivating factor for the individuals willing to take the plunge on something unproven within their organization.

Appreciation

Feeling appreciated is a fundamental human emotion that needs to be addressed before any rational thought. The universality of the need for appreciation regardless of socioeconomic status, race/ethnicity, age, or any other attribute makes this a fundamental value proposition when thinking of utilization, adoption, and engagement.

Meeting a Metric or KPI

Aside from the patient and their family/caregivers, the rest of the healthcare system is in the business of healthcare. Much of the business of healthcare attempts to instill accountability by incentivizing metrics and following key performance indicators. Because so many stakeholders have direct incentives tied to these metrics, any innovation that improves the metric can be easily quantified in value for the stakeholder who is incentivized to that metric.

Value Distribution

We have identified the identity and emotional landmines to prospectively address for each relevant entity for our new technology or innovation. We have identified the value being created for each applicable entity. Now, we need to determine how that value is captured and distributed so that the interests of the stakeholders and shareholders align. While this seems to be relatively straightforward (and one would think relatively easy), it can be rather complex in the healthcare ecosystem. I often find that promising technologies and innovations get very low utilization and ultimately are deinstalled. For example, when a patient chooses cheaper and less convenient care for an acute issue rather than the expensive emergency department, the payer benefits substantially more than the patient. Although the value was easy to capture for the payer and reduce medical claims, it is harder to distribute that value to the patient in a way that feels positive rather than less punitive.

Framework to Identify Innovation Success Early: 13 Steps for Innovation

The framework to distinguish between likely successful innovations versus those likely to fail is divided into three major categories. First is the dreaming phase, which goes from problem discovery until a minimum marketable product has been demonstrated. The next phase is the building phase, which takes that minimum marketable product and builds a business around it. Finally, the scaling phase drives operational efficiencies and continuous incremental improvements. Each step is completed in sequential order. Before proceeding to the next stage, the technology or innovation must either meet all prior goals and hypotheses for the previous steps or have a reasonable justification for skipping those steps. While this is laid out as a linear process, it often ends up being more recursive. For example, there may be a good understanding of the problem space. Still, the opportunity could be substantially larger if an overlapping or adjacent problem space were explored during the evaluation. The project would then go back to do additional problem discovery.

Dream

The dreaming phase is to increase organizational investment sequentially and resource allocation as the innovation or technology increases in its likelihood of success, finally ending and approving product-market fit with positive unit economics. This could be an external or internal product-market fit. Still, suppose there is not enough value in the solution for the stakeholders to pay for it in some meaningful way. In that case, that is likely to be a continuous uphill battle for implementation and utilization.

Step 1: Problem Discovery

    i. Goal: problem space discovery
    ii. Hypothesis: we understand problem space uniquely
    iii. Mindset: creativity, curiosity, and listening to the market
    iv. Skillset: creativity, subject matter expertise (SME), customer empathy, business understanding

Problem discovery is the most fundamental, challenging, and often poorly done step in the innovation process, leading to a lack of return on investment. I have found that using "a job to be done" framework makes it far less confusing for me to understand what problem is being addressed. A deep and thorough understanding of the problem makes it relatively easy to evaluate proposed solutions. Although a comprehensive review of this framework is outside the scope of this chapter, it is worth a summary here. Briefly, everyone is doing an action to achieve something. For example, a patient may call a doctor's office and schedule an appointment to see the doctor. We can ask, "what is the patient trying to accomplish at a more profound level?" In this example, the patient may have pain in their knee, and they want to find out whether or not it is cancer because their uncle had osteosarcoma and is wanting to figure out how to eliminate the pain so they can get back to their morning run routine. I find the "jobs to be done" framework very helpful because if we are evaluating a technology that increases the call center's capacity in this example, it would be beneficial to find an efficient way to address the patient's health concerns about their knee.

Suppose we continue with this example of a patient who wants to resolve their health concern about their knee. In this case, we can see that the competitors in the market may include (but are not limited to) other doctors' offices, a family member who serves the role of a medical expert, and social media or other information sources. Therefore, the innovation must be favorable to compete with all the other ways that the patient can achieve what they want to achieve, which is resolving their health concern about their knee. Through this framework, we can then assess what someone is trying to achieve, how a candidate solution helps them achieve it, how that solution compares to the other ways they are achieving their objectives currently, and quantify the difference in value between the different solutions.

This step is successful once we understand what the user is trying to accomplish, the competing ways to achieve their goals, and their frustrations with the current approach. We know this by validating our hypothesis with the customers we intend to target. For the palpation device, we interviewed multiple physicians, potential patients, health system operators, and payers. We discovered that patients are trying to avoid unnecessary trips to the office or hospital, physicians wanted to be able to care for patients with abdominal pain safely via telehealth, and operators and payers wanted to provide safe care for lower cost.

Step 2: Valuate

    i. Goal: determine the potential of the idea
   ii. Hypothesis: solving the problem is worth the likely investment
  iii. Mindset: rapid customer and market discovery
  iv. Skillset: market/competitive research, customer feedback, understanding of the feasibility of the idea/product, financial

Once we understand what the different stakeholders are trying to accomplish and have validated that hypothesis in step one, we are now ready to determine the potential value of the proposed idea or solution. When I'm evaluating startups, I often see this step done very poorly. For example, a startup may present a solution to reduce the cost of caring for patients with congestive heart failure. As a physician, I see many efforts to target congestive heart failure because it is a costly problem for society. However, it is expensive because it is a terrible condition where treatments only bend the cost curve modestly. The startup may propose a new method to engage patients to get their daily weights, which helps manage their disease. The potential value of this intervention is often presented as the cost of caring for congestive heart failure. Sometimes they are more accurate and report the potential value as the cost reduction associated with people who check their weights daily versus those who do not. But even this is inaccurate. Their approach will not perfectly engage people to check their daily weights, and other people already check their daily weights without this new approach. The actual value created by the startup is the difference in cost for those who check their daily weights versus those who do not because they were induced to do so through the startup's approach. The actual valuation of those solutions is the difference between the current cost of care and the new cost of care once the modestly effective treatment is implemented over the current standard. It is no wonder that the vast majority of these fail to take on or scale broadly.

We have satisfied the step two hypothesis when we have determined that there will be value created to stakeholders over the current state that will be sufficient to warrant implementation and change management. For the palpation device, we learned through market research that abdominal pain is the most common complaint in emergency and urgent care settings and the majority do not need advanced imaging. The value of safely shifting the care of the abdominal pain patients to the home/virtual care is the difference in cost for in-person versus virtual care minus the cost of the solution.

Step 3: Rapid Low-Fidelity Prototyping

    i. Goal: shape idea based on customer and market feedback
   ii. Hypothesis: problem can be solved with favorable unit economics
  iii. Mindset: rapid prototyping for customer development and business plan creation
  iv. Skillset: product development, customer discovery, business planning

Armed with a deep understanding of what the stakeholders are trying to accomplish and an appreciation for the approximate value a solution could create, it is time to start trying different solutions. At this step, the goal is to see if you can create a solution by testing the proposed solution's most significant assumption.

For example, as a physician doing telehealth, I wanted to push on a patient's abdomen remotely and know whether or not they had a surgical abdomen, such as appendicitis, or just had gastroenteritis or food poisoning. The big assumption is that a patient pushing on their abdomen does so differently when they have a surgical abdomen than when they don't have a surgical abdomen. Instead of creating a consumer-grade and high-technology device to test this assumption, we went with a faster and cheaper approach. In the case of the palpation device, my son, who was 16 years old and in high school, was obsessed with 3D printing and CAD design. He hacked together pressure sensors, a massage head, and a 3D printed device plugged into a laptop, which allowed us to test the device on real patients and see if the basic assumption could be met. In terms of time and expense, it took about four weeks and $2,000 worth of investment for my son to create the device, and then approximately three months later, the device and IRB application were approved. If the results were positive, we could build a higher-grade prototype in the next step of this process.

On the other hand, if the device could not distinguish the need for a patient to get advanced imaging, then for a mere $2,000, we avoided a considerable investment of time and resources. In this case, the device could not distinguish who needed surgery from those who did not. We then went back to step 1 to determine what we were trying to do and learned that we were actually trying to determine when a physician would want advanced imaging after examining the patient. We reexamined the data and the device distinguished between those who did and those who did not get advanced imaging with nearly perfect accuracy.

We know we have completed this step successfully when we are satisfied that the most significant assumptions about our proposed solution are correct when tested with our target customers. It is important to note that this is a very iterative process, and this is a step at which many pivots frequently happen. The problem I thought I was addressing with a proposed solution is different but may have significant value. When that happens, we take a new understanding and go back to step one with our increased expertise. Similarly, I have also found that the proposed solution that I thought would work was completely and utterly wrong. We needed to develop a very different approach to solve the problem. The point is to do that now when it is early, and pivoting is easy and inexpensive.

Step 4: Minimum Viable Feature (MVF)

    i. Goal: define the key features and value proposition
    ii. Hypothesis: we can make the key features, effectively distribute, and implement the solution
    iii. Mindset: operational/product architecture problem-solving
    iv. Skillset: SME, product development, operations, and implementation

Based on the previous step, we now understand how we can solve the riskiest assumptions to create a solution that will create value for the stakeholders. We are now in a position to create a product that can reliably deliver on that most critical value proposition that would ultimately make the stakeholders interested in purchasing or using the solution. Keeping a laser focus on the most significant value proposition helps inform the early architecture when it is still early enough for pivots and redesigns. At this stage, we are generally testing and iterating the development within a couple of reasonably small and controlled environments for rapid feedback and engineering improvements. This is the phase where we want to ensure that we are not in one isolated environment because we risk creating something that has low generalizability. I once built a robust and techno-logically sophisticated analytics tool for a multisystem anesthesia practice, which seemed diverse and representative of the anesthesia practice space. It was a huge technological step forward in the industry, embraced as a substantial competitive advantage, and I was thrilled with how it was being used. When I went to deploy our fantastic tool to other practices with different management teams, I suddenly learned that I'd created a unique vocabulary and operational approach that was overly specific to my development customer. If I had employed one or two additional develop-ment locations at this phase, I would've created a much better product and saved substantial time, money, and resources.

We are finished with this phase when we have a solution with well-defined key features, a compelling value proposition, and a feasible approach to distribute and implement to our entire target stakeholders. In the case of the palpation device, we identified that the device must be less than $10 to manufacture, be able to connect to a phone or tablet via a powered wire or wireless connection, be intuitive to use by a novice in distress, and provide high-reliability data. Pilot testing the device in settings where it is not used (like in the emergency department) will allow usability and reliability issues to be addressed. This phase will also generate the performance data needed to market the device in the next step and get FDA approval.

Step 5: Develop a Minimum Viable Product (MVP)

   i. Goal: create tangible, working MVP
  ii. Hypothesis: we can build a solution that will be used
 iii. Mindset: rapid changes, managing resistance and challenges
 iv. Skillset: innovation project management, customer discovery and new product management

Now that we have established an approach to creating our main value proposition that resonates with the stakeholders, it is time to build around that core aspect of our product to make a mini-mum viable product. While the product we created as a minimum viable feature delivers on the core value proposition, it often does not include all the necessary integrations and user experience to be broadly deployed in the mainstream workflow. The minimum viable product is the stage where we figure out how to integrate the value of our minimum viable feature into a solution that the stakeholders will want to use and deploy. Therefore, the focus is on developing a product that delights the stakeholders.

We know that we have successfully achieved this step when the market is willing to pay something for our solution and implement it. This is generally at the stage where we are doing pilots that continue to have unfavorable unit economics and rapid feedback and iteration. Like the lessons learned with the minimum viable future, the pilots must take place in a diversity of contexts that capture the spectrum of the intended target stakeholders. In the case of the palpation device, we will take the device that meets our minimum viable features, deploy it internally to our virtual care programs, and pilot it externally with virtual care providers. This will include significant support resources to ensure a positive experience, which means the unit economics will be negative at this point.

Step 6: Develop Minimum Marketable Product (MMP)

   i. Goal: Create a market proof of product-market fit and positive unit economics
  ii. Hypothesis: The market will pay for the product with positive unit economics

    iii. Mindset: Focus on customer pipeline and path to limit unit costs
    iv. Skillset: Innovation project management and adapting rapidly to new customer and market insights

We work to create a minimum marketable product at the final stage of dreaming. This builds on the minimum viable product outside of pilot arrangements with positive unit economics. While the minimum viable feature demonstrated our ability to create value, and the minimum viable product established that we could create a product that the market would pay for, the minimum marketable product demonstrates that we can sell a product with positive unit economics. Therefore, the focus is on delivering the product that delights the stakeholders while reducing our internal unit costs. We succeeded in this step when we have positive unit economics outside of customer acquisition costs. For the palpation device, we will market the device broadly based on the demonstrated value in the prior step while focusing on reducing unit costs through greater automation and less reliance on human intervention. The initial customers will become definitional customers and we will want to have as diverse a representation as possible, but will not be able to grow beyond our ability to provide high-touch service.

Build

Step 7: Integration Assessment

    i. Goal: determine the potential to integrate product within the user ecosystem
    ii. Hypothesis: we can deploy this product with positive unit economics
    iii. Mindset: navigating the relevant enterprises/organizations
    iv. Skillset: deep understanding of enterprise, partnership structures, and the product

Integration into existing systems, architecture, and workflow is often the most challenging aspect of healthcare technology innovation. It is unlikely to be successfully adopted and utilized outside of early enthusiasts if the solution cannot be integrated successfully and seamlessly into existing workflow and processes. Building on our understanding of what is a minimum marketable product, this step focuses on decreasing the customer barrier to purchase and adoption by creating integrations into the existing workflow, technology, process, and culture. While the minimum marketable product may have made the integration in the background using non-generalizable processes, such as robotic process automation or humans in the loop, at this step, the focus should be on creating robust and generalizable integration. Furthermore, the time, resources, and money integration cost should be less than the stakeholder's willingness to pay for the minimum marketable product. That is to say, the integration solution needs to have long-term positive unit economics.

    For example, one of the products we worked on collected information from patients and then made it usable within the electronic health record before the patient's visit. Initially, this function was performed by humans in the loop who pulled the information from one system and manually entered it into the electronic health record. Given that the information collected and populated changed during the development process, this approach allowed for maximum flexibility and rapid evolution. This was subsequently replaced by robotic process automation, where the computer system pulled information from the product and then entered it into the electronic health record. Finally, this was replaced by a fully integrated system that used APIs to automatically and near instantaneously move the information from one system to the other. For the palpation device, the data from the self-exam by the patient and the results of the exam will need to integrate seamlessly into the virtual care visit physical workflow (what the patient and physician are doing) and the information workflow (what information is collected and where it is available).

Step 8: Priority Assessment

    i. Goal: determine the importance of offering the product
   ii. Hypothesis: the value is greater than alternative projects
  iii. Mindset: determining value (both strategic and monetary)
  iv. Skillset: deep understanding of enterprise ops/strategy and the product

At this stage, the minimum marketable product, the value creation, the change management, deployment cost, and the long-term utilization have been well established. Identifying how this solution aligns with the purchaser's goals and the alternative ways to use resources shapes the go-to-market strategy and accelerates sales. When approaching a solution as a consumer or customer to a vendor with a novel innovation, we need to compare this overall value proposition with how it aligns with our organizational strategic plan. We know we've completed this step when we can favorably position the innovation in value proposition and required resources within the organizational or purchaser's strategic roadmap. For the palpation device, we need to identify stakeholders who want to provide expanded virtual care capabilities that can reduce emergency and urgent care utilization for abdominal pain and can either provide a palpation device prior to need or distribute it within a short time frame, such as a same-day delivery model.

Step 9: Product Definition

    i. Goal: based on prior data and feedback, define the exact solution to be generally offered
   ii. Hypothesis: solution can be delivered within operational constraints at a positive value
  iii. Mindset: managing enterprise complexities and the "new"
  iv. Skillset: deep understanding of the enterprise and the product

At this stage of the product journey, we generally know a lot more than when we started. This is where we fine-tune precisely what we are offering based on the market feedback from our deployments and sales experience. This is essentially what we have found at this step, as we have learned the operational constraints that affect a substantial portion of our potential users. We must modify our solution to work within those constraints. We know that we have successfully navigated this step when we have a precise and constrained definition of the product that we are taking to market. Put another way: we know that we are successful when we are comfortable and confident in saying no to user/customer requests for customization or special features. For the palpation device, we need to refine the device, workflow, distribution, training, and integration into a scalable standard solution that meets the needs of our core market.

Step 10: Deploy

    i. Goal: deploy the solution broadly
   ii. Hypothesis: the solution can be deployed in a repeatable process

iii. Mindset: navigating and stretching the enterprise in new ways to deliver the product
iv. Skillset: deep understanding of the enterprise and the product

I have been humbled multiple times by how challenging it can be to successfully, efficiently, and repeatedly deploy or implement a solution across various stakeholders. Just as significant effort and attention was devoted to developing the product and marketing, more attention and creativity must be paid to developing the methodology around successful deployment/implementation. We know that we are successful in this step when we have a proven, repeatable recipe for implementing the solution. For the palpation device, we will need to iterate the deployment methodology until the deployment success is high and the cost is low.

Scale

Step 11: Scale Operations

i. Goal: create repeatable implementations and operations
ii. Hypothesis: product delivery can scale through sales strategy, operational execution, and partnership
iii. Mindset: sales strategy and operational standardization
iv. Skillset: sales, distribution, operational excellence

Now that we have a successful solution with positive unit economics, a clear value proposition, and repeatable implementation processes, we focus on scaling to rapidly reach the target population.

Step 12: Drive Efficiencies

i. Goal: optimize unit economics by driving efficiencies
ii. Hypothesis: we can optimize costs, operations, and revenue
iii. Mindset: driving efficiency and delivering customer value
iv. Skillset: sales, distribution, operational excellence

While the early steps around innovation focused on creating new ways of doing things, driving efficiencies relies on creating standardized work and processes. For the palpation device, the early experimentation and rapid iteration to customer feedback is past and the focus is now on optimizing each step in the process for greatest quality, efficiency, and profit.

Step 13: Incremental Improvement

i. Goal: constant improvement to retain/gain customers
ii. Hypothesis: we can continue to improve the solution
iii. Mindset: continuous improvement based on customer/market feedback
iv. Skillset: continuous quality improvement

At this stage, care must be taken to ensure that the solution is not frozen or fossilized but continues to have incremental improvement to retain existing customers and continue to grow the customer base by identifying new opportunities to create and capture value. For the palpation device, this means careful semi-annual product updates with vigorous testing prior to deployment.

Conclusion

Innovation is challenging because there is so much uncertainty about what will or will not work. This uncertainty leads to an inappropriate allocation of resources. The first uncertainty is whether or not the idea, the problem being solved, can both create value and return that value in a meaningful way. I have introduced a methodology to ensure that an innovation produces value by paying attention to the identity, emotions, stakeholders, type of value, and method of distributing that value. The next big chunk of uncertainty is whether or not the solution will solve the problem and be successfully deployed. I then introduced a step-by-step framework that tests one specific hypothesis at a time. The innovation should be resourced appropriately to test that specific hypothesis. While this should be a linear process so that resources are sufficient but limited to testing the focused hypothesis for the step, the steps may need to be in a different order than those laid out here. For example, it may make sense to have a priority assessment prior to an integration assessment. I hope that adopting this approach will lead to a common language, assessment, and resourcing of innovation projects across a diverse portfolio. Creating a new and better future – one that did not exist until one jumped into the arena – is to participate in the miracle of creation. It is a deeply satisfying and humbling purpose (Figures 12.1 and 12.2).

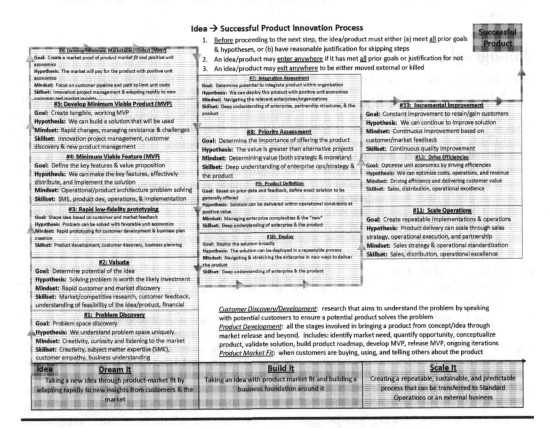

Figure 12.1   Innovation framework. Source: Tamayo-Sarver, Vituity (2022).

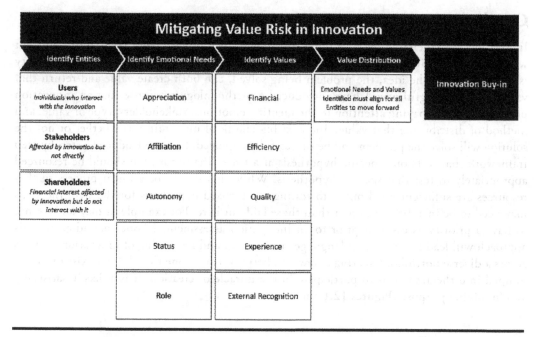

Figure 12.2 Mitigating value risk in innovation. Source: Tamayo-Sarver, Vituity (2022).

Bibliography

Alexander, Christopher S., Paul Miesing, and Amy L. Parsons, 2005. "How Important are Stakeholder Relationships." *Academy of Strategic Management Journal* 4: 1. Accessed 4 27, 2022. https://questia.com/library/journal/1g1-166751810/how-important-are-stakeholder-relationships.

Allen, Larry A., and Christopher M. O'Connor, 2007. "Management of Acute Decompensated Heart Failure." *Canadian Medical Association Journal* 176(6): 797–805. Accessed 4 27, 2022. https://ncbi.nlm.nih.gov/pmc/articles/pmc1808524.

Altman, Drew E., and Douglas H. Morgan, 1983. "The Role of State and Local Government in Health." *Health Affairs* 2(4): 7–31. Accessed 4 27, 2022. https://healthaffairs.org/doi/full/10.1377/hlthaff.2.4.7.

Arrow, Kenneth J., 1962. Economic Welfare and the Allocation of Resources for Invention. Accessed 4 26, 2022. https://nber.org/chapters/c2144.pdf.

Batalden, Paul B., and Frank Davidoff, 2007. "What is 'Quality Improvement' and how can it Transform Healthcare?" *Quality and Safety in Health Care* 16(1): 2–3. Accessed 5 3, 2022. https://ncbi.nlm.nih.gov/pmc/articles/pmc2464920.

Becher, Elise C., and Mark R. Chassin, 2001. "Improving the Quality of Health Care: Who Will Lead?" *Health Affairs* 20(5): 164–179. Accessed 5 3, 2022. https://healthaffairs.org/doi/full/10.1377/hlthaff.20.5.164.

Beck, David E., and David A. Margolin, 2007. "Physician Coding and Reimbursement." *The Ochsner Journal* 7(1): 8–15. Accessed 4 27, 2022. https://ncbi.nlm.nih.gov/pmc/articles/pmc3096340.

Briganti, Suzan E., and Alain Samson, 2019. "Innovation Talent as a Predictor of Business Growth." *International Journal of Innovation Science* 11(2): 261–277. Accessed 4 26, 2022. https://emerald.com/insight/content/doi/10.1108/ijis-10-2018-0102/full/html.

Carrière, Benoit, Robert Gagnon, Bernard Charlin, Steven M. Downing, and Georges Bordage, 2009. "Assessing Clinical Reasoning in Pediatric Emergency Medicine: Validity Evidence for a Script Concordance Test." *Annals of Emergency Medicine* 53(5): 647–652. Accessed 4 27, 2022. https://sciencedirect.com/science/article/pii/s0196064408015096.

Chaudoir, Stephenie R., Stephenie R. Chaudoir, Alicia G. Dugan, and Colin Hi Barr, 2013. "Measuring Factors Affecting Implementation of Health Innovations: A Systematic Review of Structural, Organizational, Provider, Patient, and Innovation Level Measures." *Implementation Science* 8(1): 22–22. Accessed 5 3, 2022. https://implementationscience.biomedcentral.com/articles/10.1186/1748 -5908-8-22.

Claffey, Thomas F., Joseph V. Agostini, Elizabeth N. Collet, Lonny Reisman, and Randall Krakauer, 2012. "Payer-Provider Collaboration in Accountable Care Reduced Use and Improved Quality in Maine Medicare Advantage Plan." *Health Affairs* 31(9): 2074–2083. Accessed 4 27, 2022. https://healthaf-fairs.org/doi/full/10.1377/hlthaff.2011.1141.

Conklin, Thomas P., 2002. "Health Care in the United States: An Evolving System." *Michigan Family Review* 7(1): 5. Accessed 4 27, 2022. https://quod.lib.umich.edu/m/mfr/4919087.0007.102/--health -care-in-the-united-states-an-evolving-system?rgn=main;view=fulltext.

Cutler, David M., Elizabeth Wikler, and Peter Basch, 2012. "Reducing Administrative Costs and Improving the Health Care System." *The New England Journal of Medicine* 367(20): 1875–1878. Accessed 5 3, 2022. https://nejm.org/doi/full/10.1056/nejmp1209711.

Devers, Kelly J., Linda R. Brewster, and Lawrence P. Casalino, 2003. "Changes in Hospital Competitive Strategy: A New Medical Arms Race?" *Health Services Research* 38(1 Pt 2): 447–469. Accessed 5 3, 2022. https://ncbi.nlm.nih.gov/pmc/articles/pmc1360894.

Epstein, Ronald M., and Richard L. Street, 2011. "The Values and Value of Patient-Centered Care." *Annals of Family Medicine* 9(2): 100–103. Accessed 5 3, 2022. https://ncbi.nlm.nih.gov/pmc/articles /pmc3056855.

Feldman, Debra S., Dennis H. Novack, and Edward J. Gracely, 1998. "Effects of Managed Care on Physician-Patient Relationships, Quality of Care, and the Ethical Practice of Medicine: A Physician Survey." *JAMA Internal Medicine* 158(15): 1626–1632. Accessed 5 3, 2022. https://jamanetwork.com /journals/jamainternalmedicine/fullarticle/1105592.

Field, Robert I., 2008. "Why is Health Care Regulation so Complex?" *P and T: A Peer-Reviewed Journal for Formulary Management* 33(10): 607–608. Accessed 4 27, 2022. https://ncbi.nlm.nih.gov/pmc/articles /pmc2730786.

Forrest, Christopher B., Paul A. Nutting, Sarah von Schrader, Charles A. Rohde, and Barbara Starfield, 2006. "Primary Care Physician Specialty Referral Decision Making: Patient, Physician, and Health Care System Determinants." *Medical Decision Making* 26(1): 76–85. Accessed 5 3, 2022. https://jhu .pure.elsevier.com/en/publications/primary-care-physician-specialty-referral-decision-making-patient -5.

Gardner, Rebekah, Qijuan Li, Rosa R. Baier, Kristen Butterfield, Eric A. Coleman, and Stefan Gravenstein, 2014. "Is Implementation of the Care Transitions Intervention Associated with Cost Avoidance after Hospital Discharge?" *Journal of General Internal Medicine* 29(6): 878–884. Accessed 5 3, 2022. https://ncbi.nlm.nih.gov/pmc/articles/pmc4026506.

Glied, Sherry. 2008. "Health Care Financing, Efficiency, and Equity." *National Bureau of Economic Research.* Accessed 5 3, 2022. https://nber.org/papers/w13881.pdf.

Hansen, Erik G., Friedrich Grosse-Dunker, and Ralf Reichwald, 2009. "Sustainability Innovation Cube – A Framework to Evaluate Sustainability-Oriented Innovations." *International Journal of Innovation Management* 13(4): 683–713. Accessed 5 3, 2022. https://worldscientific.com/doi/abs/10.1142/ s1363919609002479.

Hussey, Peter S., Han de Vries, John A. Romley, Margaret C. Wang, Susan S. Chen, Paul G. Shekelle, and Elizabeth A. McGlynn, 2009. "A Systematic Review of Health Care Efficiency Measures." *Health Services Research* 44(3): 784–805. Accessed 5 3, 2022. https://ncbi.nlm.nih.gov/pmc/articles/ pmc2699907.

Howard, K. B., and D. S. Pathak, 1999. "Determining the Differences among Cost Savings, Cost Avoidance, and Cost Reduction." *Pharmacy Practice Management Quarterly* 19(3): 1–7. Accessed 5 3, 2022. https://ncbi.nlm.nih.gov/pubmed/10747680.

Kim, W. Chan, and Renée Mauborgne, 2005. "Value Innovation: A Leap into the Blue Ocean." *Journal of Business Strategy* 26(4): 22–28. Accessed 5 3, 2022. https://emerald.com/insight/content/doi/10.1108 /02756660510608521/full/html.

Krueger, Hans, Patrice Lindsay, Robert Côté, Moira K. Kapral, Janusz Kaczorowski, and Michael D. Hill, 2012. "Cost Avoidance Associated with Optimal Stroke Care in Canada." *Stroke* 43(8): 2198–2206. Accessed 5 3, 2022. https://ahajournals.org/doi/full/10.1161/strokeaha.111.646091.

Needleman, Jack, and Susan Hassmiller, 2009. "The Role of Nurses in Improving Hospital Quality and Efficiency: Real-World Results." *Health Affairs* 28(4). Accessed 5 3, 2022. https://healthaffairs.org/doi/full/10.1377/hlthaff.28.4.w625.

Nguyen, H. Bryant, Win Sen Kuan, Michael Batech, Pinak Shrikhande, Malcolm Mahadevan, Chih-Huang Li, Sumit Ray, and Anna Dengel, 2011. "Outcome Effectiveness of the Severe Sepsis Resuscitation Bundle with Addition of Lactate Clearance as a Bundle Item: A Multi-national Evaluation." *Critical Care* 15(5): 1–10. Accessed 4 27, 2022. https://ncbi.nlm.nih.gov/pmc/articles/pmc3334775.

Omachonu, Vincent K., 2010. "Innovation in Healthcare Delivery Systems: A Conceptual Framework." *The Innovation Journal* 15(1): 1–12. Accessed 4 27, 2022. http://dphu.org/uploads/attachements/books/books_1028_0.pdf.

Porter, Michael E., 2010. "What is Value in Health Care?" *The New England Journal of Medicine* 363(26): 2477–2481. Accessed 4 27, 2022. https://nejm.org/doi/full/10.1056/nejmp1011024.

Rivers, Patrick A., and Saundra H. Glover, 2008. "Health Care Competition, Strategic Mission, and Patient Satisfaction: Research Model and Propositions." *Journal of Health Organization and Management* 22(6): 627–641. Accessed 4 27, 2022. https://ncbi.nlm.nih.gov/pmc/articles/pmc2865678.

Schaik, Paul van, Darren Flynn, Anna van Wersch, Andrew Douglass, and Paul Cann, 2005. "Influence of Illness Script Components and Medical Practice on Medical Decision Making." *Journal of Experimental Psychology: Applied* 11(3): 187–199. Accessed 4 27, 2022. https://eric.ed.gov/?id=ej734256.

Scott, Shannon D., Ronald C. Plotnikoff, Nandini Karunamuni, Raphaël Bize, and Wendy M. Rodgers, 2008. "Factors Influencing the Adoption of an Innovation: An Examination of the Uptake of the Canadian Heart Health Kit (HHK)." *Implementation Science* 3(1): 41–41. Accessed 5 3, 2022. https://implementationscience.biomedcentral.com/articles/10.1186/1748-5908-3-41.

Shapiro, Daniel. 2017. *Negotiating the Nonnegotiable: How to Resolve Your Most Emotionally Charged Conflicts.* Penguin.

———. n.d. *Negotiating the Nonnegotiable: How to Resolve Your Most Emotionally Charged Conflicts.* Penguin. Accessed 4 12, 2022. https://books.google.de/books?id=nmdYCgAAQBAJ&pg=PT83.

Smith, Robert C., 2002. "The Biopsychosocial Revolution: Interviewing and Provider-Patient Relationships Becoming Key Issues for Primary Care." *Journal of General Internal Medicine* 17(4): 309–310. Accessed 5 3, 2022. https://ncbi.nlm.nih.gov/pmc/articles/pmc1495036.

Stilgoe, Jack, Richard Owen, Phil Macnaghten, and Phil Macnaghten, 2013. "Developing a Framework for Responsible Innovation." *Research Policy* 42(9): 1568–1580. Accessed 5 3, 2022. https://sciencedirect.com/science/article/pii/s0048733313000930.

Tang, Ning, Ning Tang, John M. Eisenberg, and Gregg S. Meyer, 2004. "The Roles of Government in Improving Health Care Quality and Safety." *The Joint Commission Journal on Quality and Patient Safety* 30(1): 47–55. Accessed 4 27, 2022. https://ncbi.nlm.nih.gov/pubmed/14738036.

———. n.d. *The Discipline of Innovation.* Accessed 4 27, 2022. http://hbr.org/2002/08/the-discipline-of-innovation/ar/1.

Tian, Xuan, and Tracy Yue Wang, 2014. "Tolerance for Failure and Corporate Innovation." *Review of Financial Studies* 27(1): 211–255. Accessed 5 3, 2022. https://academic.oup.com/rfs/article-abstract/27/1/211/1571535.

Zarabozo, Carlos, and Scott Harrison, 2008. "Payment Policy and the Growth of Medicare Advantage." *Health Affairs* 28(1). Accessed 4 27, 2022. https://healthaffairs.org/doi/10.1377/hlthaff.28.1.w55.

Chapter 13

Playing with FHIR: The Path to Ensuring We Bring the Power of Supercomputing to How We Understand Healthcare in Medicine

Lucia C. Savage

Contents

Introduction

The first automated teller machine was launched in London, UK, in 1967. In 1969, the first ATM was installed at the Chemical Bank branch in Rockville Center, New York. By the end of 1971, there were 1,000 ATMs installed worldwide. By 2018, there were more than 410,000 ATMs in the United States (Lian An, 2014). Most of us probably use ATMs occasionally even now, although

DOI: 10.4324/9781003348603-17

other electronic exchange systems, like debit cards, credit cards, and e-banking through apps like Venmo, are more and more common. They are convenient. And the vast majority of us just let these exchange methods, based on financial industry standards, work. We don't stop to think about how they work.

The changes being implemented in 2022 and 2023 to exchange of electronic health information have the same promise of changing many aspects of healthcare as did the invention of the ATM. This chapter will explain how we got to this point, and why the US government has been on a 25-year policy campaign to fully digitize health information. If this latest iteration succeeds, then a few years, perhaps for the most common healthcare information, like lab results, prescription lists/current meds, and allergy/problem lists, doctors and hospitals who need this historical information on a patient will be able to get it as easily as we use ATM and electronic financial transactions.

Start at the Beginning

The First Wave: 1974–2003

The government's first substantial attempt to formally digitize the health information for learning was inked in 1996, in the Health Insurance Portability and Accountability Act, now known by its acronym, HIPAA. That's right, in 1996, Congress in HIPAA's Accountability provisions required that if providers and hospitals who were seeking reimbursement from the US health insurance for the elderly, Medicare ("CMS") to bill the government electronically, then they had to do so under standards for the data (Solove, 2013). The Accountability part of the HIPAA statute aimed to take advantage of advances in computing technology and nascent machine learning/algorithmic data science, to evaluate claims data for signs of fraud, waste, abuse, or lack of adherence to evidence-based medicine. In this effort, Congress largely succeeded, with approximately 97% of healthcare providers billing their claims electronically by the early 2000s. In fact, digitizing the claims data led to our ability to find and remediate life-saving gaps in care, such as whether women had their mammograms, or whether 50-year-olds had their baseline colonoscopy. Savvy readers will recognize HEDIS® measures, and it is no accident that HEDIS measures took root on a wider scale in 1997 (Schauffler, 1996).

Even before 1996, however, the federal government recognized the power of computing to improve biomedical and health services research. Belmont Report, published in 1977 by the Department of Health, Education and Welfare (now HHS), was commissioned to study and make recommendations on how to fairly use health information from humans in a new computer age, back then being the age of punch cards and Fortran, not spreadsheets or EHRs. The Belmont Report stated principles for fairly using health information collected from humans in research, including privacy and transparency principles that not only undergird US and state privacy law today but worldwide.

Once the HIPAA statute was enacted, the implementation of its "accountability" requirements came in three phases:

1. Finalizing first iteration of regulations to protect the privacy of individual's information in the claims data (December 2000; later updated in August 2002).
2. Finalizing the "transactional code sets," or the coding standards for how to submit a claim for payment electronically, in a standard format that could be processed more efficiently *and* analyzed (fall 2002).

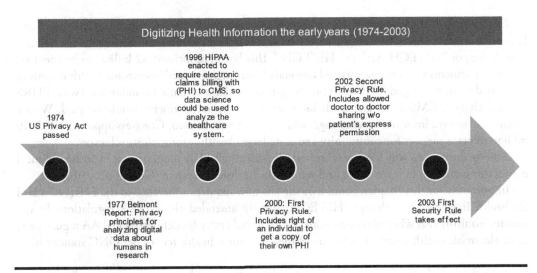

Figure 13.1 Digitizing health information, the early years 1974–2003. Source: Savage, Omaha Health (2022).

3. Finalizing standards for ensuring data subject to HIPAA was transmitted and handled with baseline security standards (fall 2003).

Figure 13.1 illustrates a timeline from 1974 to 2003 of federal efforts to use digital health information from humans fairly and for the purpose of learning.

Did digitization work in the first wave? I think so. Because we digitized claims data, we were able to use computing technology, itself evolving as fast as Moore's Law, to understand what was working and not working in healthcare as delivered to individuals. We could, for example, study the downstream comparative impact of different devices implanted in a hospital, as occurs with SharedClarity, a 2014 collaboration among Baylor Health Care System, San Francisco-based Dignity Health, and Illinois-based Advocate Health Care. Even today, we use analysis of digitized claims data to identify and remediate not just HEDIS gaps in care, but also health disparities. But, looking at claims data, as helpful as it was, was not enough.

The Second Wave: Digitizing Clinical Data

Even while the healthcare sector was busy implementing the digital requirements for claims data, other innovators were hard at work figuring out how to digitize the clinical data itself. Epic Systems was founded in 1977. Cerner Corporation was founded in 1979, and Allscripts was founded in 1986, just to give a sample, and acquired a prior EHR company, Eclipsys. The Veterans Administration home-grown VISTA system was also an early effort (Parker, 2019). The agency that ultimately became responsible for establishing standards for electronic clinical health information, the Office of the National Coordinator for Health IT, was established by Executive Order in 2004, just a year or so after the first wave of HIPAA Security regulations took effect. Prior to 2009, this small agency was exhortative; that is, it lacked authority to do more than coordinate other federal policy efforts on digital health information. But the technology was advancing rapidly. In 2006, with funds appropriated in the Medicare Modernization Act, the Agency for Healthcare Research and Quality (AHRQ) awarded the first grants to build community-based health information exchange (Dullabh, 2011).

Finally, in 2009, as part of the stimulus package called the American Restoration and Reinvestment Act (ARRA), Congress enacted the Health Information Technology for Clinical Health Act, or "HITECH Act" or "HITECH." This law appropriated $2 billion to be spent on incentive payments when physicians and hospitals "meaningfully used" electronic health records in prescribed ways. It charged ONC with promulgating regulations that set standards for what EHRS did and charged CMS to set standards for when the incentive payments would be paid. With a specific eye toward information exchange, where this chapter started, Congress appropriated $600 million in grant support for community and state-based health information exchanges.

Among the reasons why Congress created the Meaningful Use program was to create a clinical digital infrastructure to learn about and reform the healthcare system (Gold et al., 2016).

Importantly for this chapter, about *why* the law now requires adherence to the FHIR standard for health information exchange, HITECH directly amended the HIPAA regulations in the statute, requiring that where data was stored by a covered entity (as defined by HIPAA regulations) in an electronic health record (not just a certified electronic health record per ONC standards),

> (1) the individual shall have a right to obtain from such covered entity a copy of such information in an electronic format and, *if the individual chooses, to direct the covered entity to transmit such copy directly to an entity or person designated by the individual, provided that any such choice is clear, conspicuous, and specific*; and (2) notwithstanding paragraph (c)(4) of such section, any fee that the covered entity may impose for providing such individual with a copy of such information (or a summary or explanation of such information) if such copy (or summary or explanation) is in an electronic form shall not be greater than the entity's labor costs in responding to the request for the copy (or summary or explanation). (emphasis added)

More later about the above provision.

Following the enactment of HITECH, ONC did its work of implementing the Meaningful Use program. The HHS Office for Civil Rights (OCR), which enforces the HIPAA Privacy, Security and Breach Notification Rules, updated its regulations as required by HITECH. And in 2014, the federal regulations that apply to clinical labs that bill CMS were updated to require that labs release lab results to patients who request them without waiting for physician review, per HITECH.

Figure 13.2 shows the government's further policy initiatives from 2003 to 2014.

The Third Wave: Exchange "Without Special Effort"

Despite the billions spent on the Meaningful Use program, six years later, by 2015, Congress remained dissatisfied that information exchange was occurring too infrequently. After several rounds of Congressional hearings and official government analyses (report to Congress), in 2016, by overwhelming majority, Congress passed, and President Obama signed the 21st Century Cures Act. Most of the text of this law concerned updating FDA authorities and funding new large-scale biomedical research initiatives. But it also covered clinical health information exchange. The Cures Act prohibited "information blocking" by developers of Certified EHR technology and by healthcare providers. It required that developers and healthcare providers use technology that

> (A) enables the secure exchange of electronic health information with, and use of electronic health information from, other health information technology *without special effort on the part of the user*;

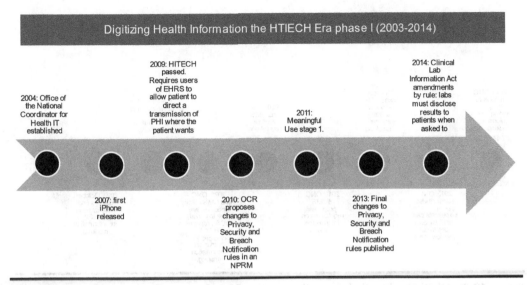

Figure 13.2   Digitizing health information, the HITECH era phase 2003–2014. Source: Savage, Omaha Health (2022).

    (B) allows for complete access, exchange, and use of all electronically accessible health information for authorized use under applicable State or Federal law; and

    (C) does not constitute information blocking as defined in section 3022(a).

(Cures Act 4003(A), amending Public Health Service Act section 3000; emphasis added)

It also required the developers of certified EHRs to implement and allow data calls on "open specification" application programming interfaces. Remember the ATMs, how they've been surpassed in many respects by electronic financial transactions via the internet? It is "application programming interfaces" or APIs that make those electronic transactions possible. In 2016, ONC's panel of experts published a report on the privacy and security of API technology in healthcare, and those resources plus others published since then will explain the technology. In summary, APIs should ideally require users' usernames and passwords. The Open Authorization (OAuth) standard can also be used with HL7 FHIR®-based Application Programmable Interfaces, APIs; enable token-based access, can be deleted or revoked any time for any reason– breach, misuse, etc. Access to tokens, built on appropriate identity proofing and authorization, can restrict permissions for specific <u>API</u> access purposes or limit access within an app.

    Let's pause for a moment on "open specification." In 21st Century Cures, the term means one that the technical specification for how the computer talks to each other *is not proprietary*. It does *not* mean that anyone can get the data that is exchanged via that specification. Congress aligned on this term in direct response to complaints from health systems, providers, and hospital executives that while they wanted to exchange health information to benefit shared patients and to reduce waste or harmful care, the cost was too high as each EHR system had its own *proprietary* technical specifications which made engineering the connections between organizations idiosyncratic and very expensive (Savage et al., 2019).

    Third, 21st Century Cures established substantial fines ($1 million per occurrence) to be paid by the developers of certified EHRs if they were found to have unlawfully blocked legitimate information exchange.

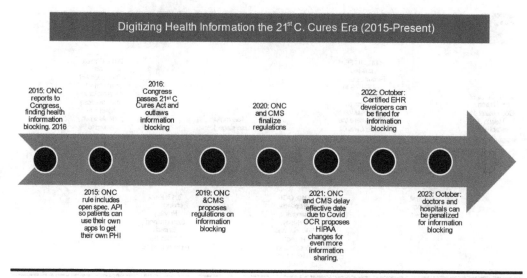

Figure 13.3 Digitizing health information in the 21st Century Cures era. Source: Savage, Omaha Health (2022).

Finally, 21st Century Cures required ONC to double-down on standards. Having been in the room where it happened, I know this is because Congress had lost patience with innovation that did not interoperate. In one meeting, an important Congressional advisor said to me words to the effect that "if I can call your Android phone with my iPhone, why can't we make this work?" Furthermore, 21st Century Cures charged ONC to try to solve the problem of a lack of standards for when health information exchange should occur and who was entitled to receive health information from a discloser by requiring ONC to develop benchmark contracts and standards on these thorny issues. This effort resulted in ONC's "Trusted Exchange Framework" and "Common Agreement."

This chapter will *not* provide the specific details of this law. Excellent government-approved infographics, explainers, standards lists, and FAQs can be found on ONC's website. URLs are in the appendix.

Figure 13.3 summarizes 2015 to the present.

I am an optimist. I actually think appropriate health information exchange will become easier, less costly from an implementation perspective, and more widespread given once the new ONC rules take effect. Already, we see early signs of the beneficial impact.

For example, through APIs, patients who want to participate in research can acquire their clinical health information and contribute it to research without time-consuming and complex inter-company data agreements (Landi, 2021).

But Why?

So, here we are, 25 years into a federal policy effort to have a learning healthcare system. The Agency for Health Care Research and Quality defines a learning health system as

> a health system in which internal data and experience are systematically integrated with external evidence, and that knowledge is put into practice. As a result, patients

get higher quality, safer, more efficient care, and health care delivery organizations become better places to work.

If we had the level of health information exchange Congress keeps passing laws to obtain, then we might truly have a learning health system.

In fact, in some respects, the speed with which the US healthcare system has been able to respond to COVID, for example, developing the vaccines and rapidly iterating on treatment methods for hospitalized COVID patients, is evidence that we've reached an inflection point in our goal of a learning health system (Middleton, 2021).

But a second important reason why is that the Congressional standard of "no special effort" (21st C. Cures sec 4003) will allow healthcare organizations to move beyond long discussions about technology, what is better, X EHR or Y EHR, leaving those resources free to focus on delivering healthcare and improving outcomes, including saving lives. Like our use of ATMs by the mid-1980s, when they were ubiquitous, we can stop wondering how they work and just use them and get on with our lives.

What Is Next?

Widespread Easier Health Information Exchange between Institutions

In January 2022, the ONC released its Trusted Exchange Framework. The framework answers Congress' charge in 21st Century Cures section 4003(b) that ONC section establish a framework which healthcare stakeholders could "build consensus" (quoting Cures Act) about when was health information exchange not legally permissible and thus not exchanging would be evidence of the prohibited "information blocking." The framework in fact documents the baseline of what is allowed, and prohibited, found in HIPAA. ONC has said it expects to fully implement the framework by the end of 2023.

In conjunction with the framework, ONC published the "Common Agreement." This is a template contract available for adoption that, if widely enough adopted, would remove the idiosyncrasies out of the health information exchange contracting process. It is hoped that the Common Agreement will be adopted by most national exchange organizations and their business partners. For example, if the physicians in your health system participate in Commonwell/Carequality, a major national exchange network, the Common Agreement is expected to impact how that exchange works.

And then there is FHIR itself, of the Fast Healthcare Interoperability Resources. Without getting into the details of how FHIR was developed, it was described to me by one of the original advocates, former US Chief Technology Officer Aneesh Chopra, as a system for naming and structuring the billions of points of data that make health information. In that respect, it is like a library cataloging system, although we are cataloging and identifying individual data points, rather than books of data. As a non-technologist, I have found this analogy helpful, and it helps me understand how the open-sourced card catalog is getting built out.

It is being built with a data specification ONC calls the "U.S Core Data for Interoperability," or USCDI, which is updated frequently with new versions. As an example, and knowing that by the time this is published, we will have another version of USCDI, version 1 was published in July 2020. Version 2 came along a year later, in July 2021, and ONC is already at work on version 3. Readers can find links to details in the appendix. And USCDI is important because it reduces the amount of data curation needed to use the health information to save lives and improve care. Imagine the impact on health system budgets when data it gets from another health institution can be immediately integrated into the receiver's EHR and used, accurately and safely, for patient care or learning. That is the goal for the future, and contrasts sharply to the often human-/manual-driven process used right now to map the way one organization reflects a lab value, for example, blood sugar measure ("A1C"), to another organization for the test that produces the value referred to as ("HbA1c"). That is why Congress and ONC are requiring standards at the data element level for the first time. The ONC's effort to standardize how a person's address is recorded in certified EHRs is a case in point. In that project, ONC seeks to establish standards for capturing a patient's address (Posnak).

Editor's note: As explained in Chapter 15, there is a lot of work to do to get FHIR to a place where use cases are truly meaningful clinically and to support the most complex use cases whereas patients present with comorbidities, and taking into account genetic and other kinds of data.

And finally, the healthcare system will have to master the technology of identity management and authentication. Federal law allows appropriately authenticated electronic signatures for nearly all commerce, thanks to the Electronic Signatures in Global and National Commerce Act (E-Sign Act), signed into law in June 2000. This law actually predates the first HIPAA Privacy rule by 6 months! HHS is clear that any time HIPAA requires that a discloser get an "authorization" from an individual, that authorization can be signed using E-Sign Act standards: "Further, the Privacy Rule allows HIPAA authorizations to be obtained electronically from individuals, provided any electronic signature is valid under applicable law."

Nevertheless, in January 2022, I joined a new health system, one with millions of members and itself a storied early adopter of an EHR. When I contacted the relevant office to arrange for them to get my records from my prior physicians, I was given a PDF that I needed to print out, fill in, sign, and then fax or email back to the relevant office, whereupon they would in fact make an electronic query for my records via exchange technology embedded in their EHR. Needless to say, this system has a robust patient app and a rapid patient credentialing and identity-proofing system, including biometric login to go with it. So, query, if a credentialed login constitutes legal identity proofing under applicable federal standards? I think it does and so does the National Institute of Standards and Technology within the Department of Commerce. That being the case, why did I have to download and sign paper? Why did this system not just ask me to approve the collection of my health history right in the app? (And today we won't explain why under federal law, my consent for this is not necessary anyway, but ONC's Trusted Exchange Framework, approved by the HHS Office for Civil Rights before being published, does a decent job.)

True confessions? Those downloaded forms sat on my desk about six weeks before I sent them in. I am confident that had I been asked about prior health history records once logged into my system's app (and therefore having sufficiently proven my identity to US government standards for two-factor authentication), my new doctors would already have my health history instantly.

Patient-Initiated Data Calls (aka "Patient-Mediated Exchange")

We're not done yet. Yes, the Trusted Exchange Framework, the Common Agreement, the USCDI, and FHIR will change the way healthcare organizations exchange data with one another, following some very intentional policymaking by Congress and ONC. But we cannot forget the other part of Congressional intent, which is that consumers will be able to take advantage of their rights to (1) get a copy of their own PHI and (2) transmit it, or ensure it ends up in the custody of, whomever they want, "without special effort." Per federal regulations, the method of getting a copy of one's own PHI can include, but is not limited to, using an app of the patient's choosing to make a data call on their healthcare provider.[1]

This has been one of the most vocally controversial of ONC's policy efforts, even if it is merely automating a right that patients have had for 22 years. The controversy lies in the fact that once a patient's health data leave the custody of a particular institution, any one of several things could happen next:

1. The patient could use an app that monetizes the patient's health information once it is in the app's hands, and the patient may or may not know about this.
2. The patient could send their PHI to a competing institution and change their delivery system.
3. The patient could send their data to a research project their prior physician does not approve of, or a practitioner, such as a homeopath, that the prior physician does not approve of.

Yet, federal law is clear, none of these are reasons why a disclosing healthcare organization can refuse the patient's request for a copy of their data.

In fact, the conditions under which a provider can block the patient's chosen app from making a data call for the patient's own data are quite limited. There are five circumstances in which the app can be blocked entirely, and three circumstances in which the app can be limited, but not stopped. The below is a high-level summary drawn from ONC infographics and FAQs, and users are cautioned to seek their own legal advice before establishing their own policies and procedures for complying with these federal regulations. Per ONC, a healthcare provider can refuse to allow a patient's chosen app to make a data call:

1. If a federal or state privacy law prohibits disclosures (e.g., where a writing is required to release reproductive health information) or a privacy policy the provider enacted before the info blocking rules went into effect *and* which patients have come to rely upon.
2. When the provider has a reasonable good faith basis to believe that doing so would threaten the security of the *providers' system*. (Note this is not allowed to be about the security of the consumer's chosen tool once the data leaves our system.)
3. When disclosure would actually potentially harm to someone, and your patient worrying about their lab result would be unlikely to count.
4. If the request is technically infeasible, for example, a provider does not have to electronically produce data that is not collected or stored electronically.
5. If the system is closed for or scheduled maintenance and disallows data calls during that time.
   A provider can also establish some procedures:
6. Time place and manner restrictions can be imposed, but they have to apply equally to all app developers.
7. Sometimes, a fee for making the data call can be charged to the app developer (never to the patient).
8. An app developer can be required to register or sign a license to make calls on the provider's system, but the restrictions of such registration or license are quite limited.

Chapter 15 will provide more details.

As you can see, what a patient might do with their own health information is not a reason to not give it to the patient or authorized representative. A provider is also not allowed to block

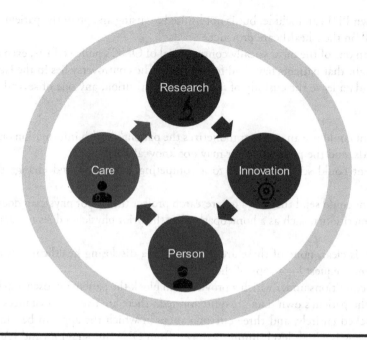

Figure 13.4   Patient needs: research, innovation, person, care. Source: Savage, Omaha Health (2022).

a patient from getting their own data because the patients might use it for something that the provider thinks is medically unsafe. Providers are encouraged to educate their patients, but in the end cannot prevent patients from getting and using their health information as the patient sees fit.

So, it is a brave new world, and maybe, just maybe, we will get where Congress has wanted to go for 25 years: to a learning health system where we (quoting AHRQ):

- Systematically gather and apply evidence in real time to guide care.
- Employ IT methods to share new evidence with clinicians to improve decision-making.
- Promote the inclusion of patients as vital members of the learning team.
- Capture and analyze data and care experiences to improve care.
- Continually assess outcomes (and) refine processes and training to create a feedback cycle for learning and improvement.

And, I would add, centered on the patient(s) and their needs, as shown in Figure 13.4.

Conclusion

We've been on a 25-year trajectory to digitize health information for learning and to save lives. Sometimes, the "learning health system" seems as if it is in reach. Other times, we drown ourselves in unnecessary paper that delays information sharing, like one's health history, that might save a life or improve health. Being optimistic, I like to think the former is more commonplace and the latter is increasingly anomalous. Let's close the chapter with a tale of two transactions for a woman who is 25 weeks pregnant when she must move across country, as shown in Figure 13.5.

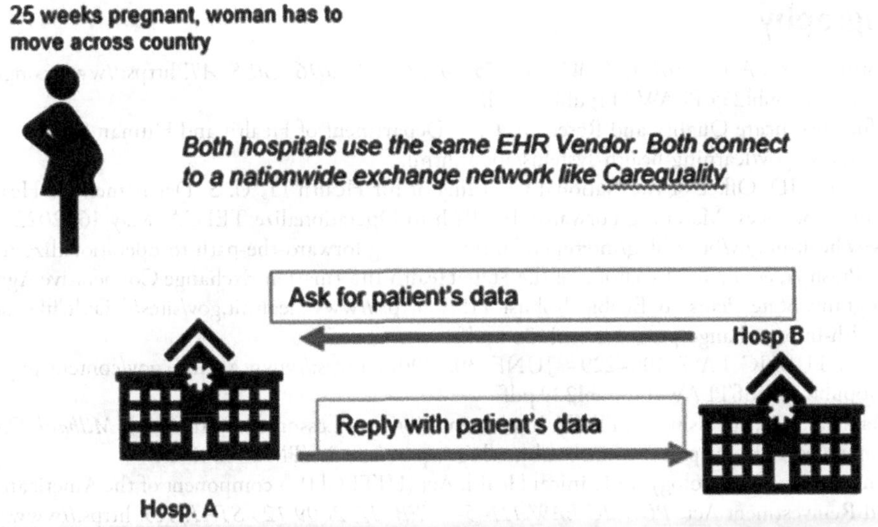

Figure 13.5   Pregnant woman health transaction. Source: Savage, Omaha Health (2022).

Paper-Based Method	*App Data Calls*
Once she finds Hosp. B, the patient asks them to go get her pregnancy history. B gives patient paper forms, but at least sends blanks by email. Patient **prints them, fills them out, signs them**, and **sends them back**. This takes three weeks. Patient is now 28-week pregnant. Because of a work holiday, the ROI offices lose a day of processing when they finally do get the forms. The forms are put in queue. It's busy. Three business days later, within the office's standard turnaround time, they process the forms. This generates a task for another person in the hospital to make a query to their networks, both with the closed network of their EHR vendor and through their exchange network. Records are found and returned, but they need to be cleaned up first, because the fields are not in a standard format. Another week passes. Finally, five weeks after she asked for the forms, the patient's data is integrated and available for her ob/gyn	Once she finds Hosp. B, the patient asks them to go get her pregnancy history from her prior doctor. The receptionist helping her schedule her first appointment issues her a credential for the hospital's patient portal app and says, "you can tell us your doctors names right in the app and we'll go get it for you." The patient finishes her app set up that night and identifies her prior physician. When the patient arrives at her first appointment a week later, her new doctor has already read her health history, and her prior prenatal tests

What's different? In an idealized app data call scenario, information delays are minimized. This means that the risk of harm to the patient from lack of complete medical history also is minimized, not to mention avoiding redundant or harmful care. Which health system would you rather receive or provide care in?

Note

1. The right to a copy of one's own PHI even in an electronic form is found at HITECH section 13004(e) and also at 45 CFR 164.524. ONCs rules for the technical standards developers of certified EHRS must implement, and for what behavior by a healthcare provider constitutes "information blocking" are at 45 CFR part 170 and part 171, respectively (published at 85 Fed Reg 25642 (May 1, 2020).

Bibliography

21st Century Cures Act, *PUBLIC LAW 114–255—DEC. 13, 2016 130 STAT.* https://www.congress.gov/114/plaws/publ255/PLAW-114publ255.pdf

Agency for Healthcare Quality and Research, U. S. Department of Health and Human Services. https://www.ahrq.gov/learning-health-systems/about.html.

Anthony, Elise, JD, Office of the National Coordinator for Health IT, U. S. Department of Health and Human Services, Marching Forward: The Path to Operationalize TEFCA, May 16, 2022. https://www.healthit.gov/buzz-blog/interoperability/marching-forward-the-path-to-operationalize-tefca.

Dullabh, Prashila, et al., The Evolution of the State Health Information Exchange Cooperative Agreement Program: State Plans to Enable Robust HIT. https://www.healthit.gov/sites/default/files/pdf/state-health-info-exchange-program-evolution.pdf.

E-Sign Act, PUBLIC LAW 106–229—JUNE 30, 2000. https://www.govinfo.gov/content/pkg/PLAW-106publ229/pdf/PLAW-106publ229.pdf

Gold, Marsha, et al., Assessing HITECH Implementation and Lessons: 5 Years Later, *Milbank Quarterly*, September, 2016, https://www.ncbi.nlm.nih.gov/pmc/articles/PMC5020152/.

Health Information Technology for Clinical Health Act (HITECH), a component of the American Rescue and Reinvestment Act, *PUBLIC LAW 111–5—FEB. 17, 2009 123 STAT.* 115. https://www.govinfo.gov/content/pkg/PLAW-111publ5/pdf/PLAW-111publ5.pdf.

Landi, Heather, Healthcare Giants Show Feasibility of Cross-Country Data Exchange, FierceHealthIT, March 31, 2021.

Lian, An, PhD, et al., Location Study of ATMS in the U. S. (2014). Available at: https://www.akleg.gov/basis/get_documents.asp?session=31&docid=22687.

Middleton, Blackford, MD, et al., Pandemic Wakes the Learning Health System: How we Developed and Continue Refining COVID-19 Treatment Guidelines, *Health Affairs Forefront*, October 21, 2021. https://www.healthaffairs.org/do/10.1377/forefront.20211019.864563/full/.

National Institute of Standards and Technology, U. S. Department of Commerce, National Institutes of Standards and Technology 800–63B. https://pages.nist.gov/800-63-3/sp800-63b.html.

Office of Human Research Protections, U. S. Department of Health and Human Services, 1974. https://www.hhs.gov/ohrp/regulations-and-policy/belmont-report/read-the-belmont-report/index.html.

Office of the National Coordinator for Health IT, U. S. Department of Health and Human Services, Report to Congress on Info Blocking, 2015. https://www.healthit.gov/sites/default/files/reports/info_blocking_040915.pdf.

Office of the National Coordinator for Health IT, U. S. Department of Health and Human Services, Joint Health Information Advisory Committee API Task Force, 2016. https://www.healthit.gov/sites/default/files/facas/HITJC_APITF_Recommendations.pdf

Office for Civil Rights and the Office of the National Coordinator for Health IT, U. S. Department of Health and Human Services. https://www.healthit.gov/sites/default/files/factsheets/exchange_treatment.pdf.

Office for Civil Rights, U. S. Department of Health and Human Services. https://www.hhs.gov/hipaa/for-professionals/faq/554/how-do-hipaa-authorizations-apply-to-electronic-health-information/index.html.

Office of the National Coordinator for Health IT, U. S. Department of Health and Human Services. https://www.healthit.gov/topic/information-blocking; and https://www.healthit.gov/isa/united-states-core-data-interoperability-uscdi; and https://www.hhs.gov/sites/default/files/combined-onc.pdf, and https://www.healthit.gov/cures/sites/default/files/cures/2020-03/InformationBlockingExceptions.pdf; and ONC FAQ IB FAQ 27.1.2020Nov, found at: https://www.healthit.gov/curesrule/resources/information-blocking-faqs.

Parker, Mitchell, Ransomware is a Leadership Problem, 2019. https://authory.com/MitchParker/Ransomware-is-a-Leadership-Problem-a33f7f92a60144a0cb339c0a33d7d188d.

Posnak, Steve, Principal Deputy, Office of the National Coordinator for Health IT, U. S. Department of Health and Human Services, "Say Yes to Project @US", 2021. https://www.healthit.gov/buzz-blog/health-it/say-hey-to-project-us-a-unified-specification-for-address-in-health-care.

Savage, Lucia, JD, et al., Digital Health Data and Information Sharing: A New Frontier for Health Care Competition? American Bar Association. *Antitrust Law Journal*, 82(2), 2019. available on Research Gate at https://www.researchgate.net/publication/332530889_Digital_Health_Data_and _Information_Sharing_A_New_Frontier_for_Health_Care_Competition#fullTextFileContent.

Schauffler, H., and Rodriguez, T., Exercising Purchasing Power for Preventive Care. *Health Affairs (Millwood)*, 15(1 Spring), 1996: 73–85. http://content.healthaffairs.org/cgi/reprint/15/1/73.pdf

Solove, D. J., HIPAA Turns 10: Analyzing the Past, Present, and Future Impact. *Journal of AHIMA*, 84, 2013: 22–28. GWU Legal Studies Research Paper No. 2013-75, GWU Law School Public Law Research Paper No. 2013-75, https://library.ahima.org/doc?oid=106325#.YolPE5PMKWA.

Chapter 14

Technology and the Engaged Patient: Evidence-Based Patient Engagement for Improved Patient Outcomes

Lucia C. Savage

Contents

Introduction

In this chapter, an activist engaged patient who has an expert-level understanding of what technology can, and cannot, accomplish in strengthening clinician–patient (family) shared decision-making and teamwork for health will summarize three examples of where patient engagement with the assistance of technology has improved patient care, improved health care quality, and either improved the clinical outcome or avoided a worse outcomes. All references here to "patient" should be read to mean the patient and their personal care partners, whether family members or close friends.

I am that activist, engaged patient. And if you read to the end of this brief chapter, you will hear at least one personal story that, in my experience, is likely emblematic.

Let's concede at the outset, though, that the clinical evidence is not definitive that a patient-centric, collaborative approach using technology *will* improve clinical outcomes. For an excellent synopsis of the complexities of research, attitudes, and outcomes, see Ross Baker (2014). Indeed, in research for this chapter, I found more articles about the lack of definitive evidence on this topic than articles I could summarize here. But I believe that *because* of significant improvements in technology's ability to help people collaborate across time, space, and levels of information sophistication, we can replicate the successes of early studies that used technology to connect patients and clinicians. We can also therefore feel confident that as we scale patient engagement to full-scale collaboration for health, we will see more and more beneficial results. And, as a patient, I want to connect with my trusted healthcare professional, not just look something up with "Dr. Google." So, let's dig into a couple of examples and see if I can persuade you.[1]

A brief word about scope: This chapter is *not* about how to organize the pieces of technology to connect clinicians to patients. For that, the whole book, and our prior book, *Mobile Medicine: Overcoming People, Culture and Governance*, should be your guide. Nor is this chapter about ensuring appropriate security and privacy of technology that connects patients and clinicians; see Chapter 8 of *Mobile Medicine* and the resource appendix for that information (Douville, 2022). Further management and risk management for remote care tips are covered in this book's, Chapter 11, Hospital at Home: Managing the Risks of Delivering Acute Care in the Patient's Home

> Rather, the scope of this chapter is limited to providing three examples of successful instances using technology to connect clinicians and patients to improve health outcomes or avoid worse outcomes. It will also illustrate the technological advances we've made over the last nine years.

Example 1: The first example is from 2016. In this study, researchers tested both patient satisfaction with and, more importantly, improved health outcomes from a remote monitoring system for infants in between a series of surgeries when those infants were born with a single heart ventricle. In 2016, the standard of care necessary for at-home monitoring in between a series of surgeries was a scale, a pulse oximeter, and a notebook for parents to record daily health data. Parents then phoned the data into the care team once weekly. The researchers developed

> a solution that was based on remote monitoring and ensured continuous, connected care for patients and their families. Our vision was that instead of a binder, the family would enter data (pulse oximetry, weight, and other information) into a mobile device with near-instant connection to an enterprise database and the ability to perform automated analytics and graded notifications to the care team.

> **(Shirali et al., 2016)**

(This study is also informative on challenges in 2016 that, through policy changes, should be decreasing, from a pulse oximeter manufacturer's refusal to connect their data directly to the study institution's EHR to non-standard transmission specifications, but full discussion of such issues is beyond the scope of this article.)

This study of infant health monitoring at home showed promise of lower infant mortality rates due to real-time information available from home for the care team. And 90% of the families included in the study preferred the electronic tools, even though they had to manually enter the data in a hospital-issued tablet, to a traditional written notebook. If you fast-forward this study to 2020 technology, you might even skip the hospital-issued tablet *and* the manual entry into the tablet; in other words, the technology in 2020 would work more like our second example.

Example 2: The second example occurred in 2019–2020 and was published in 2021. This study tested whether, with the right technology that automatically connected patients' health post-chemotherapy status at home with their oncology care team, significant adverse effects from chemotherapy were ameliorated or avoided (Pavlos Masoul, MD, MPT, et. al, *JAMA* August 30, 2021). While the disease is different from the infants in the first study, the concept is similar: can adverse events be avoided by effective home monitoring based on collaboration between the clinical professionals and the patients?

In this study, patients with genitourinary cancers were asked to use a simple app the research team developed to log particular symptoms of dizziness, nausea and vomiting, and shortness of breath post-chemotherapy. According to the study's authors: "an acceptable and fiscally sound method can be developed to create a dynamic learning system to detect and manage immune-related toxic effects" (id.)

From my patient's view of the result, the study showed that a technology-enabled virtual communications loop could be established so that patients could log potentially dangerous side effects, and clinical professionals would have real-time access to the reported symptoms and were able to intervene and successfully abate symptomology where appropriate.

Contrasting Example 1 to Example 2, let us see the evolution of technology for connecting clinicians and patients over the last five years. While Example 1 did effectively connect the collected data directly to the clinical team's EHR (Figure 14.2), it did not use the patient's own smartphone. That did not occur until Example 2, conducted in 2020. This evolution illustrates how, due to advances in technology and smartphone adoption, there are even more opportunities to have patients and clinicians partner in care for better outcomes. I illustrate this evolution by contrasting Figures 14.1, 14.2, and 14.3.

Figure 14.1 illustrates a likely commonplace data flow circa 2013, when studies hypothesizing about connecting clinicians and patients to improve outcomes first started appearing (Baker et al, supra). In 2013, according to Pew Research (https://www.pewresearch.org/internet/fact-sheet/

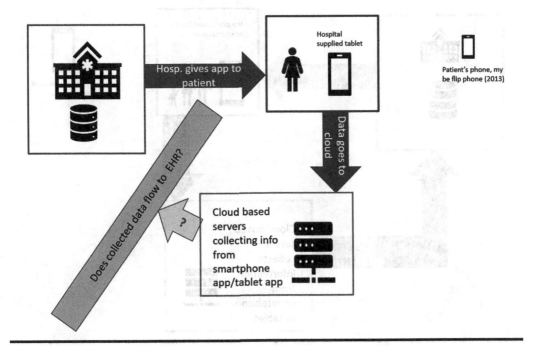

Figure 14.1 Commonplace data flow. Source: Savage, Omaha Health (2022).

mobile/), only 50% of US adults had a smartphone. Further, in 2013, only 59% of general acute care hospitals had even a basic EHR, let alone one that could connect with data inbound from other sources, however legitimate (Office of the National Coordinator for Health IT dashboard, www.https://www.healthit.gov/data/quickstats/non-federal-acute-care-hospital-electronic-health -record-adoption).

What is missing:

1. Reliable, secure connectivity to the hospital's EHR
2. Multiple pieces of equipment for the patient to manage and remember
3. Symantec, open-source data standards for easy communication, computer to computer
4. Complete information loop from clinical team to patient and back again

Figure 14.2 illustrates the infant heart study. In that study, the research team solved for problem 1, but not 2. And, if you read the study in detail, you'll see that the health information technology ecosystem in 2016 had not solved for problem 3 or 4.

Figure 14.2 illustrates the study by Shirali et al. They solved for the connectivity to the EHR, but their study required manual entry of the relevant health information in a hospital-issued tablet versus calling it in once a week from handwritten notes in a notebook. Study 2 solved for the EHR connectivity and work flow problems, but not patient convenience, and problems 3 and 4 remained unsolved across the healthcare ecosystem.

In Figure 14.3, data collection is made easy because it occurs on the patient's smartphone, not another piece of equipment the patient has to remember to use. The utility of the information for the clinical staff occurs because the data collection app integrated fully and seamlessly with their EHR. Figure 14.3 solves for problems 1, 2, and 4.

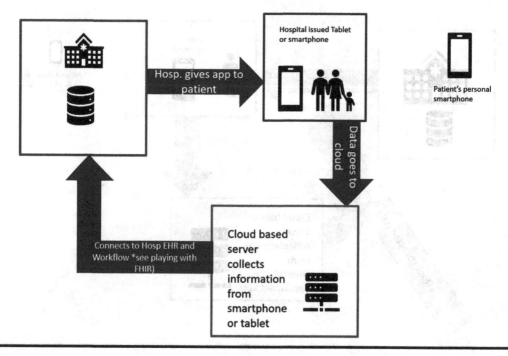

Figure 14.2   Infant heart study. Source: Savage, Omaha Health (2022).

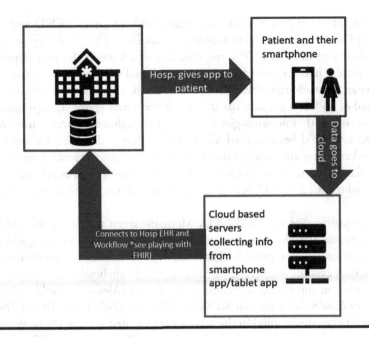

Figure 14.3   Data collection via Smartphone.

The figures illustrate how over the past nine years, technological changes have improved the likelihood that what Don Berwick wrote about in 2009 (Berwick, 2009) might actually become possible.

Here is what has changed: Compared to 51% smartphone usage in 2013, in 2021, Pew Research reports 85% of adults use a smartphone regularly. Compared to 59% of hospitals with a basic EHR in 2013, in 2020, ONC reports that 96% of general acute care hospitals have a full (not just basic) EHR, and that EHR functionality has expanded, especially its required ability to connect "without special effort" to other sources of data (quoting 21st Century Cures Act). Thus, it is only now, with ubiquitous smartphones *and* 96% adoption of full EHRs by general hospitals (ONC, id.), that we have the technology that makes Masoul's study, and its positive results, possible.

Plus, there is a third change that is unfolding that will make replicating the complete information loop of Figure 14.2 possible, and that is the required adoption of application program interface (API) technology using the internationally recognized Health Level 7 standard "fast healthcare interoperability resource" (see Chapters 10 and 15 for details). The FHIR standard, by standardizing the data elements, security engineering, and authentication protocols to enable the APIs of an app developer or a patient and an institution to effectuate health information exchange. This would provide complete information loops to support early intervention in avoiding bad health outcomes, will become easier and a more expected part of healthcare.

But all along, without the supercomputing power of the smartphone and the required connectivity "without special effort," there has been one more way in which the technology underlying mobile medicine has been available to improve care and avoid adverse outcomes. And, no, this is not another story about adverse drug interactions and e-prescribing. Rather, this is a simple story of how shared access to health history helped a caregiver and a hospitalist ultimately reach a better result for the caregiver's elderly mother.

The patient in this story is an 88-year-old woman with advanced COPD and lifelong bipolar disorder. Her COPD treatment plan was made complicated by an allergy to many steroids, which have a history of inducing 50 manic episodes that have in the past required inpatient treatment. The patient had for more than 15 years received care in a fully vertically integrated system that was an early adopter of EHRs. Her care team, all physicians within the integrated system, consisted of a PCP, a psychiatrist (for medication management), a pulmonologist, and once a year, a consult with a nephrologist for helping to manage her medication regimen with her kidney function. She'd been weaned off of lithium due to declining kidney function, and she took some other strong anti-anxiety medications not necessarily just for emotional balance, but sometimes for sleep, on the orders and with the oversight of her psychiatrist. This patient was my mother, who passed in 2017, but whose permission I have to tell stories about her healthcare experience.

Despite high-quality care for her COPD, Mom managed to get a cold which, naturally, turned into pneumonia, for which hospitalization was required. Prior to being admitted, as part of her daily medication regimen prescribed and actively managed by her psychiatrist, Mom took 15 mg of Ativan each evening for sleep. (The science of the beneficial impact of the right kind of sleep on people with mental illness is becoming definitive.) (Alexandra Gold and Louisa Silvia, *The Role of Sleep in Bipolar Disorder*, Nat Sci Sleep. 2016; 8: 207–214. doi: 10.2147/NSS.S85754.) Upon being admitted to the hospital in the vertically integrated system, the hospitalist naturally reviewed medications. The hospitalist prescribed 5 mg of Ativan (not the 15 mg prescribed by Mom's psychiatrist), since that was the standard protocol. Mom insisted she need the higher dose to sleep, but her pleas were ignored. The on-call psychiatrist agreed with the hospitalist's medication decision.

The next day, Mom called me, desperate. (I lived across the country.) She told me she had not slept since being admitted and that they wouldn't let her take her normal sleep medication for sleeping in her own bed, even though now, she was in the hospital with the lights, and the buzzers, and the middle of the night blood-pressure checks. As Mom's care partner, I said I would talk to the doctor.

I did connect by one of the oldest technologies, telephone, to the hospitalist. I pointed out that per the medical record, she and I both had access to the EHR, the inpatient medication was 67% lower than outpatient, although the ability to sleep well in a hospital was far worse. The hospitalist and I argued for 45 minutes. Never once in that time did she offer to check the EHR. Nor did the hospitalist offer to message Mom's regular psychiatrist, even though I begged her to. In frustration, I angrily said that if Mom ended up inpatient in the mental health ward due to lack of sleep, it would be on the hospitalist's shoulders. Finally, she reluctantly agreed to change the prescription back to Mom's outpatient rate, so Mom maybe could sleep. After that change, Mom slept 20 hours. As a result, she regained her emotional equilibrium, which, in turn, may have actually helped the hospital care team do its job. Eventually, Mom was discharged to a SNF then to home.

Why Does This Story Belong in This Chapter?

First, had the hospitalist checked the EHR outpatient medication history and simply messaged her psychiatrist colleague who managed Mom's medication, my phone call to the hospitalist could have been avoided entirely. A core purpose of the EHR is to create an accurate, whole-picture, longitudinal record of the patient's history and current health status, especially for use by a physician seeing the patient for the first time. Here, the EHR was not used that way.

Second, even if the hospitalist did not check the EHR immediately, once Mom insisted, that too could have been confirmed by checking the EHR and checking with Mom's psychiatrist, instead of dismissing the patient's account of their own health history

Third, failing all that, simply listening to the patient's caregiver, who was asking the hospitalist to check the EHR and consult with the regular psychiatrist, would have saved 40 minutes of hospitalist time, and thus hospital resources.

Finally, having actually listened to the caregiver's insistence, at least the hospital avoided a worse health outcome of an inpatient mental illness episode.

The technology was an important part of this story: everyone involved had access to the same EHR records, if only the hospitalist would have looked. But it was not rocket science. As we implement more technology, this story illustrates why common sense and collaboration may not need randomized controlled trials to prove their value.

Healthcare executives today are trying to take full advantage of technology to improve health outcomes, use resources wisely, and deliver high-quality care. In this environment, the very idea of engaging patients and then studying the outcomes may not be at the top of your to-do list. Moreover, anyone would be intimidated by the items in the research literature (Shaller & Darby, 2009). The following are the hallmarks of success of clinician–patient collaboration:

- Visionary leadership
- Dedicated champion
- Partnerships with patients and their care partners
- Effective communication across the organization
- Performance measuring and monitoring

It is no wonder there is more work to be done. We have come a very long way technologically since 2013, and now we have the tools to do that work.

Note

1. With much gratitude to Grace Cordovano, John Wilbanks, and Jan Oldenburg for their help in navigating the diverse research on technology and patient engagement.

Douville, Sherri. *Mobile Medicine: Overcoming People, Culture, and Governance*. New York: Routledge, 2022.

References

Berwick, Don. *What 'Patient-Centered' Should Mean: Confessions Of An Extremist*, Health Affairs Blog, Vol 28, Supp 1. 2009 doi.org/10.1377/hlthaff.28.4.w555

Ross, Baker, G. *Evidence Boost: A Review of Research Highlighting How Patient Engagement Contributes to Improved Care.* https://www.cfhi-fcass.ca/docs/default-source/itr/tools-and-resources/evidenceboost-rossbaker-peimprovedcare-e.pdf

Shaller, D., and Darby, C. . *High Performing Patient and Family Academic Medical Centers: Cross Site Summaries of Six Case Studies.* Picker Institute. 2009

Chapter 15

Interoperability and Information Blocking: How to Enable Data Sharing and Keep HHS Happy

Peter McLaughlin

Contents

HIPAA and Sharing

Any number of mainstream publications and pundits over the years have offered pithy insights as to the value of data to an organization regardless of size or sector. Healthcare is no different. Frankly, patient data at the individual and the aggregate levels could be a significant asset for a healthcare organization. This is not only because more data about an individual should improve one's ability to diagnose and treat the person, but also because more data about similarly situated patients may assist in treatment of the one. Circling back to the business side of healthcare for a moment, the logic continues that when an organization – whether a healthcare provider or a health technology provider – can retain close hold of patient data, that information asset is valuable in part because it is not shared.

DOI: 10.4324/9781003348603-19

This should not be surprising. In fact, the practice of taking conspicuous and inconspicuous steps to retain data as a competitive asset – some would say hoarding – has been publicly acknowledged and criticized (Savage, Gaynor, and Adler-Milstein 2019, 598; "SPEECH: Remarks by Administrator Seema Verma at the ONC Interoperability Forum in Washington, DC | CMS" 2018).

Contrary to this behavior, federal health information policy has for years encouraged the sharing of patient data when appropriate (Savage, Gaynor, and Adler-Milstein 2019, 598). HIPAA – the Health Insurance Portability and Accountability Act of 1996 – and the ensuing Privacy Rule have mandated for two decades now that when a patient requests a copy of their health information from a covered entity, that organization is obliged to deliver, regardless of why the patient has made the request ("45 CFR Part 164 Subpart E – Privacy of Individually Identifiable Health Information" 2022, sec. 164.524).

In 2009, when Congress passed the HITECH Act, one might think they were trying to drive that very same point home – as if the regulation were insufficient. Specifically, the language from HITECH reads:

> In applying section 164.524 of title 45, Code of Federal Regulations, in the case that a covered entity uses or maintains an electronic health record with respect to protected health information of an individual –
>
> (1) the individual shall have a right to obtain from such covered entity a copy of such information in an electronic format and, if the individual chooses, to direct the covered entity to transmit such copy directly to an entity or person designated by the individual …; and
>
> (2) … any fee that the covered entity may impose for providing such individual with a copy of such information (or a summary or explanation of such information) if such copy (or summary or explanation) is in an electronic form shall not be greater than the entity's labor costs in responding to the request for the copy (or summary or explanation).

(Health Information Technology for Clinical Health Act 2009, sec. 13405(e))

Nonetheless, as Savage et al. point out, in 2014 and 2015 data suggested that data sharing was not occurring near the levels HHS anticipated, and that in 2014 only 25% of healthcare providers routinely engaged in the "four dimensions of interoperability – finding, sending, receiving, and integrating data from outside providers" (Savage, Gaynor, and Adler-Milstein 2019, 601). Was it simply because health systems were not requesting data on behalf of their patients? A subsequent survey demonstrated that "60 percent of respondents reported that hospitals and health systems routinely or occasionally engage in information blocking, while 85 percent of respondents reported that EHR vendors do so" (Savage, Gaynor, and Adler-Milstein 2019, 603). Not to belabor the point, in March 2022 the HHS Office of Civil Rights announced four new enforcement actions against healthcare providers failing to honor patients' requests for health information. These four cases brought the number of enforcement cases OCR had won in its HIPAA Right of Access Initiative to 27 over a 30 month period (Office of Civil Rights 2022).

So here we are, after a span of years in which Congress and HHS thought they were providing clarity on patient access to data, health organizations continued to pursue the logical business incentives of maximizing health data and often sharing as little as possible. A cynic might say that those covered entities and business associates (the EHR vendors) were not honoring patient access requests have brought upon themselves the information blocking and interoperability rules discussed in this chapter. Or, in the words of the philosopher Pogo, "we have met the enemy and he is us" (Glynn, 2019).

No More Information Blocking for You

The 21st Century Cures Act, passed in 2016, prohibits practices now categorized by law as information blocking, and this prohibition applies to certain actors, particularly healthcare providers, developers of certified EHR technology, and health information exchanges and networks (21st Century Cures Act, sec. 4004). The statute essentially describes information blocking as business, technical, and organizational practices that either prevent or significantly discourage the access, exchange, or use of electronic health information when the actor knows or in some cases should know that these practices are likely to interfere with a request from or on behalf of a patient. And while the initial rollout of the final rules from ONC and CMS applied to a subset of electronic protected health information, that period has expired, so the mandate against information blocking behavior applies to the entirety of a patient's electronic health information (as a practical matter, their ePHI or their designated record set).

What might information blocking look like? Perhaps a hospital configures its EHR to block sending referrals and electronic health information to an unaffiliated provider or system? Remember that blocking refers to practices, and thus not simply "technology" hurdles. If a patient must provide to staff a written consent before that organization shares electronic health information with an unaffiliated provider or system, that could well be seen as interfering with the patient's right to move a copy of such data. Or your contract with a business associate EHR vendor prohibits their enabling access by anyone outside your organization. And if your health IT vendor fails to configure your system or discourages inclusion of data integration capabilities, both vendor and provider will likely violate the prohibition. Finally, simply the passage of time – taking a week to facilitate access for an unaffiliated provider when same-day capability exists – will likely be considered an information blocking practice.

Like the child in the sandbox who has been told to share toys and play nicely, the list of things not to do will seem quite long. This is particularly so when – as emphasized throughout this book – the use of technology involves not only hardware and software, but changes to operational practices, and as often changes in culture. It is not easy, but HHS has proposed one million reasons *per violation* as an incentive to get it right.

Interoperability: The First Step

While the interoperability rules from ONC (the Office of the National Coordinator for Health Information Technology) focus on the health IT developer audience, the Centers for Medicare & Medicaid Services (CMS) final interoperability rules direct how participating providers and payers are to validate functionality from their current and future certified EHR vendors and to configure those certified EHRs properly.

The technology foundation for interoperability is the extensive use of application programming interfaces or APIs, which are most easily understood as a protocol or process for two different applications to communicate with each other using requests and responses. The technical standards will evolve but begin with the HL7 FHIR (release 4.0.1) standard for healthcare data exchange. (For an ever evolving in-depth discussion of the Fast Health Interoperability Resources specification and real-world application in your context, follow our book editor Sherri Douville's blog on the subject titled "How Data is Failing Medicine Today & When Data Can Become Actually Useful".) The blog covers the elephant in the room and the fact that FHIR is still lacking relative maturity and coverage for data elements specific to complex patients with comorbidities for which care coordination through leveraging API's would be most economically useful for all stakeholders. In her blog, you can track practical usefulness yourself through our update summaries on the API that we maintain with industry leaders.

CMS also includes several resources on its interoperability webpages on FHIR (for the communications), OpenID Connect (as an identity layer), and OAuth 2.0 (as an authentication service), among others.

In theory, a healthcare provider or system leveraging certified EHR technology will be able to rely significantly on that business partner to make all the proper interoperability requirements happen – or at least available to you. Beyond the requirement that initial and renewal certifications must conform to the interoperability criteria, developers of certified EHR technology must provide semi-annual attestations to HHS that their products and services conform. The value of this from the perspective of HHS is that a certified EHR vendor that does not submit its statements will draw scrutiny but also that the failure to submit a truthful, accurate statement would be grounds for additional enforcement action by HHS. HHS will most likely take a more cooperative than prosecutorial stand initially, as it seeks to encourage the desired behavior, but that cannot be expected to last indefinitely. A healthcare system would want assurances from any such vendor that they have and will maintain the certified status of their products and submitted required attestations to HHS.

Practices, Behavior, and Cultural Change

Technology is just one aspect – albeit an important one – in the ways healthcare systems make patient information available. HHS recognizes that there are myriad ways in which a provider might undermine or discourage patients and their representatives from requesting PHI. Operational practices, individual behavior, and the culture of an organization habituated to not sharing patient data are critical to address.

Providers should begin with a thorough review of the organization's policies that currently address requests for access, exchange, or the use of patient information. This will be especially important for requests from patients themselves and any non-clinical caregivers. These policies are not those expressly HIPAA related but will include the broader policies governing confidential information. One of these non-HIPAA policies to examine is how the organization assesses whether a decision is reasonable or not. Recall that healthcare providers will not be engaged in information blocking unless they know that the action is unreasonable and likely to interfere with or discourage access or use of patient information. It will benefit an organization if, in defending a decision not to share information, there is a policy in effect delineating what the standard for reasonableness is. Without such a policy, the organization will have to explain without the benefit of a standard why actions that appear to some to be unreasonable are indeed not. When tackling development of a reasonableness standard specific to data sharing, it will be most helpful to include FAQs and scenarios that reflect real operations. Any procedures bolstering the reasonableness standard might include detailed workflows so that each instance of non-sharing can be justified and documented.

Healthcare organizations should also identify and document where patient data is stored. Certainly, much of a provider's ePHI will be in an EHR, remember that legacy systems, picture archiving and communication systems, practice management technologies, and diagnostic tools may also hold patient information subject to a request. Hence interoperability means the ability to exchange patient information not only with a requestor but also among the organization's own systems and those of its service providers. Understanding where patient data resides is not only a pillar of complying with the HIPAA Security Rule but also simply good practice in managing information whether for the patient or as an asset in and of itself.

After a reassessment of policies that impact the sharing of patient information, it is important to drill into workflows and detailed procedures. The information blocking prohibition considers any action that increases the complexity or the burden of obtaining patient records as improper. Something as simple as failing to prioritize request responses will be under scrutiny. Similarly, an act that results in fraud, waste, or abuse fall within the gambit of information blocking. Somewhat curiously, HHS also considers actions impeding innovation and advancements in health information exchange to be information blocking. For example, deciding not to upgrade to a system that makes data sharing easier than the prior version could be questioned. The reasonableness standard would come into play to demonstrate why the failure to upgrade was in fact a reasonable choice notwithstanding presumed benefits to the organization and its patients. In configuring any system, do not overlook any defaults or options, as the decision not to implement certain data sharing functionality will be scrutinized.

A potentially heavy lift in addition to the policy and practice review will be a systematic assessment of potential information blocking provisions within existing contracts, license terms, health information exchange sharing policies, and confidentiality agreements. While an organization certainly must oblige its vendors and data recipients to protect the information at issue, reviewers should read these agreements from the perspective of a skeptic – that is, someone looking for anything that might be interpreted as interfering with the ability to transfer patient data to a proper requestor.

The list of prohibited practices in the information blocking rule is long, and organizations will be held accountable even if there is no harm to the patient. However, there are also several exceptions that a healthcare organization must incorporate into systems, policies, and practices.

An organization need not honor a request if doing so prevents harm to a patient or another person. For example, data that is inaccurate or erroneous or has been corrupted need not be shared. Similarly, if it is simply infeasible to make the data requested available, then the organization will not be deemed to be blocking, if certain conditions are met. There may be technological reasons covered in detail in Chapter 10 or legal grounds preventing an otherwise cooperating organization from fulfilling a request for information. Consider the existence of legacy systems within the organization that function on a stand-alone basis, without connectivity to an EHR. Given the cost of systems, basic financial resources may limit an organization's ability to honor a request in whole or in part.

Considering the importance of privacy and security within healthcare generally and health IT, it is not surprising that two additional exceptions consist of privacy and security concerns. If fulfilling the information request would undermine the privacy of the relevant patient or would risk the security of the electronic health information, such actions will not be deemed information blocking. Each of these exceptions would be difficult to uphold when a request comes directly from the patient, but they could be relevant in third-party requests. The HIPAA Privacy Rule itself provides exceptions from sharing electronic health information, such as information provided by a third party subject to a confidentiality promise or the obligation to respect a patient's request that information not be shared. The security exception will be challenging for many organizations to apply. HHS has stated that an organization is not responsible for the security of whatever information system, mobile app, or other technology designated by the requestor as receiving the requested information while concurrently holding that a practice directly and genuinely related to safeguarding ePHI will not be information blocking. As with other exceptions, the organization receiving the request must assess on a case-by-case basis whether a particular exception applies; it is not enough to state generally that sharing data with consumer-grade mobile applications constitutes a security risk and thus a basis for not fulfilling the request.

Conclusion

When HIPAA was enacted in the mid-1990s, part of the rationale was for individuals to be able to retain certain benefits when switching employers. And from the earliest drafts of the Privacy Rule, HHS has stated that providing patients with a copy of their medical records was a fundamental component of portability. Regardless of the rules, however, many organizations have hesitated or simply refused to make PHI available when an appropriate request is made. In the Cures Act, Congress expressed its frustration with the lack of data sharing and required developers of healthcare technology and their client healthcare systems to pivot from data hoarding to data sharing.

Beyond the technical aspects of interoperability, healthcare organizations will need to adapt policies, practices, and sometimes organizational culture to meet this most recent prescription of data accessibility. Most actions that slow or hinder the delivery of data to a requesting patient or third party will now be illegal. Many of these are likely the result of innocent choices rather than a conscious decision not to honor a data request. But if an organization knows that its behavior or its decisions probably interfere with such requests, HHS will begin serious enforcement.

Practical Next Steps

On LinkedIn, respected serial CIO David Finn said, referring to an industry conference VIVE,

> Your comment, Sherri Douville, "there's a ton of complexity and nuance" is unintentionally understated. I'm afraid many are not prepared to make some of these decisions. Thank you and James Brady, PhD, CHCIO, CISM, CRISC, CISSP, FHIMSS for bringing some simple, rational approaches to the complexity and nuance.

We had the opportunity to discuss this ill-understood topic with Vimala Devassy and James Brady, PhD, CHCIO, CISM, CRISC, CISSP, FHIMSS, at the event, VIVE. Our advice for healthcare executives' needs to respond to regulators and penalties for it has to acknowledge a ton of complexity and nuance. Though as a blunt instrument for brevity, let's say there are "two sides." The data gets exchanged as the "bridge" earlier in the chapter mentioned and called an application programmable interface (API).

Side 1 says: Make all the patient data free upon patient's request, quality of requestor, privacy, cybersecurity be damned!
> Side 2 says: We aren't just going to let "anyone" connect to our mission-critical EHR? This is nuts!

The table stake as we have described in this chapter is that you have to release the data; it's the law.

It is difficult that there's little quality control for consumer facing apps. You can mitigate that by instituting a disclaimer that you essentially "wash your hands" of the consumer-grade potentially insecure application that the patient is requesting access to. Health system executives should seek sophisticated counsel for a disclaimer example.

1) Plan to peg your program to your business model. There are three hypothetical manifestations of common iterations of current business models.
2) Mind the maturation of FHIR for specific data elements. Because most of the health system costs stem from complex patient types whose data elements are years, potentially decades from going live in FHIR; true data liquidity and interoperability for health data will be limited. Choose your FHIR use cases accordingly (Douville, 2019). Read her cited blog for practical and fine details which are continually updated.

Scenario A: A wealthy academic center that has a nationally leading informatics center of distinction. Organizational objectives are beyond a basic financial business model and extend to include research and teaching plus clinical care; therefore, they may already have started on a roadmap to fully managing patients' requests through a Software as a Service, SaaS product solution management paradigm. This would include the building of capabilities for managing software applications development lifecycle risks. Outside counsel will have undoubtedly reviewed related policies.
Scenario B: A competent county health system with an Accountable Care Organization (ACO) risk contracts. In this scenario, they are motivated to experiment with their own FHIR apps. This would mean that they possess capabilities for managing software applications development lifecycle risks.
Scenario C: Laggard system within a region that is currently and squarely stuck in a fee for service model that will grudgingly release the API with a disclaimer. This profile of health system will likely lack depth of capabilities for managing software applications development lifecycle, specifically enterprise-grade risks. If you are a health system executive in this scenario and you need an example of a good disclaimer, you want a reference to excellent outside counsel.

We hope this chapter helps you to understand the importance of addressing information blocking and helps you to recognize where your system is. Further, we hope you can now see how you might approach your info blocking roadmap and strategy based on your system's business model and technical capabilities.

References

21st Century Cures Act, Pub. L. No. 114–255, 130 Stat. 1033 (2016).

Douville, Sherri. "How Data is Failing Medicine Today & When Data Can Become Actually Useful," (August 19, 2019). https://sherridouville.medium.com/how-data-is-failing-healthcare-today-when -data-become-actually-useful-e8e91325e431.

"Final Rule: 21st Century Cures Act: Interoperability, Information Blocking, and the ONC Health IT Certification Program," *Federal Register* 85, 25642 (May 1, 2020) https://www.federalregister.gov /documents/2020/05/01/2020-07419/21st-century-cures-act-interoperability-information-blocking -and-the-onc-health-it-certification.

HHS Office of Civil Rights, "Four HIPAA Enforcement Actions Hold Healthcare Providers Accountable with Compliance," March 28, 2022. https://www.hhs.gov/about/news/2022/03/28/four-hipaa -enforcement-actions-hold-healthcare-providers-accountable-with-compliance.html.

"Privacy of Individually Identifiable Health Information," 45 CFR Part 164, Subpart E (2021) https://www .ecfr.gov/current/title-45/subtitle-A/subchapter-C/part-164/subpart-E

Savage, Lucia, Martin Gaynor, and Julia Adler-Milstein, "Digital Health Data and Information Sharing: A New Frontier for Health Care Competition?" *Antitrust Law Journal* 82, 529–621 (2019). https:// www.researchgate.net/publication/332530889_Digital_Health_Data_and_Information_Sharing _A_New_Frontier_for_Health_Care_Competition.

U.S Department of Health and Human Services, "Remarks by Administrator Seema Verma at the ONC Interoperability Forum in Washington, DC," August 6, 2018. https://www.cms.gov/newsroom/press -releases/speech-remarks-administrator-seema-verma-onc-interoperability-forum-washington-dc.

Wilson, Glynn. "We Have Met the Enemy and He is Us," *The New American Journal* (July 3, 2019). https:// www.newamericanjournal.net/2019/07/we-have-met-the-enemy-and-he-is-us/.

Chapter 16

The Role of Standards in the Responsibility and Rewards of Medical Technology Industry Leadership: Identify Your Objectives to Build Your Path in Leading the Future of Medical Technology

William C. Harding, Ken Fuchs, Mitch Parker, and David Rotenberg

Contents

DOI: 10.4324/9781003348603-20

Introduction

In medical software, you must understand how regulators perceive the field today and how they currently plan to regulate, for example, software as a medical device. The coauthors of this chapter have experience in uniquely helping to build and lead a working group body as technical and industry leaders, thereby bringing a perspective of practical and applied understanding around the standards process that underpins regulatory processes. The problem right now is that multiple segments of the tech sector are uninformed related to regulatory processes as well as associated standards. However, in order to lead in medical software, there must exist the capability to convene and drive the evolution of an ecosystem of informed contributing partners. When this is done well, progress and the introduction and successful adoption of advanced technologies in medicine can be realized. The coauthors of this chapter will explore the current successful advancement of a technical standards working group as a model.

Industry Leadership – What Do You Need to Know about Industry Leadership and Why?

The future of medical software is likely going to align with the international medical device regulators forum, IMDRF, which manifests through a working group model that is shown to be the type of paradigm required to reconcile multiple constituents into consensus agreement on the future direction of industry. If you are going to be effective in medical software, you have to know how regulators perceive the fields from the perspective of both regulating and planning.

Why Software in Medical Technology Needs Standards

The case of the Airbus A380, while not a medical case, shows the importance of standards. Airbus formed a consortium of multiple European companies to build the A380 superjumbo aircraft. However, the wiring and wiring harnesses for the aircraft were too short. In the case of an aircraft, specifically one with a sophisticated fly-by-wire system like the A380, it involves replacing miles of wires (University of British Columbia, 2022). It also required over a thousand engineers camping out to fix it. The root cause ended up being that German and Spanish engineers used an older version of CATIA design software, version 4, to design their parts, while French and English engineers used version 5. The incompatibilities between the two versions led to all the consortium members thinking they were working well together until the final product was assembled and clearly did not.

There was a lack of standards between the organizations, which led to a US$6.1 billion cost overrun and numerous delays leading to canceled customer orders. This also led to significant public relations issues for Airbus.

Healthcare has a similar situation. There are numerous vendors and applications available to help clinicians diagnose, monitor, and treat patients' conditions. However, each of these solutions is its own stack of software and hardware. Complex interfaces need to be built between these applications and electronic health records (EHRs). A significant amount of effort in healthcare is spent on just getting systems to talk with each other to interchange data and place the data in one system of record, the EHR.

One area where standards have succeeded is Radiology/Diagnostic Imaging. The Digital Imaging and Communication in Medicine, better known as DICOM, a standard and protocol, has enabled the rapid digitization of diagnostic imaging and replaced film (MITA, 2022). Numerous imaging products from different vendors follow the DICOM standards to interchange and share images. This allows them to have one Picture Archiving and Communication System (PACS) for numerous image types (PeekMed, 2022).

While diagnostic imaging has been standardized, other areas of medicine have not. It is often very difficult to get non-Radiology systems to interoperate. This means that either the EHR vendors, providers themselves, or third parties must do the hard work of stitching everything together to present the data to care providers so that they can make informed decisions. Please refer to this book's interoperability chapters for references to resources that characterize cybersecurity requirements and internal organizational policy requirements and challenges to interoperability.

In the Great Baltimore Fire of 1904, which burned for more than a day and destroyed 1,500 buildings, the fire spread quickly and overwhelmed the city's ability to fight it alone. Fire companies from New York, Philadelphia, Wilmington, Harrisburg, and elsewhere swiftly rushed in to help. They had more than enough water and people to fight the fire, but there was a problem: most of their fire hoses wouldn't fit the hydrants, prolonging the fire to 30 hours and damaging an area the size of 70 blocks in the city's business district (Stein, 2022).

There is a similar analog to the fire hose connector in medical technology – the Luer lock connector system defined by ISO 80369-7 (ISO 2021). This connector system is the standard way of attaching syringes, catheters, hub needles, IV tubes, and other applications whenever medical fluid connections need to be made. Imagine the chaos if different manufacturers of infusion fluid bags used different connectors.

While the need for standards for physical things is well accepted, standards for medical software technology should be just as obvious. You can search online for the list of standards consensus organizations recognized by the FDA and are therefore legitimate in healthcare (i.e., "Recognized Consensus Standards"). In many cases, medical software directly replaces functionality implemented in hardware and therefore should meet the same performance and safety standards. The development of the software should follow rigorous and well-documented processes which also meet regulatory requirements. Software applications or mobile apps which communicate with each other or with cyber-physical medical devices should follow standards-based protocols to ease the task of systems integration and interoperability. Similarly, these software-based technologies are vulnerable to various types of cyberattacks and need to be analyzed for potential threats with appropriate mitigations implemented. This is another area where standards can provide guidance and/or requirements.

How Standards Inform Regulatory Processes

The development of medical devices is a world governed by standards. The use of these standards is driven by the need for the medical devices to be cleared or approved for use in the countries in which they will be marketed. Standards are typically developed by groups of "experts" in the field coming to a consensus on the current state of the art. They are developed under the auspices of Standards Development Organizations (SDOs) such as ISO, IEC, UL, AAMI, IEEE, etc. typically for global use. The SDOs make sure that proper processes are followed and protect the participants, who in many cases are competing manufacturers, from antitrust issues.

Standards tend to be classified as either process standards or technical standards. Process standards tend to specify requirements related to how a product is developed or how it is manufactured. An example would be ISO 14971 which is a safety-related standard that specifies how to analyze potential safety issues related to a product and make sure that they are mitigated to an acceptably low level.

Another process standard example would be IEC 62304 – Software Life Cycle Processes, which defines a software design development process with a particular focus on SaMD (Software as a Medical Device) and SiMD (Software in a Medical Device). This standard defines development steps such as System Requirements, Software Requirements, Software Design, and Software Coding as well as associated testing and traceability steps such as Code Reviews, Unit Tests, Verification Tests, and Design Validation. This approach is typically referred to as the V-Model of system development.

Technical standards are more prescriptive. They can range from something as simple as grading eggs by size (what is a large egg versus an extra-large egg) to providing technical requirements for how much RF energy that a device is able to radiate, different levels of water resistance, how high can you drop the device, medical device alarm behavior, minimum clinical performance requirements for a pulse oximeter, etc.

For example, in order to receive a European CE Mark, the device must show that it complies with an appropriate list of standards that have been mutually agreed upon by the manufacturer with their chosen Notified Body. The Notified Body will inspect the documentation prepared by the manufacturer which shows compliance with the chosen standards. In most cases, the standards come from international standards bodies such as IEC, ISO, or could be Europe-specific marked as EN standards like in the case of KN95 masks approved to be marketed in Europe. To earn CE mark, they must conform to American Society for Testing and Materials (ASTM) standards, and associated performance by three designated ASTM levels and related measurements tell the consumer that that specific mask with a CE mark is validated in its capability to protect against a specified percentage and size of infectious particles, for example. This can be contrasted with an N95 respirator, which must meet a parallel process of technical standards and regulation, the National Institute for Occupational Safety and Health (NIOSH) standard. The latter hinges on again specific tests to be performed and validated and outlined in the code of federal regulations; this specific standard means that the N95 mask filters out at least 95% of particles as small as 0.3 micrometers in diameter.

In the United States, we have similar processes as the EU; however, the government in the form of the FDA takes a more active role. In addition, unlike the EU, standards compliance is voluntary in the United States rather than compulsory. The FDA maintains a database of Recognized Standards as well as a long list of over 2,600 Guidance Documents (FDA 2022a) (FDA 2022b).. These documents cover a wide range of topics, including cybersecurity and interoperability. Compliance with the relevant guidance documents is close to compulsory as they are treated as guiding documents by FDA reviewers. In terms of the recognized standards, some are more "recognized" than others. For example, following the previously mentioned ISO 14971 standard is very strongly recommended for any medical device, while there are many recognized standards that can be treated as optional. In all cases, the manufacturer can follow their own approach, but the burden is on the manufacturer that their approach is equivalent to or better than the appropriate recognized standard.

This process is essentially the same at a high level whether you are developing a simple band-aid or a complex ventilator or MRI machine. The difference comes into play in terms of the rigor of the processes that need to be followed and the specific standards that need to be conformed to be based on the type of medical device being developed. While this discussion has focused on the EU and US regulatory relationships with standards, other countries such as Canada, China, and Brazil have their own approaches, but they all rely heavily on published standards.

Why Many Stakeholders in the Tech Sector Don't Know about Standards (Using Food and Other Everyday Analogies)

In highly regulated industries, most stakeholders understand why standards need to be established and followed. Those regulated industries also realize that through standards, they can reduce duplication, create transferable solutions, reduce human errors, and reduce costs through methods and practices that drive down risks. Additionally, standards help ensure that the wheel is not constantly reinvented, and best practices are followed. That said, when thinking about the value of standards, we should consider how safe our car is, how reliable our pacemaker might be, and should I eat a fast-food burger?

Highly Regulated Industries

- Petroleum, Oil, Gas, and Coal
- Electric Power
- Vehicle Manufacturing
- Credit Intermediation
- Air and Water Transportation
- Pharmaceutical and Medical Manufacturing

With consideration for unregulated industries, one must wonder why there is a lack of knowledge regarding the existence and need for standards. Of course, many individuals might come up with reasons for not creating and following standards, where, for example, the art world could make a solid argument against standards. However, if an artist wishes to use reliable paint and medium, it is hoped that the art suppliers have followed at least some level of standards (i.e., material compositions).

Regulated Industries

- Consulting Services
- Pet Services
- Entertainment
- Babysitting Service
- Facilities Support Services
- Office Administrative Service
- Legal Services
- Publishing
- Real Estate
- Door-to-Door Solicitation

So, when we think of the positives associated with standards, why do some businesses not know about standards or want to apply them? A quick answer is that you need additional resources, time, and money to implement and follow standards. You also must submit to inspections, product testing, and approval from an applicable regulatory body. Although it might appear that applying and following standards are not good for business, that is an attitude that we would take only if we didn't want to sell much product specifically in a regulated industry. For example, though most healthcare patients might not know about standards, we can't imagine many people electing to have an unregulated medical device implanted if that device wasn't covered by insurance. And since the FDA regulates the medical device manufacturing industry and insurance providers will not cover non-FDA approved devices, we understand how standards are linked to the implanted device.

It is probably not necessary for everyone to be aware of the finer details associated with standards, but it is hoped that those organizations and individuals administering healthcare therapies are aware of and follow standards. Of course, as we continue to embrace a more mobile and even virtual existence, the ignorance of standards is coming to an end. For example, during our experience with COVID-19, many of us have sought to use telepresence and personal devices to monitor and manage our personal healthcare. And with that point in mind, we have evolved to a level where we must rely on and have confidence in our personal healthcare devices. Consequently, if we continue to depend on those devices, then do we really want to use substandard or unreliable technology that has been built without consideration for standards?

All things considered, awareness of standards may never reach fever levels across the public and some industries may choose not to apply or follow standards, but we hope that you are not one of those individuals if your goal is to work in medicine and we hope that you remember one word when thinking about the need for standards. That word is "risks" and it should be reflected upon, as you proceed to run a successful business. Specifically, even if you are not part of the regulated industry, standards must be considered if you hope to establish product manufacturing repeatability and transference of knowledge. And, unless we are in a specialized creative service (e.g., making Origami swans, one-of-a-kind painting, or jewelry) and we never make the same product twice, then standards must be defined, implemented, and followed.

The Value of Convening an Ecosystem

Once a standard is published, it would be nice to imagine that ecosystem stakeholders readily adopt them, and consumer stakeholders start to demand them. Unfortunately, this is not what normally happens. In fact, many standards sit on the shelf and are never implemented, while others are implemented but never see commercial success. Standards, like most products, must fill a need in order to see wide adoption. In addition, it is possible for multiple standards to exist that fill a similar need. While there is some cooperation among SDOs, they are not in the business of picking winners or losers; so it is always possible that competing standards may get developed. These standards may "compete" with other standards and may also compete with proprietary vendor solutions.

One example of a proliferation of similar solutions targeting the same general applications area (such as short-range, low-power wireless communication) is technologies such as Bluetooth, Zigbee, and 6LoWPAN, which are all based on IEEE 802.15.4 but are not compatible, NFC (Near Field Communication) based on ISO/IEC 18000-3, and Z-Wave which was not developed by an SDO but has an open specification.

One thing that is common among these technologies is that they also have strong consortia/ecosystems behind them to aid in their commercialization and further development. For example:

- Bluetooth Special Interest Group
- Zigbee Alliance
- Z-Wave Alliance
- NFC Forum

Besides commercial promotion, these consortia may influence the future direction of the standard, develop special profiles for specific vertical markets, and provide verification testing facilities to assure interoperability among implementations from different manufacturers. Solutions that pass verification and conformity testing can potentially be "certified" and labeled as such which provides a certain level of customer confidence. For example, in the case of a Bluetooth solution, it is about providing assurance that your Bluetooth headset will work with various phones or personal computers that support the Bluetooth protocol.

Value of Innovation and Partnering with Entrepreneurs Whether They Are of Tech Companies or a Founder of a Standards Working Group

The value in entrepreneurship comes from working with visionary leaders who have discovered innovative ways outside of existing corporate structures and obstacles to solve real problems affecting patients. Oftentimes, the structures of health delivery organizations, or lack thereof, inhibit innovation. Innovation advances the tripartite mission of clinical care, research, and education (ACC 2021). It's not a new concept, but rather an extension of the organization's existing mission.

These solutions need to have a connection back to the existing structures to be most effective and to fit into existing workflows. We must mind our P's and C's (explained in the Security Frameworks as Foil Chapter 9) and integrate these innovations into the existing subcultures and systems in a positively disruptive way that fulfills the tripartite mission. However, the major challenge is that the requirements for developing new and innovative healthcare products that meet requirements have been unclear. The Certified Health IT program requires organizations to undergo a complex process for certification and controls testing. However, it does not require organizations to undergo continual security testing. This requires partnership with entrepreneurs so that they can better understand what healthcare delivery organizations really need. An idea and product are only as good as the feedback that can be used to accomplish the tripartite mission of academic healthcare. While organizations will often discuss the quadruple aim of reducing costs, population health, healthcare team well-being, and patient experience, those are performance-oriented system goals, not mission-oriented. With the experience with electronic health records and the high amounts of physician dissatisfaction with the show, it's important to focus on giving caregivers what they need to use innovative products while facilitating integration at the workflow and technical levels.

Case Study of How a Company Successfully Leverages the Confidence of Standards

One of this chapter's coauthors, David Rotenberg, works for Telefire Ltd., an organization that is a veteran Israeli leader in the field of smart buildings management, fire protection, and safety. Telefire has been established for more than 40 years, with tens of thousands of successful installations, both in Israel and worldwide.

Telefire's product uses cutting-edge technologies to make the modern building safe, efficient, and smart. Telefire has dozens of hardware and software products, including smart fire sensors, intelligent safety systems, IoT devices, and cloud computing platforms. In addition, Telefire's system has a very high-level focus on cybersecurity, making it a choice for many building managers, owners, or security officers. Telefire is certified per the ISO 27001 standard and implements best practices to establish, implement, operate, monitor, review, maintain, and continually improve information security.

UL 2900 is a series of standards published by UL (formerly, AKA Underwriters Laboratories), a global safety consulting and certification company. The standards present general software cybersecurity requirements for network-connectable products (UL 2900-1), as well as requirements specifically for medical and healthcare systems (UL 2900-2-1), industrial control systems (UL 2900-2-2), and security and life safety signaling systems (UL 2900-2-3).

Today's world is getting more and more connected. Things and systems that used to function as stand-alone entities are now connected to the cloud and/or to each other. The Internet of Things (IoT) has or will have a major impact on every industry and on everyday life around the globe. Telefire's systems, being mission-critical and life safety essential, are installed in a huge variety of locations – from shopping malls and schools to office buildings and military sites. The systems are designed to be extremely reliable and are certified per Underwriter Labs, UL and European Standards, EN fire safety standards; however, those standards don't address cybersecurity-related issues.

NIST defines cybersecurity as

> prevention of damage to, protection of, and restoration of computers, electronic communications systems, electronic communications services, wire communication, and electronic communication, including information contained therein, to ensure its availability, integrity, authentication, confidentiality, and nonrepudiation.

(NIST SP 1800-10B under Cybersecurity from CNSSI 4009-2015, NSPD-54/ HSPD-23)

In Telefire's case, cybersecurity has an even deeper meaning, as we need to protect not only data but actual human lives which is directly applicable to medicine. Though Telefire's products already had a high cybersecurity level, Telefire decided to work with the UL 2900 team and obtain official certification per the UL 2900-2-3 standard.

David's team started the process by studying the standard and figuring out which requirements can be applied to their products and how. This process alone gave great insights, since even though there was already a good cybersecurity level, it was the first time that the team had systematically reviewed and mapped all the systems from a cybersecurity POV and pointed out possible issues that might require modifications. Later came the preparation of the submission file which was submitted to UL for a preliminary review and gap analysis. Once again, this process turned out to be very productive – UL's report confirmed that our systems are indeed protected enough to comply with the standard, aside from a few minor adjustments that needed to be done before submission. In addition, we have updated our risk management plan to include more cybersecurity risks and mitigations.

Though there is no such thing as 100% security, having our products certified per UL 2900 gave the company more proof that the products are leading the way in any cyber-related matters.

In addition, it provides customers with confidence, allowing them to use our systems without applying additional security features.[1]

There is a knowledge gap between the average user and the need for cybersecurity which is why these standards are critical. This gap might be at a technical level, but an even bigger issue is the very understanding that they must implement cybersecurity in today's ever-changing and evolving connected world. We hope that having an industry leader like Telefire working on UL 2900 will inspire other companies and organizations to consider implementing this and other similar standards to further ensure their cybersecurity.

Note

1. Obviously, Telefire is not responsible in any way for the customer's cybersecurity measures and management. The products are incorporated into the customer's IT network and it is the customer's sole responsibility to make sure that they are safe and properly maintained. It is allowed and recommended for the customer to use internal or external measures and experts to ensure the safety of their systems and devices.

References

ACC. (2021). "Health Care Innovation | Making the Case: Why Academic Medical Centers Should Invest in Clinical Innovation." American College of Cardiology (ACC), December 8, 2021. https://www.acc.org/Latest-in-Cardiology/Articles/2021/12/01/01/42/Making-the-Case-Why-Academic-Medical-Centers-Should-Invest-in-Clinical-Innovation.

FDA. (2022a). "Guidance Documents (Medical Devices and Radiation-Emitting Products)." Center for Devices and Radiological Health, U.S. Food and Drug Administration (FDA), September 15, 2022. https://www.fda.gov/medical-devices/device-advice-comprehensive-regulatory-assistance/guidance-documents-medical-devices-and-radiation-emitting-products.

FDA. (2022b). "Recognized Consensus Standards." US Food and Drug Administration (FDA), September 15, 2022. https://www.accessdata.fda.gov/scripts/cdrh/cfdocs/cfStandards/search.cfm.

ISO. (2021). "ISO 80369-7:2021: Small-Bore Connectors for Liquids and Gases in Healthcare Applications — Part 7: Connectors for Intravascular or Hypodermic Applications." International Standards Organization (ISO), May 4, 2021. https://www.iso.org/standard/79173.html.

MITA. (2022). "About DICOM-Overview." Medical Imaging Technology Association (MITA), September 15, 2022. https://www.dicomstandard.org/about.

PeekMed, Inc. (2022). "What are DICOM Images? Everything You Need to Know." PeekMed, Inc., April 11, 2022. https://blog.peekmed.com/dicom.

Stein, Ben P. (2022). "Why You Need Standards." May 23, 2022. https://www.nist.gov/feature-stories/why-you-need-standards.

University of British Columbia. (2022). "Why Do Projects Fail?" University of British Columbia, April 11, 2022. https://calleam.com/WTPF/?p=4700.

Closing Thoughts

Technology has enormous potential to help extend and eventually help alleviate an overworked health-care workforce. However, we as an industry have a long way to go to make that dream a reality. It starts with deep respect for the challenges and then taking responsibility for those areas that pertain to our scope of technology. This is in the goal to be real, reliable, and useful partners to clinicians. Many stakeholders had high hopes for one "AI in Healthcare giant" with its sizzling announcements and rollout several years ago. However, industry insiders were not surprised in the least when the parent company announced selling it off (Moore 2022). The once highly anticipated program encountered numerous hurdles that underscore the need for careful strategic planning and exquisite execution in medical technology that takes into account the practices and behaviors of all stakeholders involved, in particular the physicians. In a recent interview, that company's CEO revealed that the rationale for selling their Health division stemmed from a lack of domain expertise at the company required to make the technology ideas and the "AI Health solution" work in medicine (Kunert 2022).

Perhaps Additional Similar Announcements Exiting Medicine by Tech Companies Should Follow Shortly

Future organizations won't need to spend $4B, and hire a staff of 7,000 to be told by clinicians that their tool is "irrelevant" and another pejorative term of unhelpful in their clinical context (O'Leary 2022). Hopefully, large firms with heritage outside of medicine will learn from this example. We think the way to do this is to double down on their strengths of perfecting consumer-grade products for nonclinical use, including wellness and recreation use cases versus medical-grade products needed for clinically significant findings that clinicians use to make medical decisions. Similarly, enterprise technology companies without a foundational core of medical competency would also have the obvious growth avenues of business IT and related business apps, as well as administrative use cases for apps. All stakeholders should encourage technology organizations to stick to their historical strengths and only develop in areas where they have already held deep organization-wide domain expertise. At the same time, companies should be discouraged away from areas where their domain expertise is lacking. One area in which heritage tech companies can continue to bring a lot of value is data storage.

We Owe It to Clinicians to Bring Full Technology Lifecycle Competence to Medicine

Physicians have given and sacrificed more than most humans can even imagine doing in the course of the ongoing pandemic. Prior to that, they were already suffering from high levels of burnout.

Technology contributes to physician burnout (Jackson 2018). We as technologists need to take responsibility for not just our teams' full stack technical, compliance, privacy, and cybersecurity skills across devices, data, networking, applications, and interoperability; we must competently build and lead teams that can lead the productization, implementation, and scale of genuinely helpful solutions. This takes becoming master orchestrators of ecosystems of multiple disciplines and partnering with organizational psychologists and others to make that teamwork of teams both efficient and productive for all parties. This book is meant to provide all the required seeds and plenty of guidance for that to happen for the real gifted leader.

Technologists and Engineers Should Do Everything in Our Power Not to Contribute to Physician Burnout

As a matter of observation in my daily life being related to physicians in the family and having a number of friends who practice medicine, I observe that any technology that doesn't deliver clinically meaningful evidence-based impact has the potential to add to physician burnout and frustration. This should be widely considered an unacceptable failure of technology. Why does it happen? It's because in consumer and less so though also in business IT, workflow change management and support are not typically well thought out with robust implementation plans for the users. If patients are expected or even have the opportunity to use technology and they need help, there is no infrastructure or resources to support them today. Many companies that want to offer digital health applications to patients also have a heritage of a "no or low user support model" of technology products. What happens is all the competency gaps of technology in terms of user onboarding and use then fall into physicians' laps. Anytime a patient or their caregiver doesn't know how to use a device, computer, application, or has trouble with their internet connection related to a health app, they will ask their doctor for help. This is an additional unreasonable burden on physicians. This is especially so if the applications offered don't deliver measurable proven superior clinical outcomes.

Teams of Dedicated, Committed Multidisciplinary Teams Are the Way Forward

Industry leaders repeatedly point out that healthcare is complex, a full-time job, and does not present the opportunity to disintermediate physicians whom, for the most part, patients still trust (Padmanabhan 2021). We believe that the only solution is to partner closely with physicians; whereas they're at the same time themselves making active strides to help lead increasingly multidisciplinary teams that include technologists. This is while technologists would ideally bring their specific expertise to a team that has a credible large amount of deep medical domain expertise to navigate the complexities and truly deliver for patients, clinicians, and organizations. This field is not for the ego-driven technologist that wants to build something and imagines that people will then come. It would be too hard for many engineers and technologists to accept that the technology piece is just one of a large number of puzzle pieces that have to fall into place for medicine. In healthcare IT for advanced technologies, technology is a critical piece but can't really even be called the centerpiece, the care quality is. The software future of medical technology is a calling only for those that can be fulfilled with that. If you have found this as your calling and what we've provided in this book resonates with you, we look forward to crossing paths in this incredibly special, tight-knit industry called healthcare IT as its reach expands into medical-grade software.

Please inspire and inform us by sharing your progress, losses, lessons, and wins. We look forward to your success stories of having leapfrogged enormous amounts of promotional noise, and clinically inconsequential and at times wasteful activity. We look forward to celebrating our meaningful advanced healthcare technology future together.

Sherri Douville

Bibliography

Jackson, Wesley D. "Why does technology contribute to physician burnout?" Alberta Doctors' Digest (October 2018). https://add.albertadoctors.org/issues/september-october-2018/why-does-technology-contribute-physician-burnout/.

Kunert, Paul. "IBM CEO explains why he offloaded Watson Health: Not enough domain expertise." The A Register (June 8, 2022). https://www.theregister.com/2022/06/08/ibm_ceo_arvind_krishna_explains/.

Moore, John. "Another one bites the dust: IBM to sell watson health data assets." Chilmarkresearch.com (January 21, 2022). https://www.chilmarkresearch.com/another-one-bites-the-dust-ibm-sells-watson-health/.

O'Leary, Lizzie. "How IBM's Watson went from the future of health care to sold off for parts." Slate (January 31, 2022). https://slate.com/technology/2022/01/ibm-watson-health-failure-artificial-intelligence.html.

Padmanabhan, Paddy. "Is healthcare too hard for big tech firms?" HealthcareITNews (August 23, 2021). https://www.healthcareitnews.com/blog/healthcare-too-hard-big-tech-firms.

Appendix A: Glossary of Key Terms

Brittany Partridge

We are pleased to present you with this glossary of the following terms that are key to the success of this book and this work:

45 CFR § 164.308: The section of the Code of Federal Regulations that contains the Administrative Safeguards of the HIPAA Security Rule. This section covers areas such as security management processes, security awareness training, and contingency planning in the context of preventing the loss, theft, or unauthorized disclosure of electronic Protected Health Information (ePHI).

AI/ML (Artificial Intelligence/Machine Learning): AI and ML represent an evolution in computer science and data processing. Artificial intelligence generally refers to processes and algorithms that are intended to simulate human intelligence, including the goal to mimic cognitive functions such as perception, learning, and problem-solving. Machine learning and deep learning (DL) are subsets of AI.

Ally/Allyship: An ally is any person who actively promotes and aspires to advance the culture of inclusion through intentional, positive, and conscious efforts that benefit people as a whole.

BOM (Bill of Material): A list of the raw materials, subassemblies, intermediate assemblies, subcomponents, parts, and the quantities of each needed to manufacture an end product.

CMS: Center for Medicare and Medicaid Services, US Department of Health and Human Services, is a federal agency within the US Department of Health and Human Services (HHS) that administers the Medicare program and works in partnership with state governments to administer Medicaid, the Children's Health Insurance Program (CHIP), and health insurance portability standards. In addition to these programs, CMS has other responsibilities, including administrative duties relative to the privacy law, HIPPA, and oversight of quality metrics for labs and long-term care.

DEI: Diversity, Equality, and Inclusion.

> **Diversity:** It is the presence of differences within a given setting. In the workplace, this generally refers to psychological, physical, and social differences that occur among any and all individuals. A diverse group, a community, or an organization is one in which a variety of social and cultural characteristics exist.

Equity: It ensures everyone has access to the same treatment, opportunities, and advancement. Equity aims to identify and eliminate barriers that prevent the full participation of some groups. Specifically, barriers can come in many forms, but a prime example can be found in this study. In it, researchers asked faculty scientists to evaluate a candidate's application materials, which were randomly assigned either a male or female name. Faculty scientists rated the male applicant as significantly more competent and hirable than the identical female applicant, and offered a higher starting salary and more career mentoring to the male applicant.

Inclusion: It refers to how people with different identities feel as part of the larger group. Inclusion doesn't naturally result from diversity, and in reality, you can have a diverse team of talent, but that doesn't mean that everyone feels welcome or valued. Diversity, Equity, and Inclusion are mutually reinforcing principles within an organization. A focus on diversity alone is insufficient because an employee's sense of belonging (inclusion) and experience of fairness (equity) is critically important.

Disinformation: Unproven information that involves an intention to mislead.

Diversity of Discipline: Strategies for achieving a more diverse, equitable, and inclusive discipline, where diverse individuals will yield a diverse model system, with diverse perspectives that enable us to meet the challenges of identifying the general principles and mechanisms that generate endless forms that are the most fruitful.

EMR/EHR: (Electronic Medical Records/Electronic Healthcare Records) EHRs and EMRs are methods of healthcare record keeping that can exist in electronic or paper format, as might be found on a portable or desk-mounted piece of technology. According to Rosenbaum (2015), EHRs were created to improve methods for healthcare billing (versus improving the workflow of healthcare professionals). That point is supported by Zheng, Abraham, Novak, Reynolds, and Gettinger (2016) material that proposed that EHRs were not created by the healthcare industry to solve the issues of transforming data between dissimilar solutions.

FHIR: Fast Healthcare Interoperability Resources is a standard describing data formats and elements and an application programming interface for exchanging electronic health records.

Formula 1: Formula 1 cars were given the name because they are the cars that are the fastest, most aerodynamic, power-efficient machines that have been designed on four wheels. "Formula one" basically means the formula for the best.

HEDIS: Healthcare Effectiveness Data and Information Set, copyright NCQA.

Hubris: Excessive self-confidence; overbearing presumption of superiority of thought.

IASAM3: It is a new model that enables technology integrators to evaluate new technology through the lens of evaluating aspects of socioeconomical and technology characteristics that move beyond the assessment of pre-adoption phases and toward the examination of assessment, decision-making, integration, and sustainability of technological solutions (Aizstrauta and Ginters 2017; Ginters, Mezitis, and Aizstrauta 2018).

IEEE/UL TIPPSS: It is an IEEE/UL joint venture and standard that establishes the framework with TIPPSS principles (Trust, Identity, Privacy, Protection, Safety, Security) for Clinical Internet of Things (IoT) data and device validation and interoperability. The standard includes wearable clinical IoT and interoperability with healthcare systems, including electronic health records (EHRs), electronic medical records (EMRs), other clinical IoT devices, in-hospital devices, and future devices and connected healthcare systems.

Inequality: Centers on disparities between social types, defined by the exposure to circumstances beyond individual control.

Information Blocking: Knowingly and unreasonably interfering with the exchange of electronic health information.

Input Connectors: An input connector takes data that enters a transformational engine through a data queue and stores the data in a data stream to be processed separately by the execution layer of the engine (Ladd and Hermansson 2019). Within the scope of this discussion, an input connector is an element of a transformational engine, which facilitates the transfer of data between dissimilar technological solutions.

Intellectual Humility: Ability to revise one's viewpoint; awareness of one's cognitive fallibility.

Interoperability: It is the ability for humans, hardware, and software technology to connect and communicate using a standard protocol, where data may be transformed across dissimilar technology (Wager, Lee, and Glaser 2017).

Intraoperative: It is the ability for humans, hardware, and software technology to connect and communicate using a standard protocol, where data may be transformed across dissimilar technology (Wager, Lee, and Glaser 2017). Technology that is considered interoperable represents solutions that are able to connect with other technology and exchange data using standardized protocols through a common interface such as is characterized by a transformational engine. That proposal is supported by the material provided by Evans et al. (2017) and Slight et al. (2015), where technology cannot be considered as successfully integrated within an environment if that technology is incompatible with existing technology.

Knowmad: A knowmad is a nomadic knowledge worker who is creative, imaginative, innovative, and who can work with almost anybody, anytime, and anywhere. Knowmads are valued for their individual-level knowledge, and create new value by applying what they know, contextually, to solve problems or generate new opportunities.

Marginalized: Individuals who may not have had any experience of social resilience and therefore may not be expected to be able to engage in developing a plan to build or produce such resilience. Specifically, where the marginalized is cut off from an environment which can provide access to power resources, but are unable to demonstrate agency to access what they need through the lens of resilience.

Mentor: Mentors are perceived as individuals with invaluable assets to professional and academic individuals because of the career and psychosocial support benefits they provide. Whereas knowledge, experience, guidance, and support are desirable attributes of a mentor.

Misinformation: Unproven information which can include honest mistakes.

Narcissism: Entitlement and need for admiration; personality characterized by excessive self-interest, at times lacking empathy.

Output Connectors: Output connectors take data that is changed into a different format by the transformational engine and pushes that data into queues that pass the data to designated target locations (Ladd and Hermansson 2019). Like the input connectors, an output connector is an element within the concept of a transformational engine, which facilitates the transfer of data between dissimilar technological solutions.

Person of Color (POC): Though color is a very subjective term, the use of the term "person of color" is primarily used in the United States to include African Americans, Asian Americans, Native Americans, Pacific Islander Americans, multiracial Americans, and some Latino Americans.

Privilege: Refers to the immutable identity of being a citizen of the majority, where it is afforded to those who would never know what it feels like to be part of the minority, or the marginalized.

Relational Humility: Ability to self-regulate emotion and outward manifestation of ego; use of humility to effectively shape human relationships.

Sponsor: Within the context of this chapter, the term sponsor is synonymous with that of the term "mentor," where both terms are used interchangeably outside of the United States.

Stakeholders: They are individuals and groups who take interest in a specific initiative organization and can directly impact and influence decisions.

Transformational Engines: A transformation engine is a principal element that encapsulates other elements (e.g., input and output connectors) that are used to transform data within a technology solution from one format to another format using specific data mapping rules that are defined in the engine's configuration module (Ladd and Hermansson 2019).

Underrepresented: This term typically represents individuals who are economically disadvantaged, of color, from ethnic minorities, including potentially with limited English proficiency.

VoC (Voice of Customer): It is the component of customer experience that focuses on customer's needs, wants, expectations, and preferences. To determine the VOC, an organization analyzes indirect input or data that reflects customer behaviors as well as direct input or data that reflects what a customer says.

Index

A

AAMI, *see* Association for the Advancement of Medical Instrumentation
Acceptance and commitment therapy (ACT), 147
Accountable business owner, 220
Accountable internal resources, 220
ACEA, *see* Associate Certified Enterprise Architect
ACS, *see* American College of Surgeons
ACT, *see* Acceptance and commitment therapy
Acute care delivery, 253–254
Addiction, 127, 146–147, 154
ADKAR, *see* Awareness, Desire, Knowledge, Ability, and Reinforcement
Advocacy framework, 19
AED, *see* Automated external defibrillator
AEHIS, *see* Association for Executives in Healthcare Information Security
Affiliation, 267
Agency for Healthcare Research and Quality (AHRQ), 112, 293
AHRQ, *see* Agency for Healthcare Research and Quality
AHSP, *see* American Society of Health-System Pharmacists
AICPA, *see* American Institute of Certified Public Accountants
AID, *see* Align, Improve, and self-Determination
Alarm Fatigue, 140
ALEs, *see* Applied learning experiences
Align, Improve, and self-Determination (AID), 188
Allegiances, 266
Allies
 actions and behaviors, 59–60
 bad, 52
 benefits of alliances, 53
 broad recruitment, 54–55
 building, 75
 competency checklist, 57
 evidence, 52–53
 good, 51
 ignorance, 58–59
 leader, 54
 manage our behaviors, 60

personal experience evolution, 58
potential partner service organizations, 63–64
promotes and strengthens, 56
recruiting, 54–56
relationship alliance, 53
seeing inequality, 60–62
targeted recruitment, 55–56
Ambidextrous organizations, 99
American College of Surgeons (ACS), 186, 187
American Health Information Management Association (AHIMA), 175
American Institute of Certified Public Accountants (AICPA), 181, 194, 228–229
American National Standards Institute (ANSI), 175
American Restoration and Reinvestment Act (ARRA), 294
American Society of Health-System Pharmacists (AHSP), 175
ANAB, *see* ANSI-ASQ National Accreditation Board
ANSI, *see* American National Standards Institute
ANSI-ASQ National Accreditation Board (ANAB), 228–229
Anti-Semitic conspiracy theory, 136
API, *see* Application program interface
Application program interface (API), 216, 240, 309, 315–316
Applied learning experiences (ALEs), 20
Appreciation, 80, 101, 156, 267, 279
Architectural plan, 220
ARRA, *see* American Restoration and Reinvestment Act
Arrogance, 81, 83, 86, 89, 105, 113
Associate Certified Enterprise Architect (ACEA), 205
Association for Executives in Healthcare Information Security (AEHIS), 248
Association for the Advancement of Medical Instrumentation (AAMI), 175
Autism spectrum disorder (ASD), 145
Automated external defibrillator (AED), 20
Automated teller machine (ATM), 291
Autonomous vehicles, 42–43
Autonomy, 43, 108, 112, 113, 267–268
Autopilot, 43
Awareness, Desire, Knowledge, Ability, and Reinforcement (ADKAR), 188

Healthcare Technology Management (HTM), 24
Health Communication and Health Information
  Technology (HC/HIT), 16
Health Insurance Portability and Accountability Act
  (HIPAA), 190, 215, 292, 314
Health Sector Coordinating Council (HSCC),
  242–243, 247
Hegemonic masculinity, 131–132
High-functioning clinical IoT device security teams, 26
HIMSS, *see* Healthcare Information Management
  Systems Society
HIPAA, *see* Health Insurance Portability and
  Accountability Act
HITECH Act, 314
Home program, 255
Homophily, 131–132
Hospital at Home, 253–257
Hospital Medical Grade, 255–257
HSCC, *see* Health Sector Coordinating Council
HTM, *see* Healthcare Technology Management
Hubris
  benefits, 84–85
  business/industry/society, 87–88
  collaborators/partners/teams/organizations, 86–87
  components, 81
  defining elements, 84
  definition, 81, 97
  disrupting
    culture, 90–91
    leadership selection, 90
    person to person, 88–90
  vs. humility, 82
  leadership, 85–86
  vs. narcissism, 104
  perception scale, 82
  situational narcissism, 105–106
Human experience
  knowledge, 33, 34
  objective analysis, 34
  subjective analysis, 33–34
Humility
  care providers, 106–108
  CEO, 104
  challenge, 103–104
  clinical user, 108–109
  communication models, 101–102
  culture, 100–101
  definition, 96
  educational systems, 97
  hospital executive, 116
  integrating, 102–103
  master adaptive learning, 99–100
  medical education, 97
  medical error, 111–112
  personality factors, 113–114
  physician's relationship with identity, 98–99
  physician's relationship with knowledge, 98

physician's relationship with structure, 99
risk in clinical space, 110–111
technology as physician leader, 114–115
technology expert, 115–116
Hustle porn cultures, 138

I

IASAM3, 31, 35
IDN, *see* Integrated Delivery Network
IEEE/UL TIPPSS, 30, 35
IH, *see* Intellectual humility
IHI, *see* Institute for Healthcare Improvement
IIA, *see* Institute of Internal Auditors
Implementation framework, 18
Impostor syndrome, 146
Inclusion, 35
Industry leadership, 322
Inequality, 58, 60–62
Infant heart study, 308
Information Security, 123–125
  advancement preparation, 164
  alumni, 165
  burnout, 162–163
  career path, 163
  challenges, 128–129
  communication, 164–165
  conferences, 138
  coping techniques and resilience, 164
  distractions, 164
  economic issues, 149–153
  employee assistance programs, 165
  errors, 165
  goals and actions, 164
  gratitude, 165
  hacker subcultures, 125–127
  implicit biases and microaggressions, 165
  individual development plans, 164
  isolation, 144–145
  mentoring, 164
  operating systems, 139–140
  protection and environmental considerations,
    159–162
  required certifications, 164
  reverse mentoring, 164
  signal-to-noise ratio, 140–144
  software development, 148–149
  team members, 163
  training courses, 164
Information Security Management Systems (ISMS),
  224, 234
Innovation work
  beneficiaries of service
    employer, 270
    patient, 269
    patient caregivers, 269–270
    patient family, 269